ENVIRONMENTAL POLLUTION

ENVIRONMENTAL POLLUTION

A SURVEY EMPHASIZING PHYSICAL AND CHEMICAL PRINCIPLES

LAURENT HODGES

Department of Physics
Iowa State University

HOLT, RINEHART AND WINSTON, INC.
New York   Chicago   San Francisco   Atlanta   Dallas
Montreal   Toronto   London   Sydney

To Linda and Andrew

PREFACE

The threats of pollution are real. Their economic consequences are real. Their health consequences are real. There are sufficient data to make strong cases based on facts.—Melvin J. Josephs, *Environ. Sci. Tech.* **1**:525(1967).

This book has been written to fill the need for a one-volume scientific discussion of the major types of environmental pollution—air, water, noise, solid waste, thermal, and radiation pollution—and their effects on man and on the environment.

The book has evolved from a one-quarter course on The Physics and Chemistry of Pollution taught in the physics department at Iowa State University. It could serve as a text for a one-semester or longer course or as a supplement to any of a wide variety of courses dealing with environmental concerns.

It is hoped that the book will prove useful to students, as part of their general education or as an early step toward an environmental career; to teachers, as an aid in developing formal environmental curricula or in incorporating discussions of environmental problems in existing courses; and to educated laymen, as a scientific introduction to pollution and its control.

Because the author's training has been in the physical sciences, there is greater emphasis on physical and chemical principles than on biological and ecological principles. It should be recognized, however, that it is the health and environmental effects of pollution that are of most concern to mankind.

Historical and current statistics have been included in order to show the progress that has been made in the past in overcoming some types of pollution and yet to demonstrate the magnitude of existing problems. The discussion centers largely on the American and British experiences, but an attempt has been made to show that pollution is a global concern.

In order to permit meaningful comparisons of data, the quantitative data presented in this book have been converted into SI (metric) units from a wide variety of different units. For example, all volume units are cubic meters or its multiples or submultiples, rather than cubic feet, gallons, acre-feet, barrels, cubic miles, and other specialized units. Appendix A lists conversion factors between the SI and other units.

Appendix B includes lists of books and periodicals dealing with environmental matters and of addresses of sources of environmental information. The references at the end of each chapter are intended to provide a more specialized bibliography.

ACKNOWLEDGMENTS

A very long list of names belongs here; it would begin with acknowledgments to

John Platt for an oral presentation on "What We Must Do" which inspired the course from which this book evolved.

Dan Zaffarano and the Iowa State University physics department for encouraging and supporting Physics 490P.

The Iowa State University Library.

Robert H. Socolow for his thorough critical review of the manuscript and for hundreds of invaluable suggestions.

James C. Weaver, David L. Trauger, and Allan R. Evans for helpful reviews of the whole manuscript.

Charles L. Benn, Earl G. Hammond, William C. Moldenhauer, and Donald M. Roberts for their help on individual chapters.

The scientists whose research made this book possible—men and women of other places and other times whom I know only from their writings.

Dan Serebrakian and the staff at Holt, Rinehart and Winston.

L. H.
Ames, Iowa
December 1972

CONTENTS

ENVIRONMENTAL POLLUTION

1
INTRODUCTION

Environmental pollution is the unfavorable alteration of our surroundings, wholly or largely as a by-product of man's actions, through direct or indirect effects of changes in energy patterns, radiation levels, chemical and physical constitution and abundances of organisms. These changes may affect man directly or through his supplies of water and of agricultural and other biological products, his physical objects or possessions, or his opportunities for recreation and appreciation of nature. —U.S. President's Science Advisory Committee, Environmental Pollution Panel, *Restoring the Quality of Our Environment.* Washington, D.C.: U.S. Govt. Printing Office, 1965.

Pollutants that meet the criteria of this definition of environmental pollution are numerous: gases (such as sulfur dioxide and nitrogen oxides) and particulate matter (such as smoke particles, lead aerosols, and asbestos) in the atmosphere; pesticides and radioactive isotopes in the atmosphere and in waterways; sewage, organic chemicals, and phosphates in water; solid wastes on land; excessive heating ("thermal pollution") of rivers and lakes; and many others.

The most important and controversial question left unanswered by this definition of pollution is the question of what constitutes an "unfavorable alteration." Any man-made alteration of the environment probably has unfavorable effects, at least in the opinion of some people, and favorable effects in the opinion of others, such as those whose livelihood depends on an activity that produces pollution. The determination of the extent of the favorable versus unfavorable effects—or of benefits versus costs—is difficult just because it is ultimately subjective, even though objective data may be involved in the determination.

The affluent societies of the developed nations of the world are likely to be more concerned about the unfavorable effects than those nations in which poverty and hunger are major unsolved problems. The unfavorable effects of pesticides and fertilizer runoff are likely to be of less concern in a country in which insufficient food production is leading to malnourishment and starvation or in which insect-borne diseases are major contributors to human morbidity and mortality than in a country with agricultural surpluses, relatively pure water supplies, and a strong public health program. Disagreements of this sort have already been encountered by international consumer groups attempting to define their goals [1] and at the 1972 U.N. Conference on the Human Environment.

Poverty, starvation, and pollution all reflect mankind's failure to design social and political institutions capable of properly assessing and controlling technological innovations [2]. Serious problems of poverty and hunger exist in the United States despite the progress of the last several decades, and the progress that has been made has been accompanied by the aggravation of many existing environmental problems and the production of new ones.

The rest of this chapter is devoted to examining certain characteristics of pollution: the levels and movement of pollutants, the types of effects they have and the manner in which they can interact with one another, and the relation of pollution to other technological hazards.

LEVELS AND MOVEMENT OF POLLUTANTS

NATURAL AND MAN-GENERATED POLLUTION. Most of the pollutants that concern man occur in nature, although chlorinated hydrocarbons (such as DDT) and certain short-lived radioactive isotopes are exceptions. In some cases, the environmental levels are largely due to natural sources, while in others the levels are largely produced by man's activities. Table 1-1 gives several examples. Even when natural sources are more important on a global scale, man-generated pollutants may be more important in urban and industrial areas where the adverse effects of pollution are most severe. It is thus important to distinguish between large-scale production of pollutants and local production over a few tens or hundreds of square kilometers.

TABLE 1-1. Natural versus man-generated pollution.

Almost completely man-generated
    Chlorinated hydrocarbons (DDT, etc.)
    Lead aerosols
Substantially man-generated
    Oil on the oceans
    Phosphates in running waters
Substantial contributions from natural sources
    Hydrocarbons in the atmosphere
    Radiation
    Sulfur oxides in the atmosphere

CONCENTRATIONS OF POLLUTANTS. Concentrations of pollutants are often expressed by fractions. A concentration of one part per million (1 ppm) corresponds to one part pollutant per one million parts of the gas, liquid, or solid mixture in which the pollutant occurs. In the case of a gas mixture, the reference is generally to ppm by *volume*, whereas in the case of liquids and solids

TABLE 1-2. Fractional concentrations.

Symbol	Definition	Fraction
ppm	Parts per million	10^{-6}
pphm	Parts per hundred million	10^{-8}
ppb	Parts per billion	10^{-9}
ppt	Parts per trillion	10^{-12}

the reference is generally to ppm by *weight*. More recently, it has become customary to express gaseous pollutants and particulate matter in the atmosphere in mass density units of micrograms per cubic meter ($\mu g/m^3$), and in this case it is necessary to specify the temperature (usually 0 or 25°C) and pressure (usually 1 atm) at which the concentration is expressed. Table 1-2 lists some common fractional concentrations. At first glance, a concentration of 1 ppm seems ridiculously small and negligible. Agricultural chemical manufacturers are fond of pointing out that it corresponds to one drop of vermouth in 80 fifths of gin, or 1 oz of salt in 62,500 lb of sugar, or one mouthful of food out of all the food consumed during one's lifetime. Nevertheless, concentrations of pollutants at levels of 1 ppm or less can have serious adverse effects:

1 ppm phenol in water is lethal to some species of fish.

0.2 ppm average sulfur dioxide level in the atmosphere has been shown to lead to an increased human mortality rate.

0.02 ppm peroxybenzoyl nitrate (Figure 1-1) in smog can cause severe eye irritation in humans [3].

0.001 ppm hydrogen fluoride (HF) gas in the atmosphere can injure certain sensitive plants, such as peaches.

LONG-DISTANCE MOVEMENT OF POLLUTANTS. There is accumulating evidence that many types of pollutants can become distributed over the whole earth in relatively short periods of time. Radioactive fallout from atmospheric nuclear tests is detectable throughout the world within days or weeks, and it is a simple matter for scientists to analyze the fallout and to learn a great deal about the type of test carried out; not even the polar regions are immune from fallout. Another dramatic example is the worldwide distribution of chlorinated hydrocarbons such as DDT and its metabolites. These were first developed for pesticidal use in the late 1930's and released for civilian use in 1945; since this time they have been widely used throughout the world. They are now widely distributed throughout the marine ecosystems of the

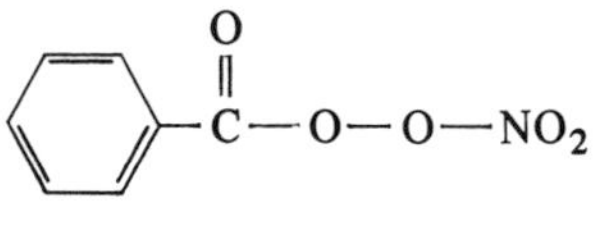

FIG. 1-1. Peroxybenzoyl nitrate, a constituent of photochemical smog and a potent eye irritant [3].

Pacific Ocean, for example, probably having been transported by air [4]. Levels of DDT and its metabolites of over 100 ppb have been found in the liver and fat of animals that never leave the Antarctic ice pack, such as Adelie penguins, while no trace could be found in an emperor penguin that had been killed in 1911 [5]. Since pesticides have not been used within thousands of kilometers of Antarctica, the DDT had clearly been transported a great distance by means not fully understood.

PERSISTENCE. Some pollutants remain dangerous indefinitely, beryllium and lead being examples. Others that are eventually broken down into harmless compounds may still persist in the environment for long periods of time. A review [6] of pesticide persistence in soils, defined as the time needed for the pesticide level to be reduced to less than 25% of the original level applied, estimated:

5 years for chlordane,
4 years for DDT,
3 years for dieldrin,
1.5 years for picloram,
5 months for 2,4,5-T, and
1 month for 2,4-D.

TABLE 1-3. Half-lives of some radioactive nuclei. The half-life is the time necessary for half the original number of nuclei to decay and thus the time over which the radioactivity will be halved.

Isotope	H^3	Be^7	Kr^{85}	Sr^{90}	I^{131}	Cs^{137}	Ra^{226}	Pu^{239}
Half-life ($t_{1/2}$)	12.3 years	53 days	10.7 years	28.9 years	8.1 days	30.2 years	1,602 years	24,400 years

It has been estimated that inorganic mercury compounds in bottom sediments of lakes and rivers may take 10 to 100 years to be converted into methylmercury, an organic form dangerous to animal life [7]. Radioactive nuclei decay exponentially with a wide range of half-lives, the half-life being the time necessary for half the original number of nuclei to decay (see Figure 1-2). Table 1-3 lists the half-lives of a number of radioactive isotopes and gives an indication of the wide range of half-lives that exist. Fallout from nuclear weapons tests occurs locally (in the immediate vicinity of the test) within the first day, tropospherically (over a large part of the world at roughly the latitude of the test) during the first month after the test, and stratospherically over a period of many years. Radioactive fission products or other debris can be carried into the stratosphere by the energy of the nuclear explosion and then only very slowly return to the troposphere and thence to the earth's surface. Even if atmospheric nuclear tests were to be ended by

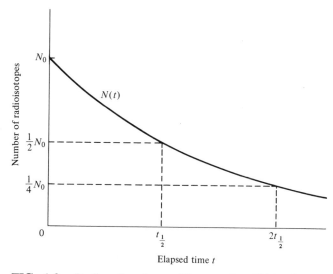

FIG. 1-2. Radioactive decay. The number $N(t)$ of un-decayed radioisotopes exhibits an exponential decrease from the initial number N_0. After a time equal to the half-life $t_{1/2}$ has elapsed, only half the initial number of radioisotopes remain; at time $2t_{1/2}$, one-fourth remain; at time $3t_{1/2}$, one-eighth remain; etc.

all nations, significant amounts of tritium (H^3) would continue to be released to the earth from the stratosphere for several more decades.

BIOLOGICAL CONCENTRATION AND DISCRIMINATION. Another important characteristic of pollutants is that they may be concentrated biologically so that levels in one part of an ecosystem are much larger than those in other parts. This typically occurs in food chains: Levels in an organism are higher than those in its food. The study [8] of one Lake Michigan ecosystem found DDT levels of:

0.014 ppm (wet weight) in mud sediments on the bottom
0.41 ppm in bottom-feeding crustacea
3 to 6 ppm in various fish (alewives, chub, whitefish)
over 2400 ppm in the body fat of fish-eating gulls

One reason for this sort of concentration is that chlorinated hydrocarbons are much more soluble in fat than in water. Another well-documented case of biological concentration through the food chain is the exceptionally high level of Cs^{137} found in humans living in the far north of Europe and North America, particularly Lapps and Eskimos. In this instance, Cs^{137} from fallout was concentrated in lichens, then in reindeer and caribou feeding on the lichens, and finally in humans eating those animals; body burdens of Cs^{137} were 10 to 100 times greater than those occurring in humans at more temperate latitudes [9].

On the other hand, pollutants may be discriminated against biologically. Sr^{90} is chemically similar to calcium, which is an essential constituent of human bones, and the two follow similar metabolic paths. The Sr^{90}, a heavier element, however, is discriminated against. Although precise figures are not known, the Sr^{90}/Ca ratio in plants may be only half that in the soil in which the plant is grown (the calcium levels are always much greater than the Sr^{90} levels), the ratio in cow's milk (the major source of dietary calcium for humans) may be only one-eighth that in the vegetation eaten by the cow, and the ratio in human bone may be only one-fourth that in the human diet (milk or plants). As a result, the Sr^{90}/Ca ratio in the bone is only a few percent of that in the soil [10]. Despite this fact, Sr^{90} levels due to fallout have been of some concern. Tritium (H^3) is believed to be treated identically with ordinary hydrogen (H^1) in all biological systems; if this is correct, then it is neither concentrated nor discriminated against.

EFFECTS OF POLLUTION

The known effects of pollution are already numerous and varied, even though there are probably many others yet to be discovered. In the next few pages some of these effects will be discussed and a number of definitions introduced but the other chapters in this book contain much more specific information.

EFFECTS ON PLANTS. Agriculture and horticulture are both affected by pollution. The characteristic types of injury produced by photochemical smog have become so severe in the Los Angeles Basin in California that it is essentially impossible to grow orchids, spinach, romaine lettuce, Swiss chard, and some other leafy plants, and over 100 km² of pine forests in the nearby San Bernardino Mountains have been severely affected [11]. Other air pollutants having severe effects on vegetation include sulfur oxides (from copper and lead smelters) and hydrogen fluoride (from such sources as fertilizer manufacturing and aluminum reduction). Even when pollution levels are not high enough to produce noticeable injury, retardation of growth may occur. Since some plants are likely to be more sensitive than others to the pollutant, whether it be an air or water pollutant or radiation, there may be complex changes in the plant ecosystem, with effects on one species leading to effects on the others [12]. Sometimes the undesirable effect of a pollutant is the increase in plant life, as when the introduction of plant nutrients such as phosphorus, nitrogen, and carbon leads to algal blooms in water bodies.

EFFECTS ON ANIMALS. Air pollutants are known to produce eye and respiratory irritation in animals as well as in humans. Hospital records show that the infamous smog episode of December 1952 in London (discussed further in Chapter 3) caused about 4000 excess deaths among humans, but at the time the only deaths noticed and reported in the newspapers were those of some of the prize cattle at the Smithfield Show. Mankind has even made use of the sensitivity of some animals to certain pollutants; an example is the use of canaries to detect poisonous gases in coal mines as well as nerve gas near trains that are carrying it. Water pollutants can also endanger aquatic life and every year millions of fish are reported killed by municipal and industrial wastes in the United States (see Chapter 10). Sewage, toxic chemicals, and disease organisms can also make water unfit for use by farm animals. There are some types of pollution known to have adverse effects on animals at levels that do not appear to affect human health. Thermal pollution—the excess heating of water in rivers or lakes—can kill fish, and pesticide levels in many species of birds have reduced reproduction rates through mechanisms such as interference in calcium metabolism.

EFFECTS ON HUMANS. Human beings are probably most concerned about the direct effects of pollutants upon their own health, even though these effects are not necessarily the most dangerous over the long run. Many substances are beneficial, even essential, to humans but become toxic at sufficiently high levels or doses. For example, the typical adult requires 10 to 15 mg of iron daily and has a body burden (i.e., a total amount in the body) of 3 g or a little less; yet each year there are a couple of thousand cases of acute iron poisoning (i.e., cases of severe illness) [13]. Occasional fatalities occur in small children from ingestion of ferrous sulfate tablets; small children need about 5 to 10 mg of iron per day but ingestion of 40 mg to 1.5 g of $FeSO_4$ has been known to produce severe effects. Similarly, copper, which is apparently needed in hemoglobin synthesis, can produce acute poisoning when 5 to 50 mg of soluble salts are ingested; the typical intake by adults is probably 1 or 2 mg daily. Thus the difference between toxic doses and actual environmental doses may not be large. Lead, which is not known to be essential to human life, can be toxic at blood concentrations above 0.8 ppm; yet the typical American has a concentration of about 0.25 ppm [14].

In determining levels that are toxic to humans, one is generally faced with the following problem. Sufficiently high levels are toxic or even lethal to everybody, although there is generally a wide range of sensitivities among the human population. For example, tartar emetic (antimony potassium tartrate) can be lethal at doses as low as 130 mg but some persons have survived doses as high as 15 g [15]. On the other hand, sufficiently low doses appear to have no toxicological effect. One possible explanation of these facts is that there exists a *threshold,* a level below which there are no effects. Another possible explanation, however, is that effects are occurring but that experimental techniques are not sophisticated enough to detect them at low doses. It is well-known that the effects of low levels of exposure are sometimes not apparent until many years have elapsed. Asbestos is known to produce increased mortality from lung cancer and asbestosis (severe scarring of the lungs by asbestos fibers) among asbestos workers but these effects take 20 to 30 years to show up [16]. There is growing evidence that many chronic diseases (such as asthma, emphysema, and bronchitis) are environmentally induced and result from long-term exposures to levels of various substances too low to produce acute effects.

A great deal of careful research will be necessary in the future to determine whether or not thresholds exist and whether or not steps can be taken to raise them artificially by aiding the body's detoxification processes. The important threshold, if exposures are limited in time, is that between irreversible effects, which cannot be undone, and reversible effects, which result in no permanent injury. Many drugs are toxic at certain doses but are of therapeutic value at lower doses, and the two doses are usually not very different. The emetic dose of tartar emetic is 30 mg, which is dangerously close to the toxic dose mentioned in the last paragraph, although if vomiting occurs, there is no danger [15]. The therapeutic value of drugs is, of course, intimately connected with their toxicity and a judgment is always necessary as to whether or not the benefits outweigh the risks.

Some effects of pollution on humans are obviously undesirable. Over the past century, as discussed in Chapter 9, it has been recognized that many diseases are waterborne and that their frequency can be checked dramatically through purification of municipal and individual water supplies. Still, the drinking water of most of the world's population is considerably less safe than that of the developed nations because of inadequacies in public health programs.

EFFECTS ON MATERIALS. Pollutants can accelerate the deterioration of materials and construction. Air pollutants, particularly sulfur dioxide gas and the sulfuric acid aerosols into which the gas is converted in the atmosphere, can corrode metals and building materials, increasing the frequency of repair and replacement. Inestimable damage is being caused to frescoes in Italy by sulfur dioxide, which converts the pigmented lime plaster to gypsum, thereby increasing its volume and reducing its cohesiveness [17]. Water pollutants, such as suspended particles or dissolved inorganic compounds, can also adversely affect pumps, industrial equipment, and bridges. With the advent of supersonic aircraft capable of producing sonic booms with overpressures of over 100 N/m^2 (newtons per square meter) (see Chapter 7), even noise can damage buildings and break windows.

SYNERGISM AND ANTAGONISM. It is impossible to discuss the effects of pollution properly merely by discussing the effects of individual pollutants. In many cases the combined effects of two or more pollutants are more severe or even qualitatively different from the individual effects of the separate pollutants—a phenomenon known as *synergism*. Widespread injury to peanut crops in Texas in 1966 and 1967 that occurred after the passage of weather fronts was eventually identified as due to synergistic action of ozone and sulfur dioxide [18]. Numerous studies have shown that some types of particulate matter, such as aerosols of soluble salts of ferrous iron, manganese, and vanadium, can increase the toxicity of sulfur dioxide [19]. Such an increase in toxicity is usually referred to as *potentiation*. Sometimes the combined effects of two pollutants are less rather than more severe, and this situation is referred to as *antagonism*. Cyanides in industrial wastes are quite poisonous to aquatic life, and in the presence of zinc or cadmium they are extremely poisonous (a synergistic effect), apparently due to the formation of complexes; in the presence of nickel, however, a nickel-cyanide complex that is not very toxic is formed [20]. The occurrence of synergistic effects makes it difficult to study the effects of pollution since so many different pollutants are present in the environment and this makes it hard to predict the effects that might occur when certain air or water quality standards are met.

GLOBAL EFFECTS. One of the most worrisome aspects of the growth of pollution is that man has now become a major factor in several of the great biogeochemical cycles and may be causing irreversible change or irreparable harm without realizing it. These biogeochemical cycles are the circulation of various elements through the biosphere (the part of the earth characterized by the existence of plant or animal life), the atmosphere, the hydrosphere (the earth's oceans and other waters), and the lithosphere (the solid part of the earth's surface). The carbon cycle, as an example, involves the steps listed in Figure 1-3. As discussed in Chapter 4, man's transfer of carbon dioxide

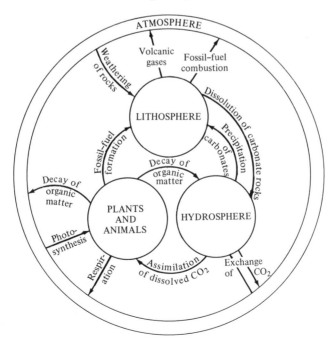

FIG. 1-3. The carbon cycle.

into the atmosphere by the combustion of fossil fuels produced during the carboniferous ages may constitute serious interference with this cycle. Fossil-fuel combustion also constitutes interference with the oxygen cycle but in this case it appears to be of negligible importance (see Chapter 4). Man's most important intervention may be in the nitrogen cycle since nitrogen fixation (the conversion of atmospheric nitrogen into organic nitrogen compounds) by man through combustion processes, fertilizer production, and the cultivation of nitrogen-fixing legumes may equal the historical nitrogen fixation from natural sources such as terrestrial and marine life and lightning [21]. The September 1970 issue of *Scientific American,* entitled "The Biosphere," discusses several of these cycles (carbon, oxygen, nitrogen, sulfur, and phosphorus) and points out our great lack of knowledge about them. Research is needed to elucidate the physical, chemical, and biological processes involved in these various pathways; the amounts transported; man's influence on the cycles; and the changes that may result from man's intervention.

TECHNOLOGICAL HAZARDS

A technological society exposes all or part of its population to many different hazards. In many cases these hazards were once occupational hazards characteristic of one or more occupations and thus of concern to only a small fraction of the population. Table 1-4 lists some examples with picturesque names. In time, many of these hazards have come to involve a much broader segment of the population. In the past, chronic mercury poisoning has generally resulted from occupational exposures but it has been learned in the past few years that some fish in the United States and Canada have levels of mercury compounds higher than the usual public health standards for human food. Chronic and fatal mercury poisoning from fish caught in waters contaminated by industrial wastes has already been reported from Minamata Bay in Japan [22]. Carbon monoxide poisoning was once primarily a problem for coal miners but is widespread today because of automobile exhausts and

TABLE 1-4. Occupational hazards of the past.

Name	Cause	Occupation
Baker's itch	Mites in flour	Bakers, grocers
Brass founder's ague	Copper and zinc fumes	Brass founders
File cutter's disease	Lead	File cutters
Grinder's disease	Fine stone and metal particles	Stone and metal grinders
Hatter's shakes	Mercury	Felt hatters
Metal fume fever	Heavy metal fumes	Metal workers
Miner's asthma	Silicate particles	Miners
Phossy jaw	White phosphorus	Match factory workers
Radium poisoning	Radium	Luminous watch painters
Silo filler's disease	Nitrogen dioxide	Agricultural workers

home furnaces. Asbestos is a clear-cut cause of occupational diseases such as asbestosis but there is evidence that it is also affecting the families of asbestos workers and persons living near asbestos plants and may be of more general concern in urban areas because of the spraying of asbestos during construction and the wearing down of brake linings in motor vehicles [16]. New occupational problems have arisen in recent years (beryllium and manganese are examples), and they too may someday affect large numbers of persons if appropriate public health measures are not taken.

There exist in the United States and other economically developed nations of the world many hazards whose presence constitutes an unfavorable alteration of our environment and may therefore be regarded as environmental pollution in the sense of the definition at the beginning of this chapter. Some of these hazards involve voluntary risks (such as those taken by smokers), while others amount to involuntary risks (such as those associated with the transportation of explosives or hazardous chemicals).

Smoking is really a personal air pollution problem but the fact that it is a voluntary activity—presumably with some benefits—is related to smokers' willingness to tolerate the high risks involved. Epidemiological studies have consistently shown that the adverse effects of smoking on health are more important than the adverse effects of air pollution, and indeed the role of smoking always needs to be considered in any epidemiological study of air pollution. Studies have shown that mortality rates among cigarette smokers will average about 70% greater than those among nonsmokers; the increase in coronary artery disease is the most important factor in this increase in mortality rate, although mortality rates due to lung cancer, bronchitis, and emphysema are about 6 to 11 times greater among smokers than among nonsmokers [23]. The great wealth of statistics accumulated since World War II regarding the health menace of tobacco has led to the recognition of tobacco use as "the greatest public health hazard in the United States today" [24].

The automobile is another important hazard. Automobile fatalities in the United States total well over 50,000 annually and serious injuries to persons in motor vehicle accidents account for another 2 million casualties annually. Motor vehicle use was clearly a voluntary luxury at the beginning of the 20th century but changes in American transportation patterns have made it a much less voluntary activity today. This change from a voluntary to a semivoluntary status has been accompanied by a large decrease in motor vehicle fatality rates per passenger-mile, but the rates are still high.

Hazardous substances expose the American public to a great many involuntary risks every day and grave accidents are not uncommon. On March 23, 1961, a barge sank in the Mississippi River near Natchez, Miss., carrying with it four long tanks containing a total of 1000 metric tons of liquid chlorine. Because of the danger of rupture or corrosion of the tanks, U.S. Army Engineers raised the tanks in October and November of 1962 while National Guardsmen stood ready to evacuate 80,000 persons in Mississippi and Louisiana from a 2000-km² area that President Kennedy had declared a major disaster area. On June 21, 1970, a dozen railroad tank cars containing liquid propane caught fire and exploded at Crescent City, Ill., destroying the downtown area of the small town and 15 individual homes—$2 million in property damage. On July 28, 1970, chlorine escaping from a chemical warehouse during a fire caused the hospitalization of a dozen firemen and policemen and led to the evacuation of 1200 persons from their homes. An explosion in the harbor at Texas City, Tex., in 1947 killed 561 persons. On March 13, 1968, the release of nerve gas at the Army's Dugway Proving Ground in Utah led to the deaths of over 6000 sheep.

It has been pointed out by Chauncey Starr [25] that American society appears to be willing to tolerate fatality rates (defined as fatalities per person-hour of exposure to risk) from various activities (auto travel, airplane travel, hunting, smoking, etc.) so long as they are comparable to the fatality rate from natural disease, which is about 10^{-6}. Greater risks are tolerated for activities providing greater benefits (defined in terms of the money people are willing to spend on them). Of particular interest is the fact that Americans are apparently willing to tolerate fatality rates for voluntary activities (hunting, skiing, smoking, general aviation) that are roughly 1000 times greater than those for activities regarded as involuntary (such as the use of electricity). This may explain the curious insensitivity of the majority of Americans to the risks of smoking and of automobile travel. The risks associated with air and water pollution can be classified as involuntary since there is no longer much that an individual can do to escape them.

If Starr's arguments are valid, society would not tolerate risks from air, water, and other pollution comparable to those from cigarette smoking or automobile usage. This would be true, at least, if society were properly informed of the risks of environmental pollution. The remaining chapters of this book detail many of these risks.

REFERENCES

1. Weinraub, Bernard, "30-Nation Consumer Unit Divided on Pollution Issue." *N.Y. Times,* July 3, 1970, p. 2.

2. National Academy of Sciences, *Technology: Processes of Assessment and Choice.* Washington, D.C.: U.S. House of Representatives, Committee on Science and Astronautics, July 1969.

3. Heuss, J. M., and W. A. Glasson, "Hydrocarbon Reactivity and Eye Irritation." *Environ. Sci. Tech.* **2:**1109–1116 (1968).

4. Risebrough, R. W., "Chlorinated Hydrocarbons in Marine Ecosystems." In Miller, M. W., and G. C. Berg (eds.), *Chemical Fallout.* Springfield, Ill.: C. C. Thomas, Pub., 1969. pp. 5–23.

5. Sladen, W. J. L., et al., "DDT Residues in Adelie Penguins and a Crabeater Seal from Antarctica." *Nature* **210:**670–673 (1966).

6. Kearney, P. C., et al., "Persistence of Pesticide Residues in Soils." In Miller,

M. W., and G. C. Berg (eds.), *Chemical Fallout*. Springfield, Ill.: C. C. Thomas, Pub., 1969. pp. 54–67.

7. Jernelov, A., "Conversion of Mercury Compounds." In Miller, M. W., and G. C. Berg (eds.), *Chemical Fallout*. Springfield, Ill.: C. C. Thomas, Pub., 1969. pp. 68–74.

8. Hickey, J. J., et al., "An Exploration of Pesticides in a Lake Michigan Ecosystem." *Jour. Appl. Ecol.* 3 (Suppl.):141–154 (1966).

9. Rechen, H. J. L., et al., "Cesium-137 Concentrations in Alaskans During the Spring of 1967." *Radiological Health Data and Reports* 9:705–717 (1968). This journal periodically reports data from Alaskan Cs^{137} monitoring studies.

10. Langham, Wright H., "Considerations of Biospheric Contamination by Radioactive Fallout." In Fowler, Eric B. (ed.), *Radioactive Fallout, Soils, Plants, Food, Man*. Amsterdam: Elsevier Pub. Co., 1965. pp. 3–18.

11. Hill, Gladwin, "1,000 Acres of Smog-Afflicted Pines to Get Ax." *N.Y. Times,* April 2, 1970, p. 15.

12. Woodwell, G. M., "Effects of Pollution on the Structure and Physiology of Ecosystems." *Science* 168:429–433 (1970).

13. Deichmann, W. B., and H. W. Gerards, *Toxicology of Drugs and Chemicals*. New York: Academic Press, 1969.

14. Patterson, C. C., "Contaminated and Natural Lead Environments of Man." *Arch. Environ. Health* 11:344–360 (1965). See also the letter about this article by R. A. Kehoe in *Arch. Environ. Health* 11:736–739 (1965).

15. Gleason, M. N., et al., *Clinical Toxicology of Commercial Products—Acute Poisoning,* 3rd ed. Baltimore: Williams & Wilkins Co., 1969.

16. Selikoff, I. J., "Asbestos." *Environment* 11(2):2–7 (March 1969).

17. Thomson, G., "Sulphur Dioxide Damage to Antiquities." *Atmospheric Environment* 3:687 (1969).

18. Applegate, H. G., and L. C. Durrant, "Synergistic Action of Ozone-Sulfur Dioxide on Peanuts." *Environ. Sci. Tech.* 3:759–760 (1969).

19. *Air Quality Criteria for Sulfur Oxides*. Washington, D.C.: U.S. Dept. of Health, Education, and Welfare, January 1969. N.A.P.C.A. Publication No. AP-50.

20. Wilbur, Charles G., *The Biological Aspects of Water Pollution*. Springfield, Ill.: C. C. Thomas, Pub., 1969.

21. Delwiche, C. C., "The Nitrogen Cycle." *Scientific American* 223(3):136–146 (September 1970).

22. Irukayama, K., "The Pollution of Minamata Bay and Minamata Disease." *Adv. Water Poll. Res.* 3:153–180 (1966). See also the photographs by W. Eugene Smith in *Life* 72(21):74–81 (June 2, 1972).

23. *Smoking and Health*. Report of the Advisory Committee to the Surgeon General of the Public Health Service. Washington, D.C.: U.S. Dept. of Health, Education, and Welfare, 1964. Public Health Service Publication No. 1103.

24. Ochsner, Alton, "The Health Menace of Tobacco." *American Scientist* 59: 246–252 (1971).

25. Starr, C., "Social Benefits versus Technological Risks." *Science* 165:1232–1238 (1969).

2

THE GROWTH OF POPULATION, PRODUCTION, AND CONSUMPTION

A chosen country, with room enough for all descendants to the hundredth and thousandth generation. —Thomas Jefferson, *First Inaugural Address,* March 4, 1801.

The founding fathers of the United States of America apparently felt that our continent could not be crowded for a very long time. In 1751, in *Observations Concerning the Increase of Mankind, Peopling of Countries, etc.,* Benjamin Franklin had written, "So vast is the territory of North America that it will require many ages to settle it fully." As the United States approaches its two-hundredth anniversary, however, there is a growing sentiment that we are already too crowded and many Americans are working for "zero population growth." Paul Ehrlich has written that "Too many cars, too many factories, too much detergent, too much pesticide, multiplying contrails, inadequate sewage treatment plants, too little water, too much carbon dioxide—all can be traced easily to *too many people*" [1].

In this chapter we shall look at world and United States population and living standards in order to demonstrate their effects on the growth of environmental problems and on the consumption of natural resources.

POPULATION GROWTH

World population growth during the past millennium has been startling. Although exact figures do not exist even for today, and population figures for several centuries ago in some countries may only be accurate to within a factor of two, it is clear that world population growth has been accelerating. Figure 2-1 plots world population data from the United Nations [2] in three different ways. Figure 2-1(a) plots the population on an ordinary arithmetic graph, on which a straight line would represent a constant annual increase; the dotted line shows how the world's population would have increased if it had continued to grow from 1800 at the rate of 5 million persons a year characteristic of the early 1800's. Figure 2-1(b) shows the population growth on a logarithmic graph, on which a straight line would indicate a constant percentage growth rate and thus larger and larger annual population increases;

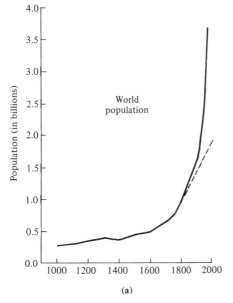

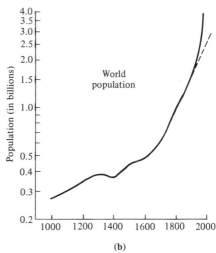

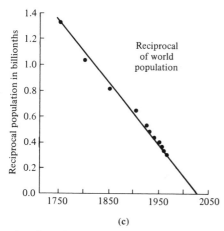

FIG. 2-1. World population plotted on three different scales. (a) Arithmetic scale; the dotted line corresponds to a constant annual increase of 5 million persons beginning in 1800. (b) Logarithmic scale; the dotted line corresponds to a constant annual growth rate of 0.5% beginning in 1800. (c) Reciprocal scale; the straight line is an approximate fit to the data that extrapolates to zero reciprocal population (infinite population) about 2026 A.D.

the dotted line plots a constant 0.5% annual increase beginning in 1800. In fact, world population is rising sharply even on a logarithmic scale, which means that the growth rate is increasing or that the time between successive doublings of the population is decreasing. Figure 2-1(c), which is included only for curiosity's sake, shows a plot of the reciprocal of world population; the plot approximately follows a straight line which has been drawn in for comparison and which predicts zero reciprocal population, or infinite population, about a quarter of the way into the 21st century. The fact that world population follows a curve of this sort was pointed out in 1960 by von Foerster, et al. [3], who pinpointed Doomsday as Friday, November 13, 2026 A.D.

This dramatic growth has coincided with the development of cheap

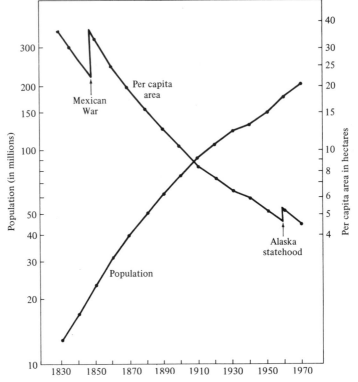

FIG. 2-2. Growth of U.S. population and decline of U.S. per capita land area plotted on a logarithmic scale.

sources of energy (coal, hydroelectric power, petroleum, natural gas, and nuclear energy) and with the industrial and scientific revolutions. The growth has not always been steady, however, as the dip in the 14th century shows. The bubonic plague pandemic of 1348 to 1350 killed over a fourth of Europe's population and other disastrous outbreaks occurred during the rest of the 14th century [4]. It is obvious that world population cannot continue forever the rapid growth experienced in the 20th century and in particular that there will be no Doomsday in 2026 A.D. It is not clear what will cause the leveling off of world population—nuclear holocaust, famine, a pandemic disease, widespread use of contraception, or other reasons. Human population has apparently made great surges in the past due to the cultural and agricultural revolutions but in each case it finally leveled off at new plateaus [5].

The growth of the U.S. population is shown on a logarithmic scale on Figure 2-2, which also indicates the trend of the per capita land area. The growth rate has moderated during the 20th century; its value dropped from 2.6% annually in the period from 1830 to 1900 to 1.4% annually in the period from 1900 to 1970. Nevertheless, the two largest *absolute* increases between decennial censuses were for 1950 to 1960 and 1960 to 1970. As the population has grown, the per capita land area has dropped steadily (except for temporary rises when the U.S. acquired land) until it is now only 4.5 hectares (about 11 acres). The breakdown of this per capita area is shown in Table 2-1.

TABLE 2-1. Approximate distribution of the per capita land area of the United States.

Land use	Area (hectares)
Grazing of livestock	1.6
Forest land	1.1
Deserts, swamps, mountains, etc.	0.8
Agricultural crops	0.7
Urban areas	0.1
Roads and airports	0.1
National and state parks and monuments	0.1
Total	4.5

For comparison, Table 2-2 lists the world's continents (with the U.S.S.R. treated separately), their population as estimated by the United Nations for 1970, and the per capita land area. The last quantity has only limited meaning because of the importance of the quality of the land. Nevertheless, it will be seen that Asia and Europe are nearly equally densely populated and both are much more densely populated than the United States.

TABLE 2-2. Population and per capita land area of the world's continents.

Continent	Population (millions)	Area per capita (hectares)
Oceania (including Australia)	19	45
South America	190	9.4
U.S.S.R.	243	9.2
Africa	344	8.8
North and Central America	321	7.6
Asia (except U.S.S.R.)	2056	1.3
Europe (except U.S.S.R.)	462	1.1
World	3635	3.7

Population growth clearly has a great deal to do with pollution growth and it causes a disproportionate negative impact on the environment [6]. Some pollution problems increase more or less in step with population growth, such as the amounts of human wastes. The *concentration* of population is a very important factor, however, since natural processes are capable of handling the wastes from a population of sufficiently low density, while higher densities put more and more of a strain on these processes. The trend toward urbanization in the United States and most of the other countries of the world is thus aggravating pollution problems, and indeed most of the problems discussed in the rest of this book are urban problems. Population concentration is also having an effect in rural areas, though, since the trend toward concentration of cattle in large feedlots is increasing the animal waste problem in the United States (see Chapter 11).

LIVING STANDARDS IN THE UNITED STATES

The rising standard of living in the United States has been accompanied by increased industrial production per capita, increased motor fuel consumption per capita, and similar per capita increases in other pollution-producing activities and is thus as important a component in the growth of pollution problems as the increase and concentration of population.

   This fact can be shown in a quantitative sense if we consider air pollution and the activities responsible for air pollution. Although there are many different types of air pollutants, the major ones are carbon monoxide (CO), particulate matter, sulfur oxides (SO_x), hydrocarbons, and nitrogen oxides (NO_x). Table 2-3 gives the total amounts of these pollutants emitted by various activities in the United States (a more detailed breakdown will be given in Table 3-2). The last two columns give the total percentage due to each activity using two different bases. The first is a mass basis, i.e., it simply treats 1 metric ton of each pollutant as equivalent to 1 metric ton of any other pollutant. On this basis, the transportation sector (largely automobiles) appears to be the chief source of air pollution, and it is because of this that one often hears comments to the effect that "the automobile is responsible for half the air pollution in the U.S." It is the emission of carbon monoxide that makes the motor vehicle show up so badly, but weight for weight, CO does not appear to be as harmful as any of the other major air pollutants. The percentage given in the last column of Table 2-3 is based on air quality criteria along the lines proposed by Babcock [8]. In constructing a combined air pollution index, Babcock has weighted the different types of air pollution in accordance with their estimated harmfulness. These weights have been used to construct the last column in the table and it should be recognized that they are no more than rough estimates that neglect such important considerations as synergistic effects. Babcock's weights regard the following as equally important in air pollution:

   40,000 parts (by weight) of carbon monoxide.
   19,300 parts of hydrocarbons.
   1430 parts of sulfur oxides.
   514 parts of nitrogen oxides.
   375 parts of particulate matter.

The notable feature of the quality percentages listed in Table 2-3 is that transportation ranks behind industry and electric power generation (and even

TABLE 2-3. Air pollutant emissions in the United States in 1968 in millions of metric tons (based on information in Reference 7).

Source	CO	Particulate Matter	SO_x	Hydrocarbons	NO_x	Percentage Mass Basis	Percentage Air Quality Basis
Transportation	58.1	1.1	0.7	15.1	7.3	42	15
Electric power generation	0.6	5.1	15.2	0.2	3.6	13	24
Industry	9.1	9.3	11.2	4.3	4.8	20	33
Solid waste disposal	7.1	1.0	0.1	1.5	0.5	5	3
Miscellaneous[a]	16.1	9.2	3.0	8.0	2.5	20	25
Total	91.0	25.7	30.2	29.1	18.7	100	100

[a] Residential heating, agricultural burning, forest fires, etc.

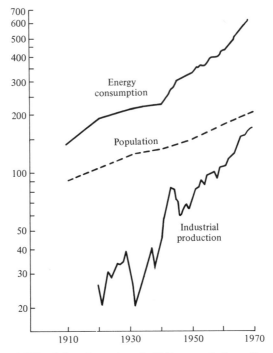

FIG. 2-3. Growth of U.S. population (in millions), industrial production (1957 to 1959 = 100), and total energy consumption (in units of 10^{17} joules) plotted on a logarithmic scale.

miscellaneous sources) as a source of air pollution when the United States as a whole is considered.

Table 2-3 suggests that in order to estimate the rate at which air pollution problems are growing in the United States it is necessary to know something about the growth of industrial activity, electric power generation, motor fuel consumption, etc., and not just the growth of population. The actual rate at which air pollution emissions are increasing will also depend on the rate at which pollution control equipment is installed, but we can ask about the rate at which the *potential* air pollution problems are increasing.

Figures 2-3, 2-4, and 2-5 plot on logarithmic scales the growth of U.S. population and the growth of certain characteristics of the American standard of living, to show that the growth rates of these characteristics are all exceeding that of the population.

Figure 2-3 plots the population according to the decennial censuses, the industrial production according to the Federal Reserve Board's index of industrial production [9], and total U.S. energy consumption according to United Nations data [10]. The energy consumption figures include the combustion of coal, petroleum, and natural gas and the generation of hydroelectricity and nuclear power, neglecting only the small contribution from the burning of wood. The data of Figure 2-3 show that industrial production and energy consumption have grown much faster than the population in recent decades, although their growth has been rather erratic, being less during depressions and recessions than during periods of rapid economic growth.

Figure 2-4 compares the growth of electric power production with the growth of the population, using the electric power statistics of the Federal Power Commission [11]. During the 20th century, electric power production in the United States has roughly doubled every 10 years and the growth rate shows no sign of abatement. In recent years several electric utility companies have found themselves with insufficient generating capacity because they under-estimated the growth in demand for electricity. It is generally estimated that electric power production will increase about sevenfold between 1970 and 2000. The implications for air, thermal, and radiation pollution of our environment are discussed further in Chapters 14, 15, and 16.

Figure 2-5 compares the growth in population with two statistics relating to transportation: motor vehicle registrations [12], which are now well above the 100 million mark, and total motor fuel (gasoline) consumption [13], which averages about 30 m³ or 750 gal per vehicle per year. In recent years the growth rate of total motor vehicle mileage has corresponded closely to that of registrations and fuel consumption, with the average vehicle traveling about 16,000 km (10,000 mi) annually. One way to demonstrate the rapid growth of automotive transportation is to point out that in 1920 there

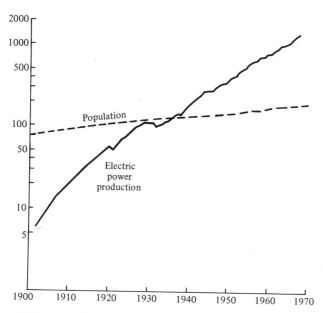

FIG. 2-4. Growth of U.S. population (in millions) and of electric power production (in billions of kilowatt-hours) plotted on a logarithmic scale.

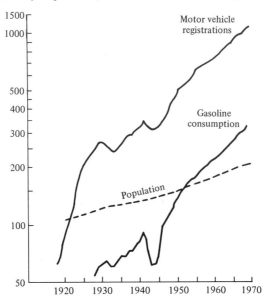

FIG. 2-5. Growth of U.S. population (in millions), motor fuel consumption (in millions of cubic meters), and motor vehicle registrations (in 100,000's) plotted on a logarithmic scale.

was one car for every 12 people; in 1950, about one car for every 3 people; and today, more than one car for every 2 people.

Table 2-4 summarizes this data and some others in numerical form, showing the annual rate of growth (compounded) for the 1950's and 1960's.

TABLE 2-4. Average annual rates of growth in the United States during the 1950's and 1960's.

Statistic	*Rate of Growth (%)*
Population	1.6
Energy consumption	3.2
Domestic crude oil demand	3.2
Industrial production	4.0
Motor vehicle registrations	4.1
Motor fuel consumption	4.5
Vehicle-miles traveled	4.5
Motor vehicles scrapped annually	5.0
Natural gas production	6.0
Electric power production	7.6
Bauxite consumption in aluminum production	7.7

A little exercise that can be carried out at this point is to assume air pollution growth rates of 4.5% (the motor fuel consumption rate) for transportation, 4.0% (the industrial production rate) for industry, 7.6% (the electric power production rate) for electric power generation, and 1.6% (the population rate) for other sources. With the help of Table 2-3, we would find that air pollution is then growing 4.1% annually on a mass basis or 4.3% annually on an air quality basis. Either way, it is clear that the air pollution problem is growing much faster than the population. (It is really the *potential* air pollution problem that we are trying to predict since the *actual* problem will depend on the extent to which control techniques are used.)

The lesson to be drawn from this example is the following: Although population control is one of the most urgent problems facing mankind today for a variety of reasons, it will not suffice to bring environmental problems to an end, and indeed our rising standard of living may be of equal or greater concern. It has been pointed out by Commoner [14] that much of the increase in pollution over and above that due to population growth is really due to the use of "faulty technology" (such as the trend to nonreturnable beverage bottles) rather than to an increase in per capita consumption.

TABLE 2-5. Annual per capita consumption or production figures for the U.S. as of the early 1970's.

Item	Amount consumed annually
Crude oil	3 m³ (800 gal)
Natural gas	3000 m³
Coal	2 metric tons
Energy in all forms	100,000 kWh
Steel	600 kg
Newsprint	40 kg
Rubber	11 kg
Tobacco	4 kg
Copper (new)	7 kg
Glass bottles	185
Nonreturnable beverage bottles	90
Packaging materials of all kinds	250 kg
Household, commercial, and municipal solid wastes	1.1 metric tons

Americans are also extraordinary consumers of natural resources. Table 2-5 lists some annual per capita consumption figures for the U.S. in the early 1970's, obtained using standard sources [9–13, 15]. These large numbers are averages over all members of the population—young and old, rich and poor—and if the reader is an adult of above average standard of living, his or her share (directly and indirectly) is likely to be much larger than this.

LIVING STANDARDS IN THE REST OF THE WORLD

Living standards in other countries are generally lower than those in the United States, at least when they are defined in terms of material goods. Comparisons of living standards between countries are very difficult to make since the relative costs of different items will not be the same in each country and since exchange rates of the countries' currencies depend on world trade rather than purchasing power.

Some information about living standards can nevertheless be gleaned from consumption statistics. One quantity of very general importance in this regard is energy consumption. The main sources of energy in the world today are solid fossil fuels (coal, lignite, etc.), liquid fossil fuels (crude and refined petroleum and natural gas liquids), gaseous fuels (natural and manufactured gas), hydroelectricity (electricity generated by water power), and nuclear power. Wood is a fuel of declining importance throughout the world and food is not generally included in computations of total energy consumption although food provides the biological energy requirements of man.

Figure 2-6 shows the per capita energy consumption for different parts of the world in megawatt-hours per year during the last 2 decades. The data are derived from United Nations statistics [10] and include the main sources of energy listed above. The U.S. average is about six times the world average and about eight times the average for the rest of the world. The averages for Western Europe and for the U.S.S.R. are about twice the world average. The underdeveloped countries are far below the world average. Japan's rapid rise from a somewhat underdeveloped status to a highly industrialized status is reflected in its energy consumption, which has approximately quadrupled on a per capita basis during the past 20 years.

The continued rise in U.S. consumption has been accompanied by a growing reliance on foreign energy sources. Until 1952 the U.S. was a net *exporter* of energy but since then it has become a consistently growing net *importer* of energy and by 1969 it was actually importing 9% of its energy requirements (largely in the form of petroleum). This contrasts with the U.S.S.R., which became a net exporter of energy in the mid-1950's and is now exporting energy equivalent to about 14% of its consumption. Western Europe imports about 60% of its energy needs.

Although the United States is a net exporter of many raw materials, it depends on the rest of the world for many others. The developing countries furnish a large portion of the natural resources used by developed countries to maintain and raise their standard of living. This exploitation has been documented by Pierre Jalée [16] using United Nations statistics. In Jalée's classification, the "Third World" of underdeveloped nations includes America apart from the U.S., Canada, and Cuba; all Africa; Asia apart from Japan, Israel, and the communist countries (China, Mongolia, North Korea, and North Vietnam); and Oceania apart from Australia and New Zealand. The Third World so defined has 47% of the world's population and 51% of its land area but it consumes only 8.8% of the world's energy even though it produces 23% (e.g., it produces 52% of the world's crude oil) and produces 23% of the iron ore, 59% of the bauxite, 44% of the copper ore, 100% of

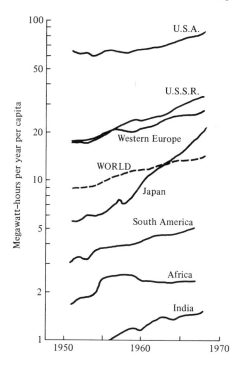

FIG. 2-6. Per capita annual energy consumption (in megawatt-hours) for different parts of the world during the 1950's and 1960's plotted on a logarithmic scale.

the natural rubber, even though its share of manufacturing is only 6.4% and its share of steel production only 4%. (These figures refer to the mid-1960's.)

Figure 2-7 shows for recent years the percentages of certain resources and commodities that are consumed by Americans and the ratio of the American per capita consumption to that of the rest of the world. Although the United States contains only about one-sixteenth (6.25%) of the world's population, it typically accounts for about 25 to 40% of the world's consumption of key resources. For cotton, the fraction is only one-sixth but cotton is not such a key good in an affluent society, especially in an age of synthetic fabrics. For natural gas, on the other hand, the fraction is about three-fifths today, and it is this high because natural gas is a relatively sophisticated (and also nonpolluting) source of energy. The fractions for phosphate, nitrogenous, and potash fertilizers (not shown on the figure) have all been close to 25% for many years.

In terms of per capita consumption, Americans consume 5 to 10 times as much as the rest of the world, even when all the other developed countries (Western Europe, U.S.S.R., etc.) are included.

In the post-World War II years, U.S. consumption relative to world consumption has generally decreased, although in some cases there was stabilization during the 1960's. The historical trends evident from Figure 2-7 are interesting. The fractions for new copper consumption were low during the depression years of the 1930's but rose sharply with the beginning of World

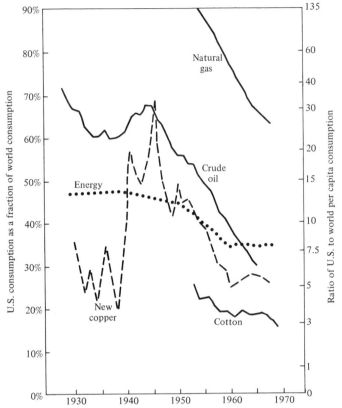

FIG. 2-7. Comparison of U.S. and world consumption of natural gas [17], crude oil [13], new copper (omitting reuse of scrap) [18, 19], cotton [20], and total energy [10, 20]. The right-hand scale indicates the ratio of per capita consumptions in the U.S. and the rest of the world assuming that the U.S. population was exactly one-sixteenth (6.25%) of the world population during the entire period.

War II and then decreased in the postwar years. Crude oil data show a decreasing trend interrupted by a wartime rise. Natural gas consumption— largely a post-World War II phenomenon—shows a steady decrease as other countries develop their natural gas resources.

NATURAL RESOURCE CONSUMPTION

A perennial question facing mankind has been, Are we running out of natural resources? This is a question that cannot be answered in any accurate fashion but there are some danger signals evident in the statistics. Over the years, new discoveries of natural resources have been made continuously throughout most of the world, increasing known or estimated reserves. In addition, technological improvements have made possible the recovery of resources from poorer reserves.

The question of natural resource depletion is generally asked with respect to nonrenewable resources, i.e., those that are not being replenished except perhaps on very long geological time scales. Minerals mined from the ground are nonrenewable resources, as are fossil fuels used for energy and chemical processes. Renewable resources are those that are constantly being replenished, such as fish in lakes and the oceans or timber in the forests. Unfortunately, commercial exploitation of renewable resources has often interfered with the natural processes of animal reproduction or reforestation and brought about the depletion or destruction of these resources. A fishing industry which lives off the aquatic "income" may continue indefinitely but one which grows too large and begins living off its "capital" may come to an end, as has the blue whaling industry. Such exhaustion of the natural wealth has hurt the seal and whaling industries in the past and biological conservation is being recognized today as an important discipline for the survival of the human species [21, 22].

With respect to nonrenewable resources, the data indicate that the United States has enormous long-lasting reserves of some materials (such as molybdenum and magnesium) but has to depend on foreign sources for other materials (such as nickel and tin). In some cases U.S. reserves are ample for a number of decades, but apparently not for centuries. Copper is such a material. Although the U.S. has been able to supply the majority of its copper needs from domestic mines over the years, the average tenor (metal content) of the copper ores mined has decreased from 1.7% in the early 1920's to 1.0% during World War II to 0.6% in the late 1960's [18, 19].

Energy resources are of particular importance because energy is vital to a technological society. In Hubbert's study of energy resources for the National Academy of Sciences' report *Resources and Man* [21], he estimated that fossil fuels (coal, petroleum, natural gas) will serve man's energy needs for only a few more centuries, forcing him to develop new energy resources such as nuclear energy. Petroleum and natural gas supplies will probably last for only a few more decades. On the basis of production and exploration data, Hubbert has estimated that the peaks of petroleum and natural gas production in the conterminous (lower 48) states will occur in approximately 1970 and 1980, respectively, with production gradually decreasing after these peaks.

Table 2-6 shows Hubbert's estimates of total original petroleum reserves (including petroleum produced to date or discovered, as well as that yet to be discovered), the current production, and the years that the total original reserves would last at current production rates and the years that would remain when one subtracts past production. If production rates continue to rise, the years of remaining production will be even less than those indicated in the last row, unless Hubbert's estimates are too low. The U.S. figures include estimated reserves of 26 billion m³ (165 billion barrels) for

TABLE 2-6. Total original petroleum reserves (estimated by Hubbert [21]), current annual production in billions of cubic meters, and years of total and remaining production at current production rates.

	U.S.	World
Total original petroleum reserves	30	215–320
Current annual production	0.55	2.5
Years of total production at current rate	55	85–130
Years of remaining production at current rate	30	70–115

the conterminous United States and 4 billion m³ (25 billion barrels) for Alaska, the latter figure being rather speculative.

These data suggest that petroleum reserves cannot last beyond a few more decades. World supplies would last only 20 to 30 years if the world's per capita consumption matched that of the United States.

Perhaps these estimates are in error, and perhaps it will become possible to extract some more petroleum from new sources such as oil shale. That the United States is soon to encounter serious problems is evident, however, from Figure 2-8, which shows the historical R/P (reserve/production) ratios for crude oil and natural gas in the United States. R/P ratios indicate the years of remaining production of proved (i.e., actually discovered) reserves at current production rates. For crude oil a reserve/consumption ratio is also shown since a sizable portion of U.S. crude oil supplies are derived from foreign sources. Crude oil R/P ratios have been decreasing in recent years, although the new Alaskan discoveries may temporarily halt this trend. The natural gas R/P ratio has been dropping precipitously (even by more than 1 year in 1 year's time) and it even dropped in 1970 when Alaskan reserves were included for the first time. Since 1968 natural gas reserves in the conterminous 48 states have actually been showing year-to-year declines. The natural gas industry has blamed this on the fact that the pricing policies of the Federal Power Commission, which regulates the industry, are discouraging exploration. Nevertheless, our natural gas supplies may be quite finite. As of December 31, 1970, proved natural gas reserves amounted to 8.3 trillion m³ and the Potential Gas Committee's estimates [23] of "probable, possible, and speculative" reserves in the U.S. totaled another 33.4 trillion m³—which implies a maximum of 67 years' supply at 1970 production rates (0.62 trillion m³/year) *if* all these reserves really do exist and *if* the natural gas can be discovered and recovered in time. The short supply of natural gas in some parts of the country has already forced suppliers to turn down new customers and occasionally even interrupt deliveries to large industrial customers.

Fossil-fuel usage will never come to an abrupt end, of course, but will decline slowly accompanied by increasing prices. Coal supplies will apparently last for several centuries, and it is likely that satisfactory methods will be developed for converting coal into liquid and gaseous fuels so that we never return to the days of extensive coal use. Other energy sources may also

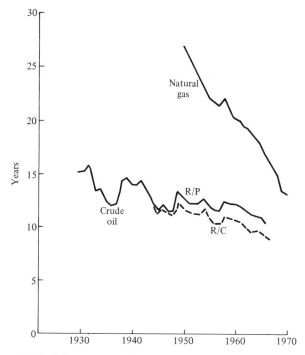

FIG. 2-8. Years of remaining production of natural gas and crude oil reserves in the U.S., expressed as reserve/production (R/P) ratios. Also shown is the reserve/consumption ratio for crude oil. Recent Alaskan discoveries of natural gas were first included in the reserves in 1970.

be developed. If controlled nuclear fusion (see Chapter 16) ever becomes feasible, the deuterium atoms in the world's oceans are theoretically capable of providing millions of times as much energy as the world's total supply of all fossil fuels.

THE LIMITS TO GROWTH

Americans have come to regard continued population and economic growth as part of the natural scheme of things. It is clear, however, that rapid growth in population and industrial production cannot continue forever on a finite earth—sooner or later we shall encounter limits to growth [24]. It is uncertain whether the controlling limit will be food production, natural resources, environmental pollution, or something else, although it is difficult to visualize worldwide growth at present rates through the 21st century [24]. The question of the limits to growth is an important one, and the idea that unchecked growth is not desirable is gaining in respectability. In 1972 the Commission on Population Growth and the American Future concluded [25]:

We have examined the effects that future growth alternatives are likely to have on our economy, society, government, resources, and environment, and we have found no convincing argument for continued national population growth. On the contrary, the plusses seem to be on the side of slowing growth and eventually stopping it altogether.

REFERENCES

1. Ehrlich, Paul, *The Population Bomb*. New York: Ballantine Books, 1968.
2. See the United Nations *Demographic Yearbook* (published annually) and the proceedings of the United Nations conferences on world population.
3. Von Foerster, H., et al., "Doomsday: Friday, 13 November, A.D. 2026." *Science* **132**:1291–1295 (1960).
4. Langer, William L., "The Black Death." *Scientific American* **210**(2):114–121 (February 1964).
5. Deevey, E. S., "The Human Population." *Scientific American* **203**(3):194–204 (September 1960).
6. Ehrlich, Paul R., and John P. Holdren, "Impact of Population Growth." *Science* **171**:1212–1217 (1971).
7. *Nationwide Inventory of Air Pollutant Emissions 1968*. Washington, D.C.: U.S. Dept. of Health, Education, and Welfare, August 1970. N.A.P.C.A. Publication No. AP-73.
8. Babcock, L. R., Jr., "A Combined Pollution Index for Measurement of Total Air Pollution." *Jour. Air Poll. Control Assn.* **20**:653–659 (1970).
9. *Statistical Abstract of the United States*. Washington, D.C.: U.S. Govt. Printing Office. Published annually.
10. *World Energy Supplies*. New York: United Nations Dept. of Economic and Social Affairs. Statistical Papers, Series J. Published annually.
11. *Electric Power Statistics*. Washington, D.C.: U.S. Federal Power Commission. Published monthly. Annual data may be found in Reference 9.
12. *Automobile Facts and Figures*. Published annually by the Automobile Manufacturers Assn., Inc., 320 New Center Building, Detroit, Mich. 48202.
13. *Petroleum Facts and Figures*. Published annually by the American Petroleum Institute, 1271 Avenue of the Americas, New York, N.Y. 10020.
14. Commoner's thesis is expounded in the following references: Commoner, Barry, Michael Corr, and Paul J. Stamler, "The Causes of Pollution." *Environment* **13**(3):2–19 (April 1971). Commoner, Barry, *The Closing Circle*. New York: Alfred A. Knopf, 1971. Commoner, Barry, "The Environmental Cost of Economic Growth." *Chemistry in Britain* **8**(2): 52–65 (February 1972). A debate in which Paul R. Ehrlich and John P. Holdren criticize Commoner's thesis and Commoner replies was published in the April 1972 issue of *Environment* and the May 1972 issue of *Bulletin of the Atomic Scientists*. Further comments by Ehrlich and Holdren and by Garrett Hardin are in the June 1972 issue of *Bulletin of the Atomic Scientists*.
15. *Glass Containers*. Published annually by the Glass Container Manufacturers Institute, Public Affairs Dept., 330 Madison Avenue, New York, N.Y. 10017.
16. Jalée, Pierre, *The Third World in World Economy*. New York: Monthly Review Press, 1969.

17. *Gas Facts.* Published annually by the American Gas Assn., 605 Third Avenue, New York, N.Y. 10016.
18. *Mineral Facts and Problems.* Published every 5 years (latest in 1970) by the U.S. Bureau of Mines.
19. *Minerals Yearbook.* Published annually by the U.S. Bureau of Mines.
20. *Statistical Yearbook.* Published annually by the United Nations Statistical Office.
21. National Academy of Sciences—National Research Council, Committee on Resources and Man, Division of Earth Sciences, *Resources and Man.* San Francisco: W. H. Freeman and Co., 1969.
22. Ehrenfeld, David W., *Biological Conservation.* New York: Holt, Rinehart and Winston, Inc., 1970.
23. *Potential Supply of Natural Gas in the United States as of December 31, 1970.* Golden, Colo.: Potential Gas Agency, Mineral Resources Institute, Colorado School of Mines Foundation, Inc., 1971.
24. Meadows, Donella H., et al., *The Limits to Growth.* New York: Universe Books, 1972.
25. *Population and the American Future.* The Report of the Commission on Population Growth and the American Future. New York: The New American Library, Inc., 1972.

3

AIR POLLUTION: INTRODUCTION

This most excellent canopy, the air, look you, this brave o'erhanging
firmament, this majestic roof fretted with golden fire, why, it appears
no other thing to me than a foul and pestilent congregation of vapours.
—William Shakespeare, *Hamlet,* Act II, Scene 2 (c. 1600 A.D.)

Air pollution is not a new problem; it has been around for centuries. Over
three centuries ago, the noted scientist and diarist John Evelyn wrote a small
tract, *Fumifugium: or, The Inconvenience of the Aer, and Smoake of London
Dissipated* [1], in which he complained about London that

> When in all other places the *Aer* is most Serene and Pure, it is here
> Eclipsed with such a Cloud of Sulphure, as the Sun itself, which gives day
> to all the World besides, is hardly able to penetrate and impart it here;
> and the weary *Traveller,* at many Miles distance, sooner smells, than sees
> the City to which he repairs.

Evelyn also described with great accuracy many of the effects of the air pollution arising from the combustion of coal: reduction in sunshine, morbidity
and mortality from respiratory ailments, dust fall, corrosion of materials.
Only in the 20th century, and especially the last few decades, have extensive
experimental and epidemiological studies been carried out to verify these
effects scientifically. In this chapter we shall consider the composition of the
atmosphere and the properties and effects of the major types of air pollution.
The next three chapters will consider these pollutants in more detail.

THE ATMOSPHERE AND ITS CONSTITUENTS

The approximate total mass of the atmosphere may be computed from the
normal atmospheric pressure $p = 1.013 \times 10^5$ N/m², the earth's mean radius
$r = 6.37 \times 10^6$ m, and the gravitational acceleration $g = 9.8$ m/s² as

$$4\pi r^2 p/g = 5.3 \times 10^{18} \text{ kg}$$

This overestimates the total mass by about 3% since the continents, which
comprise 29% of the surface area of the earth, rise above sea level by an
average of approximately 1 km, so that a volume that we have assumed to

be occupied by the atmosphere is actually occupied by land [2]. The total mass of the atmosphere is thus about 5.1×10^{18} kg.

That part of the atmosphere nearest the earth's surface, extending up to a height of about 80 km, is referred to as the homosphere because the constituent gases are very well mixed and the proportions of the principal constituents do not differ very much from those found at sea level; the homosphere contains about 99.999% of the mass of the atmosphere. The typical composition of unpolluted dry air is given in Table 3-1, together with the total masses of the individual gases in the atmosphere. The composition listed is by volume (or numbers of molecules); mass percentages can be determined by multiplying the volume percentages by the molecular weight of the gas molecule and dividing by 29, the average molecular weight of air. The volume percentages of gases present in trace amounts (under 10 ppm) are not accurately known; the SO_2 concentration has been estimated at from 0.0002 to 0.002 ppm, for example. It will be noted that the mass of the atmosphere is so great that the total mass of even a trace gas is quite large.

TABLE 3-1. The composition of unpolluted dry air and the approximate total masses of the different constituents of the atmosphere. Many trace gases are not listed.

Constituent	Molecular Formula	Volume Fraction	Total mass (millions of metric tons)
Nitrogen	N_2	78.09%	3,850,000,000
Oxygen	O_2	20.94%	1,180,000,000
Argon	Ar	0.93%	65,000,000
Carbon dioxide	CO_2	0.032%	2,500,000
Neon	Ne	18 ppm	64,000
Helium	He	5.2 ppm	3,700
Methane	CH_4	1.3 ppm	3,700
Krypton	Kr	1 ppm	15,000
Hydrogen	H_2	0.5 ppm	180
Nitrous oxide	N_2O	0.25 ppm	1,900
Carbon monoxide	CO	0.1 ppm	500
Ozone	O_3	0.02 ppm	200
Sulfur dioxide	SO_2	0.001 ppm	11
Nitrogen dioxide	NO_2	0.001 ppm	8

SOURCES AND EMISSIONS OF AIR POLLUTANTS

The estimated nationwide emissions of the five primary air pollutants—carbon monoxide (CO), particulate matter, sulfur oxides (SO_x) expressed as SO_2, hydrocarbons (HC), and nitrogen oxides (NO_x, mainly NO and NO_2) expressed as NO_2—are shown in Table 3-2, with a breakdown according to source. These are the estimates for 1968 published by the National Air Pol-

Table 3-2. 1968 U.S. emissions (in millions of metric tons) of major air pollutants [3]. CO = carbon monoxide; P.M. = particulate matter; SO_x = sulfur oxides expressed as SO_2; HC = hydrocarbons; NO_x = nitrogen oxides expressed as NO_2.

Source	CO	P.M.	SO_x	HC	NO_x
Motor vehicles (gasoline)	53.5	0.5	0.2	13.8	6.0
Motor vehicles (diesel)	0.2	0.3	0.1	0.4	0.5
Aircraft	2.4	0.0	0.0	0.3	0.0
Railroads and others	2.0	0.4	0.5	0.6	0.8
Total transportation	**58.1**	**1.1**	**0.7**	**15.1**	**7.3**
Coal	0.7	7.4	18.3	0.2	3.6
Fuel oil	0.1	0.3	3.9	0.1	0.9
Natural gas	0.0	0.2	0.0	0.0	4.1[a]
Wood	0.9	0.2	0.0	0.4	0.2
Total fuel combustion	**1.7**	**8.1**	**22.2**	**0.6**	**9.1**[b]
Industrial processes	**8.8**	**6.8**	**6.6**	**4.2**	**0.2**
Solid waste disposal	**7.1**	**1.0**	**0.1**	**1.5**	**0.5**
Forest fires	6.5	6.1	0.0	2.0	1.1
Agricultural burning	7.5	2.2	0.0	1.5	0.3
Coal refuse burning	1.1	0.4	0.5	0.2	0.2
Structural fires	0.2	0.1	0.0	0.1	0.0
Total miscellaneous	**15.3**	**8.7**	**0.5**	**7.7**[c]	**1.5**
Total	**91.0**	**25.7**	**30.2**	**29.1**	**18.7**

[a] Includes gas transmission and gas pipelines.
[b] Includes kerosine and liquefied petroleum gas.
[c] Includes 3.9 million metric tons organic solvent evaporation and gasoline marketing.

lution Control Administration [3]. They are based on studies of a great many different sources but do not include some industries and other sources for which emission factors are not known. Some other sources are discussed later in this chapter.

It will be seen by comparison with Table 3-1 that the annual emissions in the United States are often significant in comparison with total amounts in the atmosphere. For example, U.S. emissions of SO_2 annually total about 2.5 times the total amount of SO_2 in the global atmosphere at any given moment. NO_x emissions are also much larger than the NO_2 mass in the atmosphere, and CO emissions are about 20% of the CO mass. When one takes into account the fact that total worldwide emissions are probably several times larger than U.S. emissions since world energy consumption and industrial production are several times the U.S. levels, it becomes apparent that man is a major contributor to global pollution. The extent of man's contribution will be estimated in the discussions of individual pollutants later in this chapter. Even more significant is the fact that in the concentrated urban areas in which much of mankind lives, all the pollutants are due largely to man's activities.

Man-made air pollution in urban areas is often referred to as "smog." The word "smog" was apparently coined about 1905 by Dr. H. A. Des Voeux, an active organizer of British smoke abatement societies, to describe the "smoke-fog" of the London pea-soupers. Today, the London-type smog is often referred to as "classical" smog, while the Los Angeles-type smog, which is quite different, is referred to as "photochemical" smog because it is formed through chemical reactions involving sunlight. Table 3-3 lists some of the characteristics of the two types of smog [4].

TABLE 3-3. Characteristics of "classical" and "photochemical" smogs.

Characteristics	"Classical"	"Photochemical"
First occurrence noted	London	Los Angeles
Principal pollutants	Sulfur oxides, particulate matter	Ozone, nitrogen oxides, hydrocarbons, carbon monoxide, free radicals
Principal sources	Industrial and household fuel combustion (coal, petroleum)	Motor vehicle fuel combustion (petroleum)
Effects on humans	Lung and throat irritation	Eye irritation
Effects on compounds	Reducing	Oxidizing
Time of occurrence of worst episodes	Winter months (especially in early morning)	Around midday of summer months

The most serious smog episodes are those in which atmospheric conditions (see Chapter 4) permit a buildup of high pollutant levels, especially if the episode lasts for several days. London has been subjected to many of these over the years but the worst such episode occurred from December 5 to 9, 1952. Figure 3-1 shows the daily deaths registered in London between

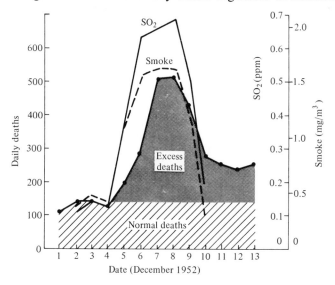

FIG. 3-1. Sulfur dioxide and smoke concentrations and daily deaths in London Administrative County before, during, and after the great smog of December 5 to 9, 1952 [5, 6].

the first and thirteenth of the month [5] and the daily average sulfur dioxide and smoke concentrations at a dozen stations between the second and the tenth of the month [6]. The very first day of the smog (December 5) was accompanied by a significant rise in mortality in both London Administrative County (for which the data are plotted) and the outer ring comprising the rest of greater London, and by the seventh and eighth the daily deaths were about four times the normal values. By December 10 the smog episode was over but excess mortality persisted for another 2 weeks. The period shown on the graph accounted for about 1800 excess deaths in London itself and about 2800 in all greater London, and it is estimated [5] that the episode produced a total excess mortality in greater London of about 4000.

Air pollution is clearly a grave danger to man if it can lead to such disastrous incidents. The various pollutants will now be considered one at a time in order to study their properties and their effects on man and his environment.

PARTICULATE MATTER

Solid and liquid aerosols suspended in the atmosphere are referred to as particulate matter. They arise either from condensation processes or from dispersion processes (erosion, grinding, spraying, etc.). Although "smoke" is popularly used to denote mixtures of particulate matter, fumes, gases, and mists, it properly refers to solid (or solid and liquid) condensation aerosols. "Dust" refers to solid dispersion aerosols and "mist" to liquid aerosols.

TABLE 3-4. Characteristics of atmospheric particles.

typical size	<0.1 μm	0.1 μm–1 μm	>1 μm
Name	Aitken particles	Large particles	Giant particles
Principal nature	Combustion aerosols	Combustion products and photochemical aerosols	Natural and industrial dust
Settling speeds[a]	Less than 8×10^{-7} m/s	Intermediate	Greater than 4×10^{-5} m/s

[a] Assuming spheres of density 1 g/cm³ and diameter as given in first row.

Table 3-4 lists some of the characteristics of these particles. Most of their mass in the atmosphere is accounted for by the "large" and "giant" particles. The giant particles arise mainly from dispersion processes, and combustion processes appear to generate most of the small particles.

Most particles are removed from the atmosphere by gravitational settling. According to Stokes' law [7] a sphere of radius R moving at speed v in a fluid of viscosity η will experience a resisting force $F = 6\pi\eta vR$. This resistance increases as the sphere's speed increases until it is exactly balanced by the gravitational force $mg = 4\pi\rho gR^3/3$ (where ρ is the density of the sphere), which means that the terminal (or maximum) settling speed is $v_0 = 2g\rho R^2/9\eta$. (Because of the buoyancy of the fluid, ρ should really be $\rho - \rho'$ where ρ' is the density of the fluid, but this is negligible when the fluid is a gas such as air.) For sufficiently small particles (size 1 μm or less) the speed exceeds that given by this formula because their motion is similar to random molecular motion, whereas for very large particles (size 100 μm or more) the speed is less than predicted by the formula because of the production of turbulence. The settling speeds are rather small, as shown in Table 3-4, since an intermediate value of 10^{-6} m/s would correspond to a descent of only 32 m (about 100 ft) in a year's time! In actual practice, of course, this figure is meaningless because the descent time of the particles will depend for the most part on the nature of the winds and the precipitation, but the point to remember is that they can remain airborne for a long time.

Since particulate matter is by definition nongaseous, concentrations in the atmosphere cannot be expressed in volume units, and the favored unit is the microgram per cubic meter (μg/m^3). The National Air Surveillance Network has been collecting samples of suspended particulate matter in the U.S. since 1957. During its first 10 years of operation, the geometric mean values in 60 center-city urban areas ranged from 38 μg/m^3 in Cheyenne, Wyo. to 185 μg/m^3 in Phoenix, Ariz. (other high values were 166 μg/m^3 in New York, N.Y., 180 μg/m^3 in East Chicago, Ind., 156 μg/m^3 in Pittsburgh, Pa., and 159 μg/m^3 in Philadelphia, Pa.), while the geometric mean values in 20 nonurban areas ranged from 10.4 μg/m^3 in White Pine County, Nev. to 57 μg/m^3 in Kent County, Del. [8]. Natural phenomena occasionally led to abnormally high values, such as 870 μg/m^3 over Ward County, N.D. in a dust storm in March 1963 and 487 μg/m^3 over Anchorage, Alaska in November 1958 due to a volcanic eruption. Nevertheless, urban areas consistently show particulate matter concentrations several times as high as nonurban areas. The data also shown a slight tendency for downtrends in urban areas over the years and uptrends in nonurban areas, perhaps due to the increasing "urban sprawl."

Atmospheric particles can scatter and absorb sunlight, thus reducing visibility (and perhaps affecting the global climate, as discussed in Chapter 4). In general, cities receive about 15 to 20% less solar radiation than rural areas and the reduction of sunlight can become as high as one-third in the summer and two-thirds in the winter [9]. The reduction in sunlight is strongly correlated with fuel combustion for industrial and household heating purposes. During the year 1932, when the depression had caused a sharp decrease in industrial activity, many cities showed higher sunlight levels than in the previous year. Particles also reduce visibility by attenuating the light from objects and illuminating the air, reducing the contrast between the objects and their background. The visual range is approximately inversely proportional to the concentration of particulate matter, with 100 μg/m^3 corresponding to a range of 12 km on the average (though sometimes it might be as much as 36 km or as little as 6 km since other factors, still unknown, are apparently important) [9]. Reduced visibility is aesthetically undesirable and it is also dangerous for aircraft and motor vehicles.

The effects of particulate matter on materials includes corrosion of metals when the air is humid; erosion and soiling of buildings, sculpture, and painted surfaces; and the soiling of clothing and draperies. Centuries of soot blackened London's buildings but many have been cleaned again in recent years, including St. Paul's Cathedral and the National Gallery, which sparkle with renewed beauty. A new problem produced by particulate matter is its corrosion and damage of electronic equipment, especially through chemical or mechanical action on electrical contacts. Many of these problems arise from particles that have settled out of the air and it used to be common to give dust-fall measurements in terms of the mass of particles settling on a given area each month or each year. In U.S. urban areas, dust fall typically ranges from 0.35 to 3.5 mg/cm^2-month (or 10 to 100 $tons/mi^2$-month [9]. Dust-fall values are not an accurate indication of corrosion or other type of damage, however, and are no longer commonly given.

The toxic effects of particulate matter on animals and humans can be classified as (1) intrinsic toxicity due to chemical or physical properties, (2) interference with clearance mechanisms in the respiratory tract, or (3) toxicity due to adsorbed toxic substances. Many toxic particles have been discovered in polluted urban atmospheres, including metal dusts, asbestos, and aromatic hydrocarbons such as the carcinogen 3, 4-benzpyrene; their concentrations are generally extremely small but they may play a role in the higher cancer rates that occur in urban areas as compared to rural areas even after correcting for the greater amount of smoking that occurs in urban areas.

Many studies have been carried out showing increased mortality and illness accompanying higher levels of particulate matter (see Chapter 11 of Reference 9). Respiratory illnesses, especially from chronic diseases such as bronchitis and emphysema, show the most pronounced association with levels of particulate matter and adverse health effects have been noted for annual geometric mean levels of as little as 80 $\mu g/m^3$. In these studies, higher particulate matter concentrations are often associated with higher sulfur dioxide levels, however, and it is not easy to separate the effects due to the two pollutants.

SULFUR OXIDES

The most important sulfur oxide emitted by pollution sources is sulfur dioxide (SO_2), although some sulfur trioxide (SO_3) is also generally produced, in amounts no more than a few percent of the SO_2 produced. SO_2 is a colorless, nonflammable gas which has an acrid taste at concentrations less than 1 ppm of air and which has a pungent, irritating odor at concentrations above about

3 ppm. The unpleasantness of this odor was noted over 2300 years ago by Theophrastus:

> Some of those [stones] that can be broken are like hot coals when they burn . . . for when they are covered with charcoal they burn as long as air is blown onto them, then they die down and afterwards can be kindled again, so that they can be used for a long time, but their odor is very harsh and disagreeable [10].

SO_2 is readily oxidized to SO_3 in the atmosphere by photochemical or catalytic processes, and in the presence of moisture the SO_3 becomes sulfuric acid or a sulfate salt and soon precipitates out of the atmosphere. The SO_2 in the atmosphere lasts only a few days at most and this is the reason that the SO_2 mass in the atmosphere (see Table 3-1) is so small compared to annual emissions by man (see Table 3-2).

Actually only about one-third of the sulfur oxides in the atmosphere are believed to be produced by man's activities. Robinson and Robbins [11] have estimated that sulfur oxides from man's activities introduce 66 million metric tons of sulfur (or 132 million metric tons of sulfur dioxide) into the atmosphere annually, largely from coal and petroleum combustion, as shown in Table 3-5. Natural sulfur sources are biologically produced hydrogen sulfide (H_2S) (arising from decay of organic matter) that is eventually oxidized to sulfur oxides and sulfates from sea spray. About 69% of the total sulfur and 93% of the man-made sulfur arise in the northern hemisphere.

TABLE 3-5. Gaseous sulfur pollutants from urban and natural sources, in millions of metric tons of sulfur annually [11].

Source	Amount of sulfur
Coal combustion	46
Petroleum combustion and refining	13
Copper smelting	6
Lead and zinc smelting	1.3
Total man-made	**66**
Biological H_2S from land	62
Biological H_2S from seas	27
Sulfates in sea spray	40
Total natural	**129**
Total	**195**

During the mid-1960's, data obtained by the Continuous Air Monitoring Project in the United States showed mean annual concentrations of SO_2 ranging from 0.01 ppm in San Francisco to 0.18 ppm in Chicago [12]. Individual SO_2 concentrations exhibited a log-normal frequency distribution (i.e., the logarithms of the SO_2 concentrations exhibited a normal, or Gaussian, distribution) and some very high values were encountered for short periods of time. The data showed that the 1-h maximum values for the year were about 10 to 20 times the annual average and 1-day maximum values for the year were about 4 to 7 times the annual average. Thus a city with an annual average of 0.10 ppm can be expected to have up to 0.4 to 0.7 ppm for the worst day of the year and 1 to 2 ppm for the worst hour of the year. Much higher concentrations can occur near single point sources of SO_2, such as coal-fired power plants, where levels of several ppm are not uncommon.

Sulfur oxides can damage materials and property, mainly through their conversion into the highly reactive sulfuric acid. Discoloration and physical deterioration are produced in building materials (limestone, marble, roofing slate, and mortar) and sculpture. The corrosion of most metals, especially iron, steel, and zinc, is accelerated by atmospheres polluted by SO_2; particulate matter, humidity, and elevated temperatures play important synergistic roles. Deterioration and fading are also produced in fabrics (such as cotton, nylon, and rayon), leather, and paper [12]. The drying time, brittleness, gloss, and even color of paints can also be affected.

SO_2 has been found to affect vegetation adversely even at concentrations below 0.03 ppm. High concentrations over short periods of time can produce acute leaf injury, such as necrotic (tissue-destroying) blotching of broad-leaved plants and grasses or brownish discoloration in the tips of pine needles. Lower concentrations over longer periods (days or weeks) leads to chronic leaf injury, such as a gradual yellowing (chlorosis) as chlorophyll production is impeded [12].

SO_2 and H_2SO_4 are both capable of irritating the respiratory system of animals and men. The levels needed to produce pathological lung change or mortality in animals are much greater than the levels encountered in urban atmospheres but the latter are capable of producing adverse health effects. In London, a rise in the daily death rate of 20% or more has been detected for SO_2 concentrations of 0.5 ppm lasting for a full day (see Figure 3-1). Daily mean concentrations of 0.2 ppm lasting for 3 or 4 days has led to increased mortality in Rotterdam, and levels of 0.1 ppm are believed to be responsible for the same result; the Rotterdam results are especially significant because particulate matter levels are low and the synergistic action of SO_2 and particulate matter does not play a role. Increased mortality is always accompanied by increased morbidity (illness) and SO_2 levels below 0.25 ppm have been associated with increased morbidity in New York as measured by hospital admissions. In all cases in which adverse health effects have been noted, the elderly and patients with heart or lung diseases have been affected most severely. An extensive review of the epidemiology of sulfur oxides may be found in Chapter 9 of Reference 12.

CARBON MONOXIDE

Carbon monoxide (CO) is a colorless, tasteless, odorless gas against which man cannot easily protect himself. It originates from the incomplete combustion of carbonaceous materials and is the air pollutant emitted in the largest quantities (except to the extent, discussed in Chapter 4, that carbon dioxide may be considered a pollutant).

Surprisingly little is known accurately about the sources and sinks of CO in the atmosphere; these are currently being actively studied. Man's activities are producing perhaps 250 million metric tons annually and there exist some biological sources of CO, although not much is known about them. It has been found, for example, that the oceans are a natural source of CO, although they probably produce only 10 million metric tons annually, and that considerable CO is present in rainwater [13]. The average concentration of CO in the atmosphere is not well-known either and the value of 0.1 ppm quoted in Table 3-1 is only an estimate, which might be several times too large or too small. Until recently it was supposed that there were no significant natural sources of CO and that the average residence time of CO in the atmosphere was approximately 2 years [14]. Radiocarbon measurements by Weinstock [15] suggested however that the CO residence time might be only a little more than 0.1 year. This suggestion has been supported [16] by recent studies at the Argonne National Laboratory, which estimate that the Northern Hemisphere alone produces well over 3 million metric tons of CO annually, probably largely from oxidation of methane in the troposphere. Urban concentrations of CO, which are often many ppm, clearly result principally from man-made sources.

CO can be oxidized to carbon dioxide (CO_2), but the rate at which this occurs seems to be very slow and mixtures of CO and O_2 exposed to sunlight for several years have remained almost unchanged. Since the residence time of CO in the atmosphere is at most only a few months, some removal process must exist. Perhaps the CO is adsorbed and oxidized on surfaces; perhaps it is removed and utilized by plants or animals; or perhaps photochemical or catalytic processes are involved in its removal [14]. Recent research indicates that soils are capable of removing large amounts of CO from the atmosphere, probably due to the activity of soil microorganisms [17].

The toxic effects of CO on human beings and animals arise from its reversible combination with hemoglobin (Hb) in the blood:

$$HbO_2 + CO \rightleftharpoons HbCO + O_2$$

Hemoglobin has a much greater affinity for CO than it does for O_2, and when O_2 and CO are present in sufficient quantities to saturate the hemoglobin, the concentrations of HbO_2 (oxyhemoglobin) and HbCO (carboxyhemoglobin) are related by the Haldane equation [18]

$$\frac{[HbCO]}{[HbO_2]} = M \frac{p(CO)}{p(O_2)}$$

where $p(CO)$ and $p(O_2)$ are the partial pressures (or volume concentrations) of the CO and O_2 gases and M is a constant that depends on the species. For man, M is about 200 to 300, while in rabbits it is less than half this value.

Since the ordinary air contains about 21% O_2, the ratio of HbCO to HbO_2 in man is approximately equal to 1/1000 of the CO concentration expressed in ppm. It generally takes a few hours to reach this equilibrium but the time is decreased by increased respiration rates [19].

The combination of hemoglobin with CO lessens the oxygen-carrying capacity of the blood so that less O_2 is available to the body cells. It also reduces the dissociation of oxyhemoglobin (HbO_2) into hemoglobin and oxygen so that anoxia (oxygen starvation) may result even though the blood is carrying several times as much O_2 as the body requires. It is thought that CO may also impair cell functioning by blocking oxidation in other ways as well.

The approximate HbCO levels (compared to total HbCO $+$ HbO_2) at which different symptoms occur are [15, 18]

> 0.0–0.1 No observable symptoms, but some evidence of physiologic stress may be detectable.
>
> 0.1–0.2 Labored respiration during exertion.
>
> 0.2–0.3 Headache.
>
> 0.3–0.4 Muscular weakness, nausea, dizziness.
>
> 0.4–0.5 Slurring of speech, tendency to collapse.
>
> 0.5–0.6 Convulsions.
>
> 0.6–0.7 Fatal coma if duration of poisoning is prolonged.
>
> 0.8   Instantaneous death.

CO levels low enough to give only 20% HbCO have been known to kill when the victim remained in the poisonous atmosphere and this corresponds to only about 250 ppm CO. The most effective treatment for CO poisoning is to place the victim in a hyperbaric (high-pressure) chamber with 2 to 2.5 atm of O_2; this speeds up elimination of the CO and, more importantly, it also corrects tissue anoxia by providing large amounts of dissolved O_2 in the blood plasma, permitting the body to bypass the hemoglobin mechanism.

In U.S. urban areas, CO concentrations of several ppm are common, maximum annual 8-h averages can be 10 to 40 ppm and occasional short-term concentrations exceed 100 ppm [15]. These levels can easily lead to HbCO concentrations in the blood of approximately 2% [20]. Actually, a nonsmoker breathing air containing no CO will have a background HbCO level of about 0.4% from biological CO production inside the body. Smokers who smoke a pack of cigarettes daily and inhale the smoke may have blood HbCO levels of 5% or more. Although these levels do not lead to clinical symptoms, they have been associated with impairment of mental performance and visual acuity and other functions, and little is known of the effects of chronic exposure to low levels of CO on human health, behavior, and performance [20]. There is some epidemiological evidence suggesting that weekly average CO concentrations of around 10 ppm may produce increased mortality among hospitalized heart patients and that blood HbCO levels above 5% in patients with heart disease is associated with physiologic stress [15].

In view of the data that exist, it is apparent that CO levels of 10 ppm are undesirably high. The U.S.S.R. has actually set a 24-h average of 1 ppm as the desired maximum.

At present, the most dangerous results of exposure to carbon monoxide are the serious intoxications and even death that occur from CO production in closed areas—from automobile exhausts in garages (or exhaust fumes leaking into the interior of the automobile), blocked furnace flues in homes, etc. Several hundred Americans die from CO poisoning each year; many of them are suicides [21].

HYDROCARBONS

Hydrocarbons are chemical compounds containing only carbon and hydrogen. Several simple examples are shown in Figure 3-2. *Open-chain hydrocarbons* contain noncyclic chains (sometimes straight, sometimes branched) of carbon atoms to which hydrogen atoms are bonded; they may be saturated (paraffinic) as are methane and propane or unsaturated (olefinic) as is ethylene. *Cyclic hydrocarbons* contain rings of carbon atoms; they may be saturated or unsaturated, and the unsaturated hydrocarbons derived from benzene are referred to as *aromatic* hydrocarbons.

The light hydrocarbons are gaseous at ordinary temperatures. Methane occurs naturally and is the principal constituent of the fuel known as natural gas; it is colorless and odorless (the odor of natural gas is due to sulfur compounds added so that humans can detect it). Ethylene and propane are also gases.

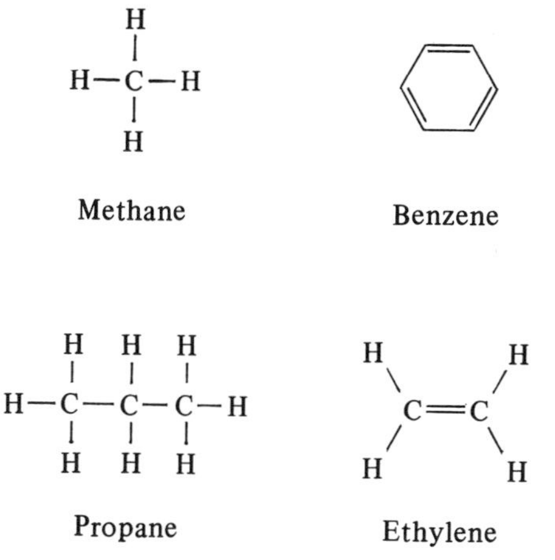

Methane   Benzene

Propane   Ethylene

FIG. 3-2. The molecular structure of several representative hydrocarbons.

Heavier hydrocarbons, such as those that occur naturally as petroleum, are liquids. They are usually separated in petroleum refining processes into various liquid fuels and solvents such as gasoline and kerosine. Some of these hydrocarbons are quite volatile. Very heavy hydrocarbons may be solids at ordinary temperatures.

The gaseous and volatile liquid hydrocarbons are the ones of particular interest as air pollutants. Hydrocarbons with more than about 12 carbon atoms are not present in the atmosphere in concentrations high enough to be of concern.

Natural sources of hydrocarbons are largely biological. Worldwide methane production probably amounts to well over 1 billion metric tons annually, mainly from anaerobic decay of organic matter. Some plants produce volatile terpenes (such as those that constitute turpentine) and isoprenes, which are complicated cyclic hydrocarbons; photochemical aerosol formation by these compounds is probably responsible for the blue haze of the Appalachian Mountain region and similar vegetation hazes. Nonurban air contains perhaps 1.0 to 1.5 ppm of methane and less than 0.1 ppm of other hydrocarbons [22].

Hydrocarbon emissions due to man total about 30 million metric tons annually in the U.S. (see Table 3-2) and probably about three times as much worldwide. A large fraction of this total consists of the higher-molecular-weight hydrocarbons, which are of more concern in the formation of photochemical smog. For this reason, it is the non-methane hydrocarbon concentrations that are of most interest in considering air pollution. The role of hydrocarbons in photochemical smog is discussed in more detail in Chapter 6.

The only pure hydrocarbon known to be capable of harming plants at concentrations that occur in or near urban areas is ethylene, whose role in inhibiting plant growth, first noted in the early 1900's, is believed to be due to interference with the activities of plant hormones. Plants are affected by exposure to as little as 0.01 ppm ethylene for several hours [23].

Open-chain hydrocarbons appear to have no effects on human beings at levels below 500 ppm, or several hundred times as much as is ever found in the atmosphere. Methane is not toxic in itself (concentrations of 50% or more can be dangerous because suffocation may result) but concentrations of a few percent may lead to explosions. Aromatic compounds can be irritating to the mucous membranes at concentrations lower than 500 ppm and injury can result but no adverse effects on health have been reported at levels below 25 ppm [22].

Hydrocarbons are of particular concern because of their involvement in the production of photochemical oxidants (see Chapter 6), which cause eye irritation and other effects. Formaldehyde (HCHO), which is a partially oxidized form of methane, has been found to produce physiological responses at concentrations of about 0.06 ppm and eye irritation beginning about 0.01 to 1.0 ppm in different individuals [22].

NITROGEN OXIDES

Although many different oxides of nitrogen are known (including laughing gas, which is nitrous oxide, N_2O), only nitric oxide (NO) and nitrogen dioxide (NO_2) are emitted to the atmosphere by man in significant quantities. They are formed by reaction of the nitrogen and oxygen in the atmosphere when combustion takes place at high temperatures (typically exceeding 1100°C) and cooling occurs fast enough to prevent decomposition:

$$N_2 + xO_2 \rightleftharpoons 2NO_x$$

Usually less than 0.5% of the NO_x is initially emitted as NO_2 [24].

Robinson and Robbins [25] have estimated that biological production of NO and N_2O amounts to about 1 billion metric tons annually, while man's combustion processes produce about 48 million metric tons of NO_x (expressed as NO_2) annually. These emissions are important parts of the environmental nitrogen cycle, although biological ammonia (NH_3) production of over 1 billion metric tons annually is also significant. The concentrations of NO and NO_2 in nonurban areas are only a few parts per billion, however, and these gases have a residence time in the atmosphere of only 3 or 4 days [24].

Man-made NO_x pollution is significant in urban areas since peak NO concentrations are often above 1 ppm and NO_2 levels occasionally exceed 0.5 ppm. These oxides both play an important role in the production of photochemical smog (see Chapter 6).

Nitrogen oxides are known to produce fading of textile dyes and additives, deterioration of cotton and nylon, and corrosion of metals due to production of particulate nitrates [24]. The concentrations at which significant effects occur have not been determined, however.

Several sensitive plants have been found to be adversely affected (by leaf injury and reduction of growth) by NO_2 concentrations of 1 ppm for a day or as little as 0.35 ppm for several months [24].

NO is not an irritant and it is not believed to have any adverse health effects on humans at concentrations that occur in the atmosphere, even in highly polluted areas. These concentrations are dangerous, however, because of the possibility of oxidation to NO_2. Hemoglobin has an extraordinary affinity for NO (about 1500 times its affinity for CO) but fortunately atmospheric NO appears to be unable to enter the bloodstream to react with the hemoglobin.

NO_2 is a reddish gas whose odor can be detected at concentrations of about 0.12 ppm. Its importance in photochemical reactions is due to its strong absorption of ultraviolet radiation. Concentrations of 100 ppm or more for a few minutes can be lethal to humans and animals, and exposures to 5 ppm for a few minutes leads to effects on the respiratory system. Exposure of monkeys to 15 to 50 ppm for 2 h damaged the lungs, heart, liver, and kidneys and produced pulmonary changes similar to those occurring in human emphysema. Long-term exposures to 0.06 ppm have been related to an increase in acute respiratory disease in humans [24]. It is thus clear that NO_2 levels in polluted urban atmospheres are associated with adverse health effects.

Serious illness and death has resulted from short exposures to NO_2. For example, it was responsible for 124 deaths in a fire at Cleveland's Crile Hospital on May 15, 1929, when X-ray film containing nitrocellulose accidentally caught fire and produced NO_2.

OTHER AIR POLLUTANTS

There are other air pollutants besides the five major types just discussed. In certain areas these may be significant contributors to the total air pollution burden. Marchesani, et al. [26] have listed a number of minor sources of emissions and estimated their quantities; some of their data are reproduced in Table 3-6. Some particulate matter is produced by abrasion (such as the wearing away of rubber tires and footwear); some organic vapors result from cosmetics and aerosol cans; and of course there is the smoke from tobacco products that is so annoying to many nonsmokers.

TABLE 3-6. Minor air pollutant emissions in the U.S. in millions of metric tons annually (adapted from Reference 26). The assumptions made in deriving these estimates may be found in the original source.

Pollutant	Amount
Ground dust from natural sources	27
Aerosols and vapors from aerosol cans	0.34
Rubber particulates from vehicle tires	0.28
Smoke from cigarettes	0.21
Footwear use	0.05
Smoke from cigars	0.04
Organic compounds from perfumes and colognes	0.03
Naphthalene from mothballs	0.005

Another class of air pollutants are the aeroallergens, the most important being ragweed pollen, although grass or tree pollen can cause allergic responses in some persons. This is an important problem since 10 or 15 million persons are adversely affected. Ragweed is crowded out by grass and other vegetation except in places where the soil has been disturbed (railroad tracks, developed land) so ragweed pollen is really more a man-made pollutant than a natural one. Some communities have found it necessary to institute programs to control ragweed.

Industrial processes, especially in the chemical industry, can produce special air pollutants—sulfuric acid, hydrogen chloride (HCl), formaldehyde, various alcohols, and many other sophisticated chemical compounds characteristic of our technological society. HCl can also be produced by the incineration of polyvinyl chloride plastic. Various fluorine compounds (HF, F_2, SiF_4, H_2SiF_6, etc.) are emitted by phosphate fertilizer manufacturing plants (since large amounts of fluoride are present in phosphate rock, typically 3%), aluminum reduction [which is accomplished by electrolysis of alumina (Al_2O_3) in a bath of molten cryolite, Na_3AlF_6], steelmaking, ceramic firing, and some chemical processing. Fluorides are readily absorbed by vegetation and can cause leaf injury and reduction of growth and yield even in very low concentrations, as low as 0.1 ppb [23]. In addition, livestock have fallen victim to fluoride poisoning from ingesting contaminated vegetation; moderate adverse effects can occur in dairy cattle whose feed contains 40 to 60 ppm fluoride (as F^-) [27].

The presence of stockpiles of nerve gases in the U.S. and their testing has alarmed many Americans. Although they cannot really be termed an air pollution problem as such, there is potential for great danger as long as these nerve gases exist. On March 13, 1968, a test of the highly toxic nerve gas VX at the U.S. Army's Dugway Proving Ground in Tooele County, Utah, apparently released some of the VX high enough in altitude for it to be carried in the atmosphere over two mountain ranges to be deposited in Skull Valley and Rush Valley, where several thousand sheep became ill and died [28]. The effect on sheep lasted over 3 weeks but scientists were unable to pinpoint the cause for several months, in part because the Army provided little information about its test. Since the Dugway incident, nerve gas supplies have been disposed of in such places as the Atlantic Ocean after being transported by rail through worried communities, fortunately without accidents.

A particulate air pollutant that has attracted special interest in recent years is asbestos. Its occupational hazards are well-known. Since 1943, Dr. I. J. Selikoff has been engaged in a longitudinal study of 632 members of the Insulation Workers Union of New York to determine their health history [29]. Through April 1967, a total of 349 had died compared to an actuarially predicted 250 while 144 had died of cancer compared to a predicted 45; there had been 27 deaths accompanied by asbestosis, a severe scarring of the lungs by asbestos fibers. All persons in urban areas are exposed to asbestos in the air from the wearing down of automobile brake linings and from the practice of spraying insulating asbestos during the construction of new buildings. In an effort to protect the public and the asbestos workers, New York has issued regulations governing asbestos spraying, requiring that the area be enclosed in tarpaulins during spraying and that workers be provided with respirators and coveralls [30].

Any given locality is likely to have other, specialized air pollution problems. In Iowa, for example, the use of anhydrous ammonia as an agricultural fertilizer has meant the presence throughout the state of anhydrous ammonia tanks that occasionally rupture or release ammonia during filling operations; many times each year nearby residents are forced to flee their homes very suddenly and remain elsewhere for many hours.

Two other air pollution problems are dealt with in later chapters: pesticides in Chapter 12 and radiation in Chapter 15.

THE COSTS OF AIR POLLUTION

What are the total costs in damages due to air pollution? What will be the cost in investment and operation for adequate controls? Neither of these questions can be answered very accurately but there have been some pioneering studies carried out to provide estimates.

The Council on Environmental Quality has estimated the total air pollution costs in the U.S. at $16 billion annually, which corresponds to $80 annually for every man, woman, and child in the country [31]. In 1970 the Council estimated some particular annual costs as:

  $100 million for painting steel structures.

  $800 million for laundering and dyeing of soiled fabrics.

  $240 million for washing cars dirtied by air pollution.

  $500 million for damage to agricultural crops and livestock.

  $40 to 80 million for adverse effects on air travel, largely due to reduced
    visibility.

Amounts were unknown for replacement and protection of precision instruments, maintenance of cleanliness in production of foods and beverages, soiling of homes and their furnishings, medical costs, the cost of absenteeism from work due to illness, and the cost of fuels wasted in incomplete combustion [32]. Lave and Seskin [33] have attempted to determine the medical costs and they estimate that a 50% reduction in air pollution levels in major urban areas would save $2.08 billion annually in terms of decreased mortality and morbidity—or 4.5% of all economic costs associated with mortality and morbidity. They also say that the total annual cost of these health effects *might* run as high as $29 billion.

The Council on Environmental Quality has estimated [31] that the total annual costs (including capital investment, interest, operation, and maintenance) of air pollution abatement would have to be $4.7 billion by 1975 in order to meet the federal air quality standards provided for by legislation (see Chapter 20). The 1970 costs were only $0.5 billion. The increase would correspond to an increase in the per capita annual cost from $2.50 to a little over $20.

In conclusion, there is evidence that investments in many areas of air pollution control are warranted and that the annual costs will be much less than 1% of our gross national product. The total cost of a clean environment is discussed further in Chapter 19.

REFERENCES

1. John Evelyn's *Fumifugium* was first published in 1661 and has been reprinted by the National Society for Clean Air (see Appendix B) as an inexpensive paperback.
2. Robert H. Socolow (private communication).
3. *Nationwide Inventory of Air Pollutant Emissions 1968.* Washington, D.C.: U.S. Dept. of Health, Education, and Welfare, August 1970. N.A.P.C.A. Publication No. AP-73.
4. Feldstein, Milton, "Toxicity of Air Pollutants." *Progress in Chemical Toxicology* 1:297–316 (1963).
5. Logan, W. P. D., "Mortality in the London Fog Incident, 1952." *Lancet* **264:** 336–338 (1953).
6. Wilkins, E. T., "Air Pollution Aspects of the London Fog of December 1952." *Quart. Jour. Roy. Met. Soc.* **80:**267–271 (1954).
7. Joos, Georg, *Theoretical Physics.* 3rd ed. New York: Hafner Pub. Co., Inc., 1958. pp. 218–222.
8. Spirtas, Robert and Howard J. Levin, *Characteristics of Particulate Patterns 1957–1966.* Washington, D.C.: U.S. Dept. of Health, Education, and Welfare, 1970. N.A.P.C.A. Publication No. AP-61.
9. *Air Quality Criteria for Particulate Matter.* Washington, D.C.: U.S. Dept. of Health, Education, and Welfare, 1969. N.A.P.C.A. Publication No. AP-49.
10. Caley, Earle R., and John F. C. Richards (eds.), *Theophrastus on Stones.* Columbus, Ohio: The Ohio State University, 1956.
11. Robinson, E., and R. C. Robbins, "Gaseous Sulfur Pollutants from Urban and Natural Sources." *Jour. Air Poll. Control Assn.* **20:**233–235 (1970). See also Kellogg, W. W., et al., "The Sulfur Cycle." *Science* **175:**587–596 (1972).
12. *Air Quality Criteria for Sulfur Oxides.* Washington, D.C.: U.S. Dept. of Health, Education, and Welfare, 1969. N.A.P.C.A. Publication No. AP-50.
13. Swinnerton, J. W., et al., "The Ocean: A Natural Source of Carbon Monoxide." *Science* **167:**984–986 (1970); Swinnerton, J. W., et al., "Carbon Monoxide in Rainwater." *Science* **171:**943–945 (1971).
14. *Air Quality Criteria for Carbon Monoxide.* Washington, D.C.: U.S. Dept. of Health, Education, and Welfare, 1970. N.A.P.C.A. Publication No. AP-62.
15. Weinstock, B., "Carbon Monoxide: Residence Time in the Atmosphere." *Science* **166:**224–225 (1969). See also Weinstock, B., and H. Niki, "Carbon Monoxide Balance in Nature." *Science* **176:**290–292 (1972).
16. Maugh, T. H., II, "Carbon Monoxide: Natural Sources Dwarf Man's Output." *Science* **177:**338–339 (1972).
17. Inman, R. E., et al., "Soil: A Natural Sink for Carbon Monoxide." *Science* **172:**1229–1231 (1971). Soil may also be a sink for other air pollutants: see

Abeles, F. B., et al., "Fate of Air Pollutants: Removal of Ethylene, Sulfur Dioxide, and Nitrogen Dioxide by Soil." *Science* **173**:914–916 (1971).

18. Bour, H. and I. McA. Ledingham (eds.), *Carbon Monoxide Poisoning.* Amsterdam: Elsevier Pub. Co., 1967.

19. Wolf, Philip C., "Carbon Monoxide. Measurement and Monitoring in Urban Air." *Environ. Sci. Tech.* **5**:212–218 (1971).

20. *Effects of Chronic Exposure to Low Levels of Carbon Monoxide on Human Health, Behavior, and Performance.* Washington, D.C.: National Academy of Sciences, 1969.

21. *Accident Facts.* Published annually by the National Safety Council, 425 N. Michigan Ave., Chicago, Ill. 60611.

22. *Air Quality Criteria for Hydrocarbons.* Washington, D.C.: U.S. Dept. of Health, Education, and Welfare, 1970. N.A.P.C.A. Publication No. AP-64.

23. Hindawi, I. J., *Air Pollution Injury to Vegetation.* Washington, D.C.: U.S. Dept. of Health, Education, and Welfare, 1970. N.A.P.C.A. Publication No. AP-71.

24. *Air Quality Criteria for Nitrogen Oxides.* Washington, D.C.: Environmental Protection Agency, 1971. Publication No. AP-84.

25. Robinson, E., and R. C. Robbins, "Gaseous Nitrogen Compound Pollutants from Urban and Natural Sources." *Jour. Air Poll. Control Assn.* **20**:303–306 (1970).

26. Marchesani, V. J., et al., "Minor Sources of Air Pollutant Emissions." *Jour. Air Poll. Control Assn.* **20**:19–22 (1970).

27. McCune, D. C., and others, "Symposium on Fluorides." *Environ. Sci Tech.* **3**:720–735 (1969).

28. Boffey, Philip M., "Nerve Gas: Dugway Accident Linked to Utah Sheep Kill." *Science* **162**:1460–1464 (1968).

29. Selikoff, I. J., "Asbestos." *Environment* **11**(2):2–7 (March 1969).

30. Knapp, Carol E., "Asbestos: Friend or Foe?" *Environ. Sci. Tech.* **4**:727–728 (1970).

31. *Environmental Quality. The Second Annual Report of the Council on Environmental Quality.* Washington, D.C.: U.S. Govt. Printing Office, August 1971.

32. *Environmental Quality. The First Annual Report of the Council on Environmental Quality.* Washington, D.C.: U.S. Govt. Printing Office, August 1970.

33. Lave, L. B., and E. P. Seskin, "Air Pollution and Human Health." *Science* **169**:723–733 (1970).

4

AIR POLLUTION: METEOROLOGY AND CLIMATOLOGY

The frost still continuing more and more severe . . . London, by reason of the excessive coldnesse of the aire, hindring the ascent of smoke, was so filld with the fuliginous steame of the Sea-Coale, that hardly could one crosse the streete, & this filling the lungs with its grosse particles exceedingly obstructed the breast, so as one could scarce breath. —John Evelyn, *Diary* (entry for January 24, 1684)

This may be the first recorded description of a thermal inversion, a meteorological condition in which air pollutants are unable to rise and be dispersed in the atmosphere, so that they accumulate and produce high concentrations of pollution. Atmospheric conditions can have profound effects on pollution and its harmfulness; this is one topic to be considered in this chapter. The reverse side of the coin is that air pollution can affect atmospheric conditions, at least on a localized scale, and may have global climatic effects; this is the other topic that will be discussed.

In order to understand these topics, it is necessary to understand the temperature distribution in the atmosphere and the radiation balance of the earth.

TEMPERATURE DISTRIBUTION IN THE ATMOSPHERE

The temperature of the atmosphere depends on the rate at which energy is being received from the sun and on the various energy transport mechanisms (electromagnetic radiation, convection, evaporation, etc.) among the atmosphere, the oceans, and the earth's land surface. As a consequence the temperature is not a constant but varies with height, latitude, season, time of day, amount of cloud cover, and many other variables. For purposes of dealing with air pollution, it is necessary to have a rough idea of the temperature distribution with height.

The approximate distribution of temperature with height is shown in Figure 4-1. Several distinct regions can be identified:

1. The *troposphere* is the region nearest the surface, in which the temperature decreases fairly steadily from the ground temperature

to a temperature of -50 to $-80°C$ at a rate of about $-6.5°C/km$ (about $-19°F/mi$), which is referred to as the *lapse rate* of the troposphere. The air in the troposphere is well mixed by convection currents and it contains approximately 90% of the total mass of the atmosphere. This layer contains most of the atmosphere's water (and clouds) and particulate matter.

2. The temperature curve changes slope rather suddenly in a narrow transitional layer known as the *tropopause*, which is usually between 10 and 20 km above the earth's surface, being highest and coldest at the equatorial latitudes.

3. Above the tropopause is the *stratosphere*, in which the temperature curve shows a warming trend with increasing height. This warming is due to absorption of solar ultraviolet radiation by ozone (O_3), whose concentration in the stratosphere ranges about 1 to 5 ppm by volume. Both the pressure and the density of the stratosphere range from about $1/10$ to $1/1000$ of the corresponding quantities near the surface of the earth. The air in the stratosphere is very dry, and clouds and convection currents from the troposphere usually do not penetrate easily into it. Some transport of air masses does take place between the two regions and their compositions are very similar.

4. The *stratopause* at the top of the stratosphere, typically 50 km high, is a transitional layer that is relatively warm and not much cooler than the earth's surface. It was first detected by its property of refracting sound waves from the earth back to the surface.

5. Above the stratosphere is the *mesosphere*, in which the temperature again decreases with height to the very coldest temperature in the atmosphere, typically about $-100°C$.

6. The region of the minimum temperature is the *mesopause*.

7. Above the mesopause, atmospheric temperatures rise very rapidly to the very high temperatures of the *thermosphere*, which is characterized by low densities (1 mg/m^3 or less) and low pressures (0.01 N/m^2 or less).

The troposphere, the stratosphere, and the mesosphere are fairly uniform in composition, with some exceptions, and are sometimes referred to as the *homosphere* for that reason, while the higher layers, in which the composition is very uneven, are referred to as the *heterosphere*.

Of particular importance to air pollution is the troposphere. The lapse rate of the troposphere, about $-6.5°C/km$, can be understood fairly simply. When some disturbance causes a parcel of air to rise or descend in the atmosphere, it generally does so adiabatically—i.e., it exchanges heat with its environment far more slowly than it loses or gains energy due to the work that it performs. A parcel of air that rises will expand (since it encounters

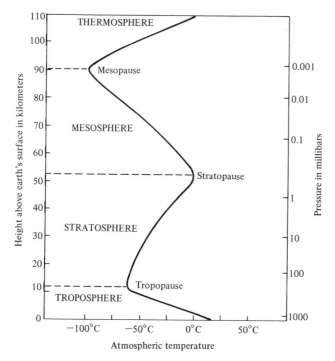

FIG. 4-1. Temperature distribution of the atmosphere. One millibar equals 100 N/m².

a lower pressure) and thereby cool, while one that descends will contract and warm. Such a parcel of air will find itself in neutral equilibrium with its environment—tending neither to increase nor decrease its rate of ascent or descent—only for one very particular temperature distribution, whose lapse rate is referred to as the *adiabatic lapse rate.*

For an ideal gas, the adiabatic lapse rate turns out to be just $-gM/C_p$, where g is the gravitational acceleration, M is the average molecular weight of the gas, and C_p is the molar specific heat of the gas at constant pressure; the derivation of this result is given in Table 4-1. Using the C_p and M values for air gives the value $-9.8\,°C/km$, which is generally referred to as the *dry adiabatic lapse rate* since it makes no allowance for the presence of moisture in the air. If the air is moist, the water present may undergo a phase change (from vapor to liquid or vice versa) so that heat in the amount of latent heat of vaporization is released or absorbed during the adiabatic process. When a moist parcel of air rises, expands, and cools, the water vapor can condense and release latent heat. Thus the actual *moist adiabatic lapse rate* is less than the dry rate and the lapse rates occurring in the troposphere are typically -6 to $-7\,°C/km$, with the higher values occurring at greater heights where the water content is less.

It is, of course, difficult for the adiabatic lapse rate to be maintained exactly because of all the energy transport processes involving the atmosphere.

TABLE 4-1. Derivation of the adiabatic lapse rate for an ideal gas.

Use: (1) The equation of state for an ideal gas: $PV = nRT$, where P is the pressure, V is the volume, n is the number of moles of the gas, R is the universal gas constant 8.314 J/mole-K, and T is the temperature.

(2) The hydrostatic equation $\partial P/\partial h = -g\rho = -gnM/V$, which gives the gradient of the pressure with height h in terms of the gravitational acceleration $g = 9.8$ m/s² (whose value does not change much in the troposphere) and the density $\rho = nM/V$, where M is the average molecular weight ($M = 28.97$ for air).

(3) The law for adiabatic expansion and contraction, which can be written as $PV^\gamma = P_0V_0^\gamma$, or using (1), $T/T_0 = (P/P_0)^{R/C_p}$, where γ is the ratio of the molar specific heats C_p and C_v at constant pressure and constant volume, respectively. For air, $\gamma = 1.40$ and C_p is $3.5R$.

Then, the adiabatic lapse rate $\partial T/\partial h = (\partial T/\partial P)(\partial P/\partial h)$. From (3), $\partial T/\partial P = T_0(R/C_p)P^{(R/C_p)-1}/P_0^{R/C_p}$, which can be simplified using (1) and (3) to $\partial T/\partial P = V nC_p$. $\partial P/\partial h$ is given by (2).

Thus, the adiabatic lapse rate $\partial T/\partial h = -gM/C_p$, which is $-9.8\,°C/km$ for air, treating air as though it were an ideal gas. The adiabatic lapse rate for a non-ideal gas includes an extra factor of $T(\partial P/\partial T)_\rho/\rho(\partial P/\partial \rho)_T$ [1].

If the actual lapse rate is greater in magnitude than the adiabatic lapse rate, a rising parcel of air will become warmer and less dense than the surrounding air so it will continue to rise. This would be desirable if the parcel contained pollutants; for example, smokestacks emit warm air that begins to rise and this air will continue to rise if the lapse rate is great enough. On the other hand, if a parcel of air begins to rise when the lapse rate is smaller in magnitude than the adiabatic lapse rate, it will become cooler and denser than the surrounding air and so will tend to return to its original level. This would be undesirable if the parcel contained pollutants. The two situations are shown in Figure 4-2. They correspond, respectively, to unstable and stable atmospheric conditions since in the first case a displacement of the parcel of air tends to produce an even greater displacement in the same direction, while in the second case a displacement tends to be opposed by a force that restores it to its original height. The situation (also indicated in Figure 4-2) in which the temperature actually increases with height for some interval of altitude is referred to as a *temperature inversion* and it clearly corresponds to extremely stable conditions that prevent dispersion. Inversions will be considered in more detail later in this chapter.

Although temperature variations in the troposphere are to be expected, there is also the possibility that long-term trends toward warmer or cooler temperatures may be occurring. The knowledge that the earth's temperature has changed in the past to produce great variations in climate leads to the question, Is the earth's average temperature changing at present and, if so, in what direction and at what rate?

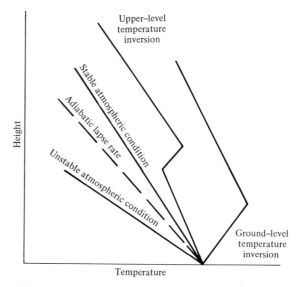

FIG. 4-2. Possible temperature distributions in the troposphere.

An answer can be given to this question only for the past century since adequate meteorological data are lacking from earlier periods. Mitchell [2] has analyzed weather records from stations distributed as uniformly as possible over the earth's surface and shown that the data indicate a worldwide warming of about 0.6°C in the period from 1880 to the mid-1940's and a subsequent cooling of about 0.3°C in the quarter-century since then, as shown in Figure 4-3. This fluctuation may seem rather small in comparison to daily and seasonal temperature variations but it is not insignificant climatologically. It has been accompanied by important changes in sea levels, glacier positions, and desert margins. It is also believed that the world's average temperature changed only 6°C between the glacial and interglacial periods of the Pleistocene Ice Age [2]. It is of great interest to mankind to know if the cooling trend will continue and perhaps lead to another ice age or if a warming trend sufficient to melt the polar ice caps could occur. The fluctuation may be a natural fluctuation over which man has little control or, as discussed later in this chapter, it may be a result of man's activities.

THE RADIATION BALANCE OF THE EARTH

A physical object of a given temperature T will radiate energy in the form of electromagnetic waves. The higher T is, the more energy is radiated; according to the Stefan-Boltzmann law, the total amount radiated is proportional to the surface area and to the fourth power of the temperature (T^4). The

FIG. 4-3. Trends of mean annual temperature between 0° N and 80° N latitude from 1870 to the late 1960's [2]. (*Courtesy of Springer-Verlag New York, Inc.*)

energy radiated by the sun, which is of crucial importance to life on earth, arises from various thermonuclear fusion processes deep within the sun and is radiated at wavelengths largely in the visible portion of the electromagnetic spectrum.

A "black body"—an object capable of 100% absorption of any electromagnetic radiation that falls on it—radiates energy in accordance with Planck's radiation law: The total energy radiated per unit time per unit surface area per unit wavelength interval at the wavelength λ is [3]

$$E(\lambda) = \frac{2\pi hc^2}{\lambda^5} \frac{1}{\exp(hc/\lambda kT) - 1}$$

where c is the speed of light (3.0×10^8 m/s), h is Planck's constant (6.63×10^{-34} J-s), T is the absolute temperature, and k is Boltzmann's constant (1.38×10^{-23} J/K). If $T = 6000$ K, the curve of $E(\lambda)$ is that shown in Figure 4-4(a). This curve, which peaks in the visible part of the electromagnetic spectrum (for which λ is between 0.4 and 0.7 μm), is closely identical to that of the sun, which radiates very nearly (but not perfectly) as a blackbody of temperature 6000 K.

The earth also radiates energy but since its average temperature is much lower (its effective radiative temperature is about 250 K), its radiation is much less and tends to occur at longer wavelengths, as shown in Figure 4-4(b). Since the earth's average temperature is not changing very fast, we may suppose that the earth is in approximate radiative equilibrium; i.e., the amount of solar radiation falling on the earth approximately equals the amount of radiation emitted by the earth. The two radiation distributions are shown in Figure 4-4(c), drawn so that the areas under the two curves are approximately equal. The sun radiates vastly more energy than the earth, of course, but only a small fraction of it actually falls on the earth.

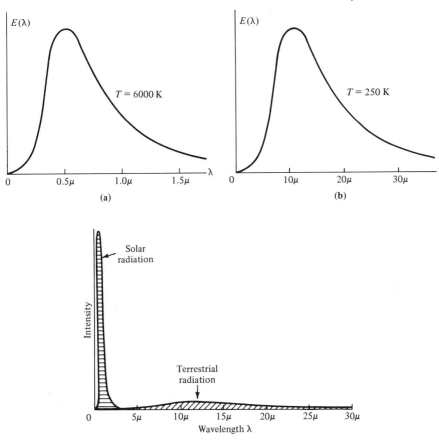

FIG. 4-4. Electromagnetic energy radiation curves as a function of wavelength λ. (a) Blackbody radiation for $T = 6000$ K. (b) Blackbody radiation for $T = 250$ K. (c) Approximate relative intensities for total solar radiation as received by the earth and total terrestrial radiation.

Since the curves for solar radiation and terrestrial radiation overlap very little, it is possible to separate them by wavelength:

$\lambda < 4$ μm: solar or shortwave radiation.

$\lambda > 4$ μm: terrestrial or longwave radiation.

It is of particular interest to consider the detailed radiation balance at this point. The solar constant, the amount of solar radiation falling on the earth at normal incidence, averages 1370 W/m². This is not a fixed quantity. It changes with variations in solar activity and with the variations in the distance from the earth to the sun. The average distance is about 149.5 million km but it is about 5 million km closer at perihelion (first week in January) than at aphelion (first week in July), and the maximum solar constant for the year is about 7% greater than the minimum value.

The total amount of solar power falling on the earth is that intercepted by a circle of radius equal to the mean radius of the earth, which is about 6370 km, and it amounts to 1.76×10^{17} watts, a truly staggering quantity. This is the main source of energy received at the surface of the earth, being about 5000 times greater than the geothermal flux from beneath the surface, which amounts to about 3×10^{13} watts [4].

Only about half the solar radiation actually reaches the surface of the earth, the rest being reflected or scattered back into space or absorbed by atmospheric gases. Since the earth is in radiation balance, the total radiation leaving the earth, either as solar shortwave radiation reflected or scattered back into space or as terrestrial longwave radiation going into space, must equal the solar constant and total the same 1.76×10^{17} watts. The balancing does not necessarily occur at any given instant but it must occur over the year as a whole.

The actual processes involved in the radiation balance (or "heat budget") of the earth are shown in Figure 4-5; percentages are used to represent the average energy fluxes, with 100% corresponding to the solar constant.

Of the incoming solar radiation (100%),

1. 49% is intercepted by clouds, with 24% reflected back into space, 2% absorbed by the clouds, and 23% diffused through the clouds and absorbed at the earth's surface.

2. 17% is absorbed by the atmosphere.

3. 20% is absorbed by the earth's surface directly.

4. 2% is reflected back into space from the earth's surface.

5. 12% is scattered by the atmosphere, with 7% being scattered back out into space and 5% scattered toward the earth and absorbed by the earth's surface.

The albedo of the earth, or the total part of the incoming solar radiation that is reflected or scattered back into space, is about 33% (24% from clouds, 2% from earth, and 7% from the atmosphere). The total absorption by the atmosphere is 19% (2% by clouds and 17% by atmospheric gases), and the total absorption by the earth's surface is 48% (20% directly, 5% after scattering by the atmosphere, and 23% after diffusion through the clouds).

The balancing in the long wavelength region, with which we shall include the transport of latent and sensible heat, is as follows:

1. The atmosphere radiates about 165% of the solar radiation, 61% into space and 104% back to the earth. Of this 165%, 19% was

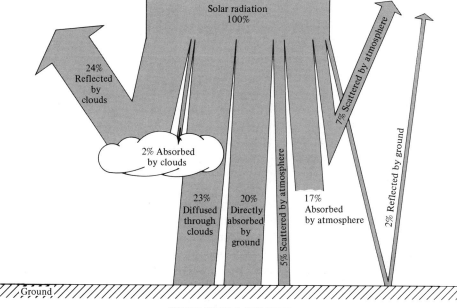

SHORT–WAVE RADIATION

(a)

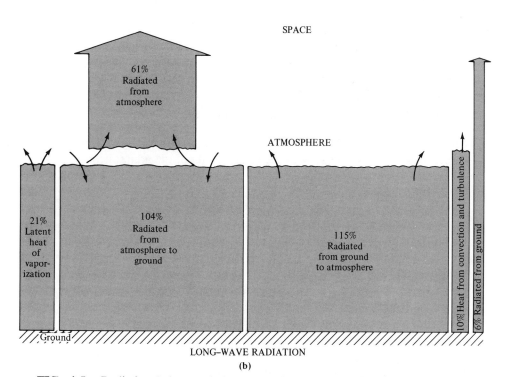

LONG–WAVE RADIATION

(b)

FIG. 4-5. Radiation balance of the earth-atmosphere system in the shortwave ($\lambda < 4$ μm) and longwave ($\lambda > 4$ μm) regions.

originally absorbed in the shortwave region, 115% is long wavelength radiation from the earth to the atmosphere (mainly absorbed
by the water vapor in the atmosphere), 21% is latent heat arriving
from the earth through evaporation, and 10% is sensible heat
arriving from the earth through convection and turbulence.

2. The earth radiates 6% directly into space and conveys 146% to
the atmosphere through radiation, evaporation, convection, and
turbulence as indicated above. Of this total of 152%, 48% was
originally absorbed as shortwave radiation, and 104% is long wavelength radiation absorbed from the atmosphere.

The occurrence of radiation fluxes amounting to more than 100% of the solar
radiation (or more precisely, greater than the 67% not reflected or scattered
back into space) may seem surprising at first, but it simply results from the
fact that radiation energy is absorbed and radiated back and forth by the
atmosphere and the earth's surface before being permanently radiated away
from the planet.

None of these percentages are accurately known. For many years the
earth's albedo was thought to be well above 40% and recent data from
meteorological satellites [5] suggest that the mean annual albedo may be
only 29%. A recent survey of these questions may be found in Reference 6.

The absorption of shortwave and longwave radiation that occurs in the
atmosphere is due to several important constituents, especially water vapor,
ozone, oxygen, and carbon dioxide. Figure 4-6 shows the absorption for a
solar beam reaching ground level after passing through the atmosphere [7].
Below 0.3 μm, in the ultraviolet region of the electromagnetic spectrum, there
is almost complete absorption by oxygen and ozone; the warm layer in the
vicinity of 50 km in the upper atmosphere (see Figure 4-1) is due to ultraviolet absorption by ozone. Between 0.3 and 0.8 μm, which includes all the
visible region and the majority of the solar radiation, the atmosphere is essen-

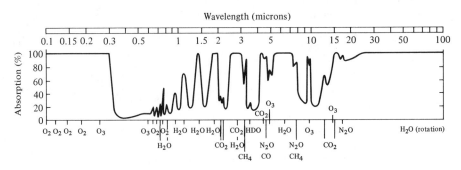

FIG. 4-6. Atmospheric gaseous absorption spectrum for a solar beam reaching
ground level. Molecules responsible for the different peaks are identified. (From
Goody, R. M., *Atmospheric Radiation* [7]. *Courtesy of The Clarendon Press,
Oxford.*)

tially transparent. In the region above 0.8 μm there are a number of absorption bands due to water vapor and carbon dioxide and several other gases. The atmosphere is again moderately transparent in the region around 10 μm, where the longwave terrestrial radiation is concentrated, and a "window" is provided for this radiation to escape back into space.

TEMPERATURE INVERSIONS

A temperature inversion is said to exist when the normal lapse rate in the lower atmosphere is inverted and the temperature actually increases with height; examples are shown in Figure 4-2. Exhaust gases and pollutants will rise only a certain distance under these conditions and persistent inversions that last several days can be very dangerous. All the noted smog episodes of London, Los Angeles, New York, and other great cities have resulted when inversions were present.

Temperature inversions can be produced in a number of ways:

1. The ground surface, and thus the lower layers of air adjacent to it, may be cooled by its radiation of heat. During the day, solar radiation will reach the ground and be absorbed, leading to warming of the surface, but at night the radiation of energy by the surface will lead to its cooling. An inversion may be produced but it will generally disappear during the next day. This behavior is shown in Figure 4-7, which sketches the typical temperature distribution at different times of day and night and compares it to the adiabatic lapse rate. Pollutants may accumulate during the night beneath a very stable atmospheric level a few hundred meters high and then be carried down to the ground in the morning when the surface begins to warm and thermal convection leads to stirring of the air; this sudden increase in low-level pollution in the morning is called *fumigation*.

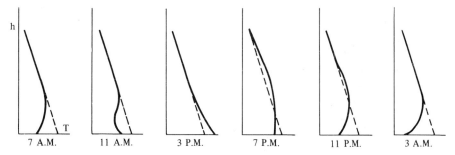

FIG. 4-7. Typical temperature distributions in the troposphere as a function of height at different times of the day. The dashed line in each case represents the adiabatic lapse rate.

2. Cold air, by virtue of its greater density, often collects in the bottom of a valley, creating a localized inversion. Farmers know that plants grow better on the valley slopes than on the valley floor for this reason. This type of inversion may be only a few meters high.

3. Inversions can be created by the meeting of two air masses of different temperatures since the colder mass, being denser, will generally push under the warmer mass. These inversions are called *frontal inversions* since they occur at the fronts or boundaries of the air masses.

4. Horizontal motion of warm air over a cold surface will lead to cooling of the lower layers of air just above the surface, producing an inversion.

5. A subsidence inversion results when upper layers of air descend (subside) during a developing anticyclone (high pressure center). The subsiding air warms as it contracts but the warming will be greater at the upper part than at the lower part near the ground. If the anticyclone persists several days over an area, the inversion may also persist and lead to pollutant buildup. This type of inversion produced the great smog episode in London in December 1952, in which pollutants were only able to rise about 100 m.

The higher an inversion, the lower the pollution concentrations. The Los Angeles Basin in California is enclosed on three sides by mountains and is subject to cool ocean breezes from the fourth side; inversions occur there about 320 days of the year. At times the inversion ceiling is below the tops of skyscrapers, while other times it is much higher, as shown in Figure 4-8. A noted subsidence inversion was produced over much of the eastern United States during Thanksgiving week 1966 by a stagnant high-pressure area [8]. On the afternoon of November 26 the atmospheric mixing depth (the height over which mixing of the air was occurring) was only 164 m in New York and 250 m in Boston. Peak pollutant concentrations occurred a few days earlier, however, with the hourly SO_2 average concentration reaching 0.97 ppm in New York on November 24 when the daily mean was over 0.5 ppm. The average concentration of particulate matter reached 390 $\mu g/m^3$ in Philadelphia on November 25 [8]. As might be guessed from the data presented in Chapter 3, increased mortality resulted from this episode, with excess deaths in New York alone totaling 168 according to Glasser, et al. [9].

The first smog episode to attract widespread attention was one that occurred along the Meuse River (between Liege and Huy) in Belgium during December 1 to 5, 1930. About 6000 persons became ill and 60 deaths were attributed to the fog, the death rate being over 10 times normal. The inversion was about 80 m high, lower than the hills bordering the valley. Well before the 1952 London smog episode, Firket [10] pointed out that a comparable disaster in London might kill 3200 persons.

FIG. 4-8. Three views of the Los Angeles Civic Center. (Upper) a clear day. (Middle) Smog engulfs the Civic Center on a day when the base of the temperature inversion is approximately 460 m (1500 ft) above ground level. (Bottom) Smog is trapped by a temperature inversion at approximately 90 m (300 ft) above the ground. The upper portion of the Los Angeles City Hall is visible in the clear air above the base of the temperature inversion. (*Photographs courtesy of the Los Angeles Air Pollution Control District.*)

Another noted smog episode occurred in Donora, Pa. in October 1948. Some 6000 persons (out of 14,000) became ill and there were about 20 deaths [11].

Although these were the only two smog incidents attracting international attention before the London smog of 1952, it is now clear that there have been many others in major industrial cities over at least the past century. Some prominent smog episodes are listed in Table 4-2 (see Reference 12 for a discussion of some of these). The quotation from John Evelyn's *Diary* at the beginning of this chapter would seem to describe a similar incident, for Evelyn remarked about the respiratory difficulties produced.

TABLE 4-2. Notable smog incidents and estimated number of excess deaths.

Location	Year	Excess Deaths
London	1873	268
London	1880	692
London	1891	572
Glasgow	1909	592
Meuse Valley, Belgium	1930	60
Donora, Pa.	1948	20
London	1952	4000
New York	1953	200
London	1956	1000
London	1962	750
New York	1963	200–400
New York	1966	168

There is essentially nothing that can be done to prevent a temperature inversion. The energy that would be required to stir the polluted air artificially to permit dispersion of the pollution is too great. Tall smokestacks may enable plumes to rise enough to go above the inversion ceiling, and smokestacks as tall as 381 m have been built. The most effective measure during a smog episode is to reduce pollution—using natural gas or low-sulfur fuel oil and coal instead of high-sulfur fuels, reducing the use of automobiles, shutting incinerators down temporarily, etc.

LOCAL EFFECTS OF POLLUTION ON CLIMATES

Pollution in the atmosphere is capable of very important effects on the climate in localized areas, especially urban areas [13, 14]. The effects in urban areas include the following:

1. *Higher temperatures.* Minimum daily temperatures in urban areas are often 5 to 10°C higher than those in the surrounding rural areas and annual mean temperatures are typically 0.5 to 1.3°C

higher [13]. In summer, this is due to the fact that the tall buildings and pavements of cities absorb more solar radiation (and reflect less) than the vegetation and soil of the rural areas in daytime and release more heat in nighttime because of their higher specific heat. In addition, less heat goes into evaporation (see entry 2 below). Increased fuel consumption, especially in wintertime, is also responsible for artificial heat production. These effects result in the production of an "urban heat island" that has been the object of many scientific studies [13]. Increased convection above cities is one notable example of how meteorological conditions are affected.

2. *More rapid runoff of water.* Cities are characterized by more rapid runoff of water, which in a rural area would normally be largely absorbed in the soil. Well-documented consequences are a reduction in evaporation of water in the city and lower relative humidity (typically 2 to 8% lower).

3. *Attenuation of solar radiation.* As discussed in Chapter 3, particulate matter in the atmosphere, whose urban concentrations are typically 10 times those of rural areas, is capable of reducing the amount of solar radiation falling on the city by 15 to 20%. These concentrations are also accompanied by a reduction in visibility.

4. *Lowering of wind speeds.* The presence of urban construction leads to increased turbulence because of the increase in surface roughness and wind speeds near the surface of the earth are reduced [14].

5. *Increased cloudiness.* This is probably due to the updrafts produced by the urban heat island and partly to the large number of small particles produced by man's activities, which are capable of serving as condensation nuclei for water vapor in the atmosphere [14]. Urban areas appear also to have 5 to 10% more precipitation than nearby rural areas but this is just about the limit of accuracy of precipitation measurements. Some much larger precipitation effects have been reported but their existence is a matter of controversy in the scientific literature.

Even in nonurban areas, man's activities have led to climatic effects. Conversion of forests to pasture has often been followed by overgrazing, with increased soil erosion by water and wind being the end result. Similarly, conversion of grasslands to agricultural crop production has led to erosion, as in the "Dust Bowl" days of the central United States in the 1930's.

Irrigation practices also affect the water and heat balance. Large man-made reservoirs have led to reduction in temperature extremes, increase in humidity, and effects on wind patterns [14].

In some cases, man is deliberately attempting to affect the climate. Several methods of dispersing fogs at airports have been tried. Weather modification has included cloud seeding to increase precipitation and seeding of hurricanes to reduce their wind and storm damage. The results have not been too successful and cloud seeding may not increase precipitation but just redistribute it.

GLOBAL EFFECTS OF POLLUTION ON CLIMATE

The global effects of pollution on climate are less marked than the local effects but they may be of great importance in the long run [14, 15]. Several important questions that have been raised will be discussed in turn.

HEAT PRODUCTION. Is the amount of heat produced by man through fuel combustion (largely fossil fuels) of global significance? In 1968, the world's total energy consumption of fossil fuels was the equivalent of about 6×10^9 metric tons of coal [16], which would mean 1.8×10^{20} joules (see Appendix A). This would correspond to an average rate of heat production of 5.5×10^{12} watts, which is insignificantly small compared to the average power of solar radiation reaching the earth's surface, which is approximately 8×10^{16} watts (i.e., about half the 1.76×10^{17} watts of solar radiation falling on the earth and its atmosphere, discussed earlier in this chapter). The heat production can be important locally, as the existence of urban heat islands demonstrates. As an example, Los Angeles County in California has a population of 7 million in an area of 10,000 km². If the per capita energy production in the county equals the U.S. average of 100,000 kWh annually, the total energy produced will be 7×10^{11} kWh or 2.5×10^{18} joules; dividing by the number of seconds in a year (3.15×10^7) gives an energy production rate of 8×10^{10} watts, which amounts to 8 W/m² over the county. This is several percent of the average solar radiation reaching the earth's surface, which is about 170 W/m² when averaged over the whole year, including day and night.

OXYGEN CONSUMPTION. Fears have occasionally been expressed about global oxygen reserves or about depletion of oxygen in local areas. Oxygen depletion can be a serious problem in poorly ventilated buildings to the extent that incomplete fuel combustion may occur with carbon monoxide production. There is no possible danger of depleting the global oxygen supply, however [17]. The rate at which atmospheric oxygen is being depleted by the combustion of fossil fuels is insignificant. The combustion of coal to CO_2 ($C + O_2 \rightarrow CO_2$) requires 32/12 as much oxygen as carbon by mass and the combustion of hydrocarbons (petroleum and natural gas) requires somewhat more. At present, man is using less than 20,000 million metric tons of oxygen annually, or 1/60,000 of the total atmospheric oxygen (see Table 3-1). Well before we can deplete our oxygen reserves, we shall run out of fossil fuels (which are sufficient to last only a few more centuries) and encounter other serious environmental problems. Man's alteration of photosynthetic rates is also of no concern as far as oxygen reserves are concerned [17]. Recent studies [18] have found no evidence that the concentration of atmospheric oxygen is changing, nor that it has changed since 1910.

CARBON DIOXIDE. The importance of carbon dioxide and other atmospheric gases in the radiation balance of the earth has been recognized for over a century. In 1861 the noted physicist John Tyndall wrote [19]

> It is exceedingly probable that the absorption of the solar rays by the atmosphere . . . is mainly due to the watery vapor in the air . . . Every variation of this constituent must produce a change of climate. Similar remarks would apply to the carbonic acid [CO_2] diffused through the air.

The release of vast—and atmospherically significant—quantities of carbon dioxide by the combustion of fossil fuels has led to fears of significant climatic influences should the CO_2 in the atmosphere actually increase with time. The absorption bands of CO_2 in the infrared region (see Figure 4-6) would become more important with an increase of CO_2 concentration and the present radiation balance of the earth would be changed in such a manner as to increase the average surface temperature. The first calculation of this increase was by the famous chemist Svante Arrhenius in 1896 [20]; he estimated that a doubling of the atmospheric CO_2 concentration would increase the surface temperature by 5 to 6°C, with different warmings at different latitudes. Arrhenius felt that CO_2 variations due to volcanic activity might be of climatic importance.

In 1938, Callendar [21] showed that there was evidence of a gradual increase in atmospheric CO_2 and that man's activities might be responsible. In the mid-1950's, detailed studies of the CO_2 absorption by Plass [22] led him to predict a 3.6°C rise in the surface temperature from a doubling of the atmospheric CO_2 concentration and a 3.8°C decrease from its halving. More recent calculations [23, 24] incorporating the effects of convection, humidity, and cloudiness have predicted rises of from 0.8 to 2.9°C in surface temperature from a doubling of CO_2 concentration.

During the 19th century, the accepted atmospheric CO_2 concentration was 290 ppm (by volume), while the present value is about 320 ppm, or 10% more. The older data exhibit a great deal of scatter, however, and the accepted value may not have been accurate, although CO_2 production by fossil-fuel combustion could certainly have accounted for the increase. In recent years very precise measurements have been made at the Mauna Loa Observatory in Hawaii and at the South Pole, where the air is well mixed and far from CO_2 sources or vegetation. The precise data beginning in 1958 show a continuous uptrend in CO_2 concentration superimposed on an annual cycle that is due to the fact that levels are reduced in the growing season by photosynthesis [2, 25]. This trend is shown in Figure 4-9. The increase is about 0.7 ppm annually. Man's fossil-fuel combustion is adding perhaps

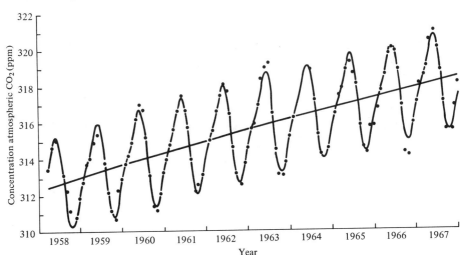

FIG. 4-9. Long-term variations in the concentration of atmospheric CO_2 at Mauna Loa Observatory in Hawaii. The dots indicate the observed monthly average concentrations. The oscillating curve is a least-squares fit of these averages based on an empirical equation containing 6- and 12-month cyclic terms and a trend function. The slowly rising curve is a plot of the trend function, chosen to contain powers of time up to the third [25]. (*Courtesy of the American Philosophical Society and C. D. Keeling.*)

9×10^3 million metric tons of CO_2 into the air each year [26], which would increase the CO_2 concentration by 1.2 ppm. It thus appears that a large fraction of man-generated CO_2 is remaining in the atmosphere but much of it is going elsewhere, probably into the oceans by dissolution.

PARTICULATE MATTER. The particulate matter in the atmosphere is believed to have the effect of cooling the earth. Small particles are effective scatterers of electromagnetic radiation, with the intensity of the scattered radiation being inversely proportional to the fourth power of the wavelength of the incident radiation as well as directly proportional to the intensity of the incident radiation. (The theory of scattering by spherical particles is usually referred to as Mie theory, and its long wavelength limit is known as Rayleigh scattering [27].) Thus blue light is scattered more than red light, producing the blue color of the sky (from above the atmosphere, the sky looks black) and the red color of a sunset, from which most of the blue light has been removed by scattering. An increase in the particulate matter in the atmosphere would thus scatter the solar shortwave radiation more than the terrestrial longwave radiation, increasing the albedo of the earth and cooling it somewhat. It has long been recognized that the dust injected into the atmosphere by great volcanic eruptions—such as Tambora in 1815, Krakatoa in 1883, and Mount Agung in 1963—can cool the earth for the following few years, mainly because much of the dust will be thrown into the stratosphere and only slowly descend back to earth [28]. Krakatoa ejected an estimated 6 to 18 km³ of rock dust into the air and the ash was soon present over the whole earth, giving rise to spectacular red sunsets.

Large volcanic eruptions greatly increase atmospheric particulate matter concentrations but man's activities may be contributing to a gradual increase in these concentrations. Data on solar radiation at Mauna Loa Observatory in Hawaii have been analyzed by Bryson, et al. [29, 30], who have argued that a long-term trend of increasing turbidity due to increasing atmospheric particulate matter can be seen from the data, and Ellis and Pueschel [31], who have argued that the turbidity shows no such trend. The Mount Agung volcanic eruption in the East Indies in March 1963 led to a dramatic increase in turbidity that remained through the late 1960's, and much of the controversy centers around the effect of this eruption.

Although volcanic dust in the stratosphere leads to cooling of the earth's surface, it is not certain that increasing atmospheric aerosol concentrations would have the same effect since particulate matter is capable of absorbing radiation as well as scattering it [14, 24]. Dust on snow and ice reduces reflection and increases absorption and can promote melting.

VAPOR TRAILS AND SUPERSONIC TRANSPORTS. In recent years the growth of the aviation industry has brought new concerns. Jet aircraft often produce persistent condensation trails that could increase the earth's albedo just as ordinary clouds do and thereby lead to a cooling of the earth. Although this is currently a matter of scientific controversy [14, 29], such effects appear to be masked by natural weather fluctuations. A more important problem seems to be the effect of the planned supersonic transports on the environment. These aircraft will fly at heights of about 20 km, well into the stratosphere, where they will consume 60 metric tons of fuel per hour and produce 75 metric tons of water, 65 metric tons of CO_2, and about 4 metric tons each of CO and NO per hour [32]. The main problem may be water since an accumulation of water vapor in the normally dry stratosphere could affect the earth's radiation balance by absorption or reflection (especially if clouds form) or lead to a decrease in the ozone concentration, permitting more ultraviolet radiation to reach the ground. At present, estimates of these effects are little more than speculation since much more research is required to determine the amounts of water already present in the stratosphere (they are only a few ppm), their fluctuation over time, and the manner in which water is removed from the stratosphere. No adverse effects or cause for alarm have yet been detected from existing supersonic flights. The development of SST's (which also cause the noise problems discussed in Chapter 7) have been opposed by environmental groups who fear that SST's might be used before their environmental effects are fully understood or even after the existence of damaging effects has been established.

THEORIES OF CLIMATIC CHANGE. The earth's climate has undergone dramatic changes in the past and may do so again. For most of recorded geological history the earth's average surface temperature has been greater than it is at present but there have also been ice ages in which it has been lower. The concern about CO_2 and particulate matter concentrations in the atmosphere arises from the possibility that their changes might affect climate by affecting the average temperature. The increase in average temperatures from 1880 to the mid-1940's mentioned earlier in this chapter has been attributed to CO_2 production and the subsequent decrease to atmospheric aerosol increase [29]. On the other hand, solar activity and volcanic activity have also been considered strong possibilities as causes. Some scientists believe the observed trends have just been natural climatic fluctuations due to large-scale and long-term air-sea interactions and are of no particular significance in discussions of air pollution [33]. Scientific research must continue if significant global effects of pollution are to be anticipated in time.

More information about the many different theories of climate change can be found in texts on climatology [34] and papers dealing with one theory or another, such as the carbon dioxide theory (which is much richer than we have hinted at in the present discussion) [35] or the volcanic theory [36]. A good recent summary is Reference 37.

REFERENCES

1. Staley, D. O., "The Adiabatic Lapse Rate in the Venus Atmosphere." *Jour. Atmos. Sciences* **27**:219–223 (1970).
2. Mitchell, J. M., Jr., "A Preliminary Evaluation of Atmospheric Pollution as a Cause of the Global Temperature Fluctuation of the Past Century." In Singer, S. F. (ed.), *Global Effects of Environmental Pollution*. New York: Springer-Verlag, Inc., 1970. pp. 139–155.
3. Derived in modern physics texts, such as Blanpied, W. A., *Modern Physics*. New York: Holt, Rinehart and Winston, Inc., 1971. pp. 335–338.
4. Stacey, Frank D., *Physics of the Earth*. New York: John Wiley & Sons, Inc., 1969.
5. Vonder Haar, T. H., and V. E. Suomi, "Satellite Observations of the Earth's Radiation Budget." *Science* **163**:667–669 (1969).
6. Kondratyev, K. Ya., *Radiation in the Atmosphere*. New York: Academic Press, 1969.
7. Goody, R. M., *Atmospheric Radiation. I. Theoretical Basis*. Oxford: Clarendon Press, 1964. p. 4.
8. Fensterstock, Jack C., and Robert K. Fankhauser, *Thanksgiving 1966 Air Pollution Episode in the Eastern United States*. Durham, N.C.: U.S. Dept. of Health, Education, and Welfare, 1968. N.A.P.C.A. Publication No. AP-45.
9. Glasser, M., L. Greenburg, and F. Field, "Mortality and Morbidity During a Period of High Levels of Air Pollution. New York, November 23–25, 1966." *Arch. Environ. Health* **15**:684–694 (1967).

10. Firket, J., "Fog Along the Meuse Valley." *Trans. Faraday Soc.* **32**:1192–1197 (1936).

11. Schrenk, H. H., et al., *Air Pollution in Donora, Pa. Epidemiology of the Unusual Smog Episode of October 1948*. Washington, D.C.: U.S. Public Health Service, 1949. Public Health Service Bulletin No. 306.

12. *Clean Air Year Book*. Published annually by the National Society for Clean Air (see Appendix B).

13. Peterson, James T., *The Climate of Cities: A Survey of Recent Literature*. Raleigh, N.C.: U.S. Dept. of Health, Education, and Welfare, 1969. N.A.P.C.A. Publication No. AP-59.

14. Landsberg, H. E., "Man-Made Climatic Changes." *Science* **170**:1265–1274 (1970).

15. Singer, S. F. (ed.), *Global Effects of Environmental Pollution*. New York: Springer-Verlag, Inc., 1970.

16. *World Energy Supplies 1965–1968*. New York: United Nations Dept. of Economic and Social Affairs, 1970.

17. Broecker, Wallace S., "Man's Oxygen Reserves." *Science* **168**:1537–1538 (1970).

18. Machta, L., and E. Hughes, "Atmospheric Oxygen in 1967 to 1970." *Science* **168**:1582–1584 (1970).

19. Tyndall, J., "On the Absorption and Radiation of Heat by Gases and Vapours, and on the Physical Connexion of Radiation, Absorption, and Conduction— The Bakerian Lecture." *Phil. Mag.* **22**:169–194, 273–285 (1861).

20. Arrhenius, S., "On the Influence of Carbonic Acid in the Air Upon the Temperature of the Ground." *Phil. Mag.* **41**:237–276 (1896).

21. Callendar, G. S., "The Artificial Production of Carbon Dioxide and Its Influence on Temperature." *Quart. Jour. Roy. Meteor. Soc.* **64**:223–237 (1938).

22. Plass, G. N., "The Carbon Dioxide Theory of Climatic Change." *Tellus* **8**: 140–154 (1956).

23. Manabe, S., and R. T. Wetherald, "Thermal Equilibrium of the Atmosphere with a Given Distribution of Relative Humidity." *Jour. Atmos. Sci.* **24**:241– 259 (1967). See also Manabe, S., "The Dependence of Atmospheric Temperature on the Concentration of Carbon Dioxide" on pp. 25–29 of Reference 15.

24. Rasool, S. I., and S. H. Schneider, "Atmospheric Carbon Dioxide and Aerosols: Effects of Large Increases on Global Climate." *Science* **173**:138–141 (1971).

25. Keeling, C. D., "Is Carbon Dioxide from Fossil Fuel Changing Man's Environment?" *Proc. Amer. Phil. Soc.* **114**:10–17 (1970).

26. Goldberg, E. D., "The Chemical Invasion of the Oceans by Man" on pp. 178–185 of Reference 15.

27. Kerker, Milton, *The Scattering of Light and Other Electromagnetic Radiation*. New York: Academic Press, 1969.

28. Lamb, H. H., "Volcanic Dust in the Atmosphere; With a Chronology and Assessment of Its Meteorological Significance." *Phil. Trans. Roy. Soc. London* **A266**:425–533 (1970).

29. Bryson, R. A., and W. M. Wendland, "Climatic Effects of Atmospheric Pollution" on pp. 130–138 of Reference 15.

30. Peterson, J. T., and R. A. Bryson, "Atmospheric Aerosols: Increased Concentrations during the Last Decade." *Science* **162:**120–121 (1968).
31. Ellis, H. T., and R. F. Pueschel, "Solar Radiation: Absence of Air Pollution Trends at Mauna Loa." *Science* **172:**845–846 (1971).
32. Nuessle, V. D., and R. W. Holcomb, "Will the SST Pollute the Stratosphere?" *Science* **168:**1562 (1970).
33. Namias, J., "Climatic Anomaly over the United States during the 1960's." *Science* **170:**741–743 (1970).
34. Critchfield, Howard J., *General Climatology*. 2nd ed. Englewood Cliffs, N.J.: Prentice-Hall, Inc., 1966.
35. Plass, G. N., "Carbon Dioxide and the Climate." *American Scientist* **44:**302–316 (1956).
36. Wexler, Harry, "Volcanoes and World Climate." *Scientific American* **186**(4): 74–80 (April 1952).
37. *Inadvertent Climate Modification*. Report of the Study of Man's Impact on Climate. Cambridge, Mass.: The MIT Press, 1971.

5

AIR POLLUTION: INDUSTRIAL EMISSIONS AND CLASSICAL SMOG

I want the sky to be filled with the smoke of American industry, and upon that cloud of smoke will rest forever the bow of perpetual promise. That is what I am for. —Robert Ingersoll, October 23, 1880 [1]

This quotation from the speech of Colonel Robert G. Ingersoll before the Cooper Institute in New York City is reported to have brought forth great cheering and cries of "Good! Good!" Today, with the perspective of another century, the same words would be more likely to elicit booing.

In this chapter we shall review the history of smoke pollution and then discuss some important industrial air pollution problems of today and the control techniques available for their abatement.

THE HISTORY OF SMOKE POLLUTION

Over the centuries, air pollution problems have been referred to as "smoke" problems, even though sulfur oxides as well as particulate matter have generally been involved. The major smoke problems in industrial towns arose from coal combustion. In the series from wood to peat to brown coal to lignite to subbituminous coal to bituminous coal to semibituminous coal to anthracite, one finds more and more carbon (50% in wood and over 90% in anthracite) in the dry nonash portion. Anthracite tends to be the least smoky and it also contains the smallest amount of sulfur, which means that it contributes less sulfur oxides [2].

Although there is evidence that coal from surface outcroppings was used in Great Britain as early as 1450 B.C., wood remained the chief fuel for many centuries. Mining of coal for nearby use occurred in the 12th century and coal from Newcastle was being shipped to London in the reign of Henry III (1216 to 1272), apparently as a fuel for dyers and brewers. As the population grew and manufacturing became more important and timber became more scarce, the need for coal increased. There are reports that some townspeople complained about the disagreeable nature of the smoke, which they correctly thought to be injurious to human health. In 1306, Edward I

73

issued a royal proclamation forbidding the use of sea coal (coal from sea-shore outcroppings) while Parliament was in session. Although one manufacturer is reputed to have been executed for violating this decree, a Commission of Inquiry had to be set up the next year to "inquire of all such who burnt sea-coal in the city or parts adjoining, and to punish them for the first offence with great fines and ransoms, and upon the second offence to demolish their furnaces" [3].

Restrictions on the use of sea coal made little headway. In fact Richard Whittington, four times (not just thrice as the old story of Dick Whittington and his cat puts it) Lord Mayor of London just before and after 1400, made his fortune hauling coal to London, and his "cat" was not a feline animal but a collier, or boat for hauling coal. Coal became a domestic fuel for the poor as well as an industrial fuel for manufacturers and its use increased during the 15th and 16th centuries. One record from 1578 stated that Queen Elizabeth I "findeth herselfe greatly greved and anoyed with the taste and smoke of the sea-cooles." The upper classes did not adopt coal in place of wood until the 17th century.

In 1661 John Evelyn wrote his antismoke tract *Fumifugium* "being extremely amazed, that where there is so great an affluence of all things which may render the People of this vast City, the most happy upon Earth; the sordid and accursed Avarice of some few particular Persons, should be suffered to prejudice the health and felicity of so many." In his complaint, addressed to King Charles II, he said of London that "Her *Inhabitants* breathe nothing but an impure and thick Mist, accompanied with a fuliginous and filthy vapour, which renders them obnoxious to a thousand inconveniences, corrupting the *Lungs,* and disordering the entire habit of their Bodies, so that *Catharrs, Phthisicks, Coughs* and *Consumptions,* rage more in this one City, than in the whole Earth besides. . . . Is there under Heaven such Coughing and Snuffing to be heard, as in the *London* Churches and Assemblies of People, where the Barking and the Spitting is uncessant and most importunate" [4].

Evelyn's solution to the problem of air pollution was to move the "*Brewers, Diers, Sope* and *Salt-boylers, Lime-burners,* and the like . . . five or six miles distant from *London* below the River of *Thames.*" He also urged the planting of trees throughout the city.

By this time London's fogs were becoming notorious for their thickness and duration and unnoticed air pollution disasters were probably occurring. During the 19th century several committees were appointed by Parliament to investigate the smoke problem and recommendations were received for abatement of smoke from furnaces and steam engines but no meaningful action resulted. The Public Health Act of 1875 contained a smoke abatement section but the air pollution reformers of the day were unable to counteract the industrialists' slogan, "Where there's muck there's money."

In time Londoners even came to speak proudly of their "pea-soupers." One that occurred in December 1873 killed some of the prize cattle at the

Smithfield Show and is now realized to have caused several hundred human deaths from bronchitis. A number of smoke abatement societies were formed in the late 19th and early 20th centuries, one by Dr. H. A. Des Voeux, who coined the word "smog." An amalgamation took place in 1929 when the National Smoke Abatement Society was formed; its name was changed to National Society for Clean Air in 1958.

Despite occasional committee reports and some new legislation, the British situation apparently worsened during the first half of the 20th century. Smog episodes in London went unnoticed and the well-publicized Meuse Valley and Donora episodes were an insufficient warning of the imminence of a major disaster. During World War II the Ministry of Home Security actually urged the use of smoky coal and the dismantling of control devices in order to conceal important targets from Nazi reconnaissance planes, but the stand was reversed in 1943 when a coal shortage developed and the conservation of fuel became an important consideration.

Major disaster struck London during the first week of December 1952 when the anticyclone center of a cold, dry, slow-moving air mass stopped for several days over the city, producing a thermal inversion that hemmed in the sulfur oxides and particulate matter from London's industries and homes. Respiratory cases increased dramatically and ambulances were unable to handle the extra load between Thursday, December 4 and Monday, December 8.

The reports on the "fog" in the *London Times* shows that there was no suspicion of the effects on human health, even though several articles mentioned the difficult breathing and even death of some of the prize cattle at the Smithfield Show. The Saturday article was headlined, "Fog Delays Air Services." The Monday article headlined "Transport Dislocated by Three Days of Fog" stated that "veterinary officials worked on Saturday and yesterday treating animals which appeared to have contracted respiratory troubles either on their long journeys or in the fog-filled hall" at the Smithfield Show. On Tuesday, the *Times* report was headlined "More Deaths at Cattle Show."

The smog finally ended and the port of London reopened after being closed for 4½ days. The medical profession realized that mortality had been abnormally high during the episode and for some days thereafter (see Figure 3-1), and the publication of the statistics in medical journals led to widespread public interest in the matter. At the prodding of the National Smoke Abatement Society, the government appointed a committee headed by Sir Hugh Beaver to study the episode. The committee's final report in November 1954 urged tighter emission controls and the establishment throughout England of smokeless zones—regions in which only certain smokeless fuels such as anthracite may be used. A Clean Air Act was passed in 1956 and extended in later years to require new furnaces to be as smokeless as possible and chimneys to be of sufficient height to provide adequate dispersal of pollutants; to control collieries, railway engines, and ships; and to provide for the estab-

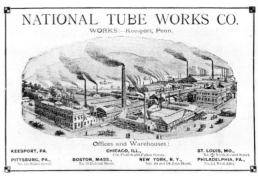

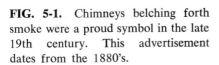

FIG. 5-1. Chimneys belching forth smoke were a proud symbol in the late 19th century. This advertisement dates from the 1880's.

lishment of smoke control areas. In 1955 London itself became a smokeless zone and the area of smoke control zones in the rest of the country is being increased each year, although the process may not be completed until 1990.

Great Britain's Clean Air Act has dramatically reduced coal smoke but sulfur oxide emissions have actually increased somewhat over the years and automobile pollution has soared. Nevertheless, London's appearance has improved with the reduction in particulate matter and songbirds absent from the city for a century have returned.

A number of American cities gained notoriety for their smokiness during the 19th and early 20th centuries. As early as 1867, St. Louis passed an ordinance requiring smokestacks to be at least 20 ft higher than adjoining buildings. In 1893, St. Louis passed a smoke ordinance stating,

> The emission into the open air of dense black or thick gray smoke within the corporate limits of the City of St. Louis is hereby declared to be a nuisance. The owners, occupants, managers, or agents of any establishment, locomotives, or premises from which dense black or thick gray smoke is emitted or discharged, shall be deemed guilty of a misdemeanor, and, upon conviction thereof, shall pay a fine of not less than ten nor more than fifty dollars. And each and every day wherein such smoke shall be emitted shall constitute a separate offense.

When the city brought suit against an obdurate offender, the Heitzeberg Packing and Provision Company, however, the Missouri Supreme Court declared the law invalid because "while it is entirely competent for the city to pass a reasonable ordinance looking to the suppression of smoke when it becomes a nuisance to property or health or annoying to the public at large, this ordinance must be held void because it exceeds the powers of the city under its charter to declare and abate nuisances and is wholly unreasonable." St. Louis did not have an enforceable smoke ordinance until 1902, a year after the state legislature of Missouri passed a law enabling communities to pass such ordinances [5].

Citizens' groups worked for smoke abatement legislation in a number of other cities, with occasional success. A group formed in Chicago in 1874 was instrumental in getting the City Council to pass a smoke ordinance in 1881, probably the first in the U.S. Many of these laws (and some still in existence today) specified violations in terms of shades of grayness of the smoke, defined by means of the Ringelmann chart (Figure 5-2). The four numbered shades correspond to 20, 40, 60, and 80% blackness. At a sufficient distance from the observer they appear as different shades of gray and are to be held in line with the smoke being observed [6]. Ringelmann charts provide a useful measure of the aesthetic insult of smoke but they are obviously rather unscientific since the grayness of smoke is not a good indication of the amounts and dangers of the emissions.

Pittsburgh, often called the "Smoky City," was one of the smokiest cities in the United States for many years because of the extensive use of coal for household and industrial heating. From 1911 to 1913 the Mellon Institute carried out an extensive investigation of the physical and physiological damage resulting from smoke. These studies documented the reduction of sunshine caused by the smoke particles, their irritating effects on the respiratory

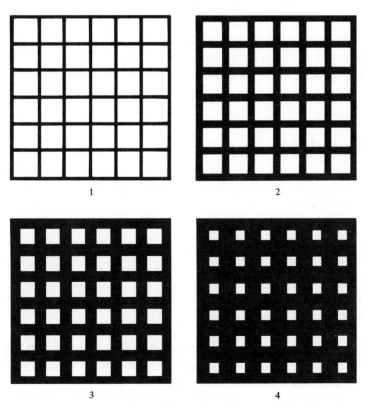

FIG. 5-2. The Ringelmann chart. Shades numbers 1, 2, 3, and 4 correspond, respectively, to 20, 40, 60, and 80% blackness. From a distance the grids merge into tints of gray that may be compared with smoke issuing from a stack [6].

and gastrointestinal tracts, the "depressing, devitalizing effect" of the constant darkness of the city, and the fact that "smoke clouds are inimical to the highest aesthetic development of urban communities" [7]. In the late 1930's some civic leaders began a crusade for clean air that encountered strong opposition from industrial and railroad corporations; some legislation was passed but not enforced during World War II. A wealthy Democratic industrialist, David Lawrence, was elected mayor in 1945 on a platform pledging smoke control but legislation permitting countywide smoke control only passed several years later after Richard King Mellon and other industrialists supported it. The increased use of smokeless fuels (especially natural gas) and control devices led to marked increases in visibility and days of sunshine and to a sharp drop in dust fall.

The great reduction in smoke in American cities in the past few decades has been due to the replacement of coal by fuel oil and natural gas and the control of particulate matter emissions on installations that still burn coal. The amount of energy produced by coal in the U.S. was almost unchanged between 1937 and 1965 but the percentage of total energy from coal dropped from 55% to 23%, while oil's percentage increased from 30 to 43% and that of natural gas from 11 to 30%; in both cases hydroelectric power supplied about 4% of the nation's needs [8]. As late as 1947, coal supplied over 50% of household and commercial energy needs; the present fraction is only a few percent.

INDUSTRIAL EMISSIONS

The data presented in Table 2-3 showed that industrial air pollution and electric power generation contribute a sizable portion of the pollutant emissions in the U.S.:

> 87% of the sulfur oxides.
>
> 56% of the particulate matter.
>
> 45% of the nitrogen oxides.
>
> 15% of the hydrocarbons, and
>
> 11% of the carbon monoxide.

These emissions arise from a great variety of sources, each with its own special problems and control technologies [9]. In this section, a number of specific sources and industries will be discussed; in the next section, the control methods appropriate to each main type of air pollution will be discussed.

FUEL COMBUSTION BY STATIONARY SOURCES. The importance of air pollution due to fuel combustion by stationary sources (as opposed to motor vehicle sources) may be seen from Table 3-2. Particulate matter, sulfur oxides, and nitrogen oxides are all major problems. The particulate matter arises mainly from coal, which typically contains about 10% noncombustible mineral ash of various sorts. Sulfur oxides arise from the sulfur content, typically a few percent, of coal and fuel oil. Nitrogen oxides, as discussed in Chapter 3, are a product of any high-temperature combustion process and thus result from the combustion of any of the fossil fuels. The particulate matter emissions would be over five times the actual current emissions if it were not for control measures; at the present time, electric power plants trap about 86.5% of their particulate emissions and industries trap about 62% of theirs [10]. The sulfur oxide emissions would be greater if it were not for the willingness of some users to pay a premium for low-sulfur coal or fuel oil, whose use is legally required in some places. In terms of heat content, it has been estimated [8] that 57.5% of U.S. coal reserves are low-sulfur (less than 1% sulfur), 17.4% are medium sulfur (1 to 3% sulfur), and 25.1% are high-sulfur (greater than 3% sulfur). Unfortunately, the reserves in the northeastern part of the country, which uses the most coal, tend to be higher in sulfur.

PETROLEUM REFINERIES. Petroleum refineries begin with crude oil and separate it into "fractions" of different boiling point ranges: gas, gasoline, kerosine, diesel fuel, fuel oil, heavy bottoms, etc. Depending on the amount of refining that takes place, these fractions may undergo catalytic or thermal "cracking" to produce hydrocarbons of lower molecular weight, or catalytic reforming to rearrange the molecular structure. In an advanced refinery many special products may be manufactured. These processes involve the production of a large number of pollutants that can often be controlled fairly easily at no great cost. Emissions include the usual products of combustion processes (CO and NO_x) plus sulfur oxides and hydrocarbons. Hydrocarbon losses may range from 0.1 to 0.6% of the crude oil input depending on the nature of the refining processes employed and the abatement facilities installed [11]. Refinery emissions of hydrocarbons tend to be the lower-molecular-weight paraffins, which are not as reactive photochemically as the olefinic and aromatic compounds and the higher-molecular-weight paraffins. During the 1950's, Los Angeles successfully instituted rigid controls on refinery emissions. By 1966, emissions per ton of crude oil had been reduced, relative to precontrol days, by about 93% for hydrocarbons, 96.5% for sulfur oxides, 94% for CO, and over 85% for particulate matter [11].

PULP AND PAPER MILLS. Pulp and paper mills are notorious sources of air pollution and obnoxious odors throughout the world. The preparation of cellulose wood pulp may involve mechanical processes but chemical processes are most common today. The leading process, and the one used by almost all the new pulp facilities, is the kraft process in which the raw wood is digested in chip form for several hours in a hot, pressurized solution of sodium sulfide and sodium hydroxide ("white liquor"). This dissolves the wood lignins, carbohydrates, organic acids, etc., leaving cellulose fibers suspended in the spent cooking solution ("black liquor"). The digestion process produces gases that must be bled off from time to time to prevent excess pressurization. These gases may contain up to 5% by volume of disagreeable-smelling mercaptans (compounds of the form RSH, where R is an alkyl hydrocarbon group), especially methyl mercaptan (CH_3SH), as well as up to a few hundred ppm of hydrogen sulfide (H_2S). When the digestion is stopped, the black liquor is evaporated until solids amount to about two-thirds of the total mass and more gases are lost to the atmosphere in the process. Burning of the concentrated liquor in a recovery furnace can lead to the release of more mercaptans and H_2S and other organic sulfur compounds, although proper oxidation will lead to SO_2 that can be retained in the residue in sulfite form. (The recovery furnace residue, or smelt, which consists mainly of Na_2CO_3 and Na_2S, is treated with lime to produce an alkaline sodium sulfide solution, which can be reused in the digestion process.)

Progress in the reduction of air pollution and odors has been marked in Scandinavia, where great efforts have been taken to recover the waste heat in the exhaust gases and use it in digestion, in order to reduce fuel costs. The cheapness of fuel in North America has not provided an incentive of this sort, although utilization of waste heat might reduce heat needs by about 30%. A great deal of effort has gone into pollution control, such as oxidation by air or chlorine to yield less smelly gases, scrubbing, absorption, and the introduction of digester and evaporator gases into the black liquor oxidation process [12]. Nevertheless, pulp and paper mills remain major pollutants of the air (and the water, as discussed in Chapter 10) in many localities.

These problems are largely avoided by the recycling of wastepaper, in which the wastepaper is cleaned, chopped, moistened, reconditioned, repulped, and made into paperboard for cereal or detergent boxes, newsprint, wallboard, or other new products.

INORGANIC CHEMICALS. The inorganic chemical industry is faced with the control of a wide range of undesirable odors, gases, and particulate matter [13]. Hydrochloric and sulfuric acid plants are faced with controlling acid fumes. Hydrofluoric acid plants, which manufacture the HF by reacting fluorspar (CaF_2) with sulfuric acid, produce HF, SO_2, H_2S, and fluosilicic acid (H_2SiF_6). Lime production plants, which burn limestone in kilns, have serious dust problems all the way from the original blasting in the quarries to the pulverization and bagging. Phosphate fertilizer plants acidify phosphate rock and produce a number of undesirable gases, including SiF_4, which subsequently hydrolyzes to HF gas or mist (phosphate rock typically contains about 4% fluoride). Chlorine and bromine vapors are problems in plants producing those gases. Tear gas emissions from a chemical plant in Great Meadows, N.J., have irritated residents, bringing on spasms of uncontrolled sneezing as well as copious tears [14].

SMELTERS. Smelting processes require high temperatures and produce a number of dusts, fumes, and gases. Some metals (such as copper, lead, and zinc) occur in nature as sulfides and the smelting process produces vast amounts of SO_2. The smelting of copper can be summarized by the chemical reaction

$$Cu_2S + O_2 \rightarrow 2Cu + SO_2$$

so that for every ton of copper produced there will be 0.49 tons of SO_2 produced (or 0.77 tons of H_2SO_4 if the SO_2 is used to make sulfuric acid). Smelters have often been located in remote areas and smoke easements have sometimes been purchased from nearby landowners. Nevertheless, extensive damage to vegetation can result over huge areas. A famous example is the Ducktown-Copperhill, Tenn., area, where copper mining has been taking place since 1850. Some 3000 hectares are completely devoid of vegetation and another 7000 hectares have no trees; injury has been detected on white pines 50 km away. In the 1920's a smelter at Trail, British Columbia, Canada, injured trees 84 km away. Today, large smelters generally pass the exit gases through a sulfuric acid plant, thus cutting down SO_2 emissions by 90% while producing a marketable product (the sulfuric acid).

AGRICULTURE AND FOOD PROCESSING. There are a number of air pollution problems caused by agriculture and the food industries based on it; some of the agricultural problems are discussed in Chapters 11 and 12. Dust can be blown from fields in the process of wind erosion; the great dust storms of the 1930's were due to removal of the grass cover on the Great Plains in the process of converting them to the growing of wheat. Agricultural pesticides can be transported by air, especially if sprayed from planes; in one incident in Texas, the herbicide 2,4-D damaged 4000 hectares of cotton 25 to 30 km away from the rice field on which it had been applied by airplanes. For many years California citrus growers used terribly smoky "smudge pots" in the belief that the smoke protected the trees and fruit from the sun after frosty nights, although it was only the heat from the pots that helped the fruit. Complaints are becoming more numerous about obnoxious odors from cattle, hog, and sheep pens and yards, especially from large feedlots. Factories which process and pack meat and fish may produce disagreeable odors due to nitrogen and sulfur compounds (such as amines) which arise from protein decomposition. The first factory successfully prosecuted under the federal Clean Air Act was the Bishop Processing Company plant in Bishop, Md., which rendered (processed) waste chicken parts (entrails, head, feet, and feathers) from chicken packing plants in the area and sent sickening odors across the state border into Delaware; the company, which had been in existence 15 years before being closed in 1970 and was doing $1.5 million of business annually, was actually making a useful product (high-protein animal food) out of a solid waste. Air pollution problems can also result from the decay of organic wastes at fruit and vegetable processing plants. The odors produced by bread bakeries, candy and cookie factories, or coffee roasters qualify as air pollution (and have elicited complaints) although many persons enjoy them [15].

CONTROL TECHNIQUES

PARTICULATE MATTER. Various types of equipment are available for removing particulate matter from exhaust gases but the selection of the appropriate method is far from an exact science [10]. The physical and chemical characteristics of the particulate matter and of the gas carrying it are very important but they vary greatly from one industry to another. Generally, however, particles larger than 50 μm may be removed satisfactorily by cyclone separators and simple wet scrubbers; particles smaller than 1 μm are handled effectively by electrostatic precipitators, high-energy scrubbers, and fabric filters.

Cyclone separators (see Figure 5-3a) are designed so the incoming gas stream produces a vortex from which particulate matter is removed by centrifugal forces and then collected. They are very simple in design and easy to maintain. They handle large particles very well but the collection efficiency of small particles is low. Efficiencies on a total weight basis may range from 50 to 95%.

Wet scrubbers (Figure 5-3b) clean the gas by wetting with water; the particles are removed in the waste water, along with some corrosive gases and mists. Efficiencies by weight range from 75 to over 99% and can be varied. There may, however, be corrosion problems associated with this process and there is the added expense of treating the waste water. The effluent gas may carry some water vapor and have a visible plume.

An electrostatic precipitator (see Figure 5-3c) removes solid and liquid particles from a gaseous stream by passing the gas between pairs of electrodes. The negative electrodes have a voltage high enough with respect to the grounded collection electrodes to produce a corona discharge. The gas ions in the corona are attracted to the collection electrode but they may collide with solid and liquid particles and transfer charge to them; the particles are then attracted to and deposited on the collection electrode. Collection efficiencies by weight are from 80 to over 99.5% but are usually designed in the range of 98 to 99.5%. The initial costs of electrostatic precipitators are relatively high but their power requirements (to produce the corona) are small and little maintenance is usually needed, although corrosive materials can be troublesome. The principle of the electrostatic precipitator dates back to the early 19th century but the first practical device was built by Cottrell in the early 1900's. This device is especially suitable for the fine particles produced in coal combustion and called *fly ash* (as opposed to the *bottom ash*

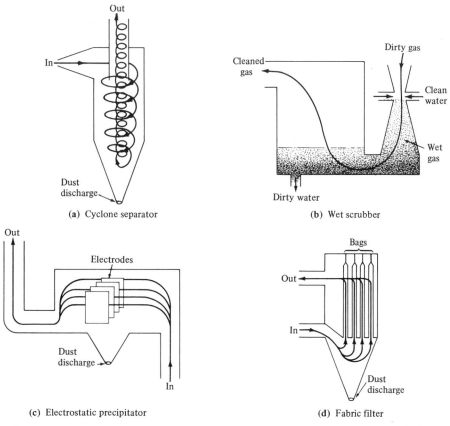

Out

In

Dust
discharge

(a) Cyclone separator

Dirty gas

Cleaned
gas

Clean
water

Wet
gas

Dirty water

(b) Wet scrubber

Out

Electrodes

Dust
discharge

In

(c) Electrostatic precipitator

Bags

Out

In

Dust
discharge

(d) Fabric filter

FIG. 5-3. Schematic drawings of four different control techniques for particulate air pollution.

residue from the combustion). At present, industrial installations in the U.S. produce about 30 million metric tons of fly and bottom ash annually—an important solid waste problem.

Fabric filters (see Figure 5-3d) extract particulate matter from gas streams by filtration with woven fabric, paper, fibrous mats, or aggregates such as sand, gravel, or coke. Large dust collectors may contain several thousand filter bags, each several meters long; individual bags can be disconnected from the gas stream for cleaning. Removal efficiencies may be 95 to 99.9% but high-temperature gases must first be cooled to 100 to 300°C, and the fabric may still be susceptible to chemical attack. These filters are widely used, especially in cement, carbon black, clay, and pharmaceutical plants.

In some cases, when the particulate matter to be removed is combustible, it is possible to burn it with direct-flame or catalytic afterburners. Afterburners are devices that complete the oxidation of combustible materials, generally without the gain of any useful energy.

SULFUR OXIDES. The control of sulfur oxide emissions is an important but not satisfactorily resolved problem [16]. A number of removal methods are being developed at the present time but the substitution of low-sulfur fuels has been of greater value to date [17].

The development of new sources of low-sulfur fuels, which command a premium price, has increased in recent years. Coking coals from western Canada have suddenly found great favor because of their low-sulfur content (about 0.5%) but most of the current production and the expansion scheduled in the early 1970's in these coalfields has been contracted for by Japanese firms. Several petroleum companies have built plants for the desulfurization of fuel oil. Standard Oil (New Jersey) has built a $120 million facility at Amuay, Venezuela, that will remove sulfur (as H_2S) from crude oil using hydrogen gas (see Figure 5-4); 1.9 million m^3 of hydrogen will be used daily to remove 335 metric tons of sulfur from 100,000 barrels (16,000 m^3) of crude oil. There are also studies underway to remove sulfur from coal and to convert coal into low-sulfur liquid and gaseous fuels.

There exist a number of methods of removing sulfur oxides from stack gases but most have been used only in small demonstration units, and electric utilities and industries have been reluctant to try them on a large scale, preferring to shift to low-sulfur fuels until one or more methods have proved their superiority. The simplest method involves the addition of limestone ($CaCO_3$) or dolomite ($CaCO_3$-$MgCO_3$ mixtures) to the furnace, with production of $CaSO_4$ or $MgSO_4$ particles upon reaction with sulfur oxides; SO_x removal efficiencies of over 80% are possible when a wet scrubber is placed on this system. This method converts a gas problem into a particulate matter problem and it leads to a large amount of waste material containing no recoverable by-products. More complicated methods are more expensive but they lead to valuable by-products (sulfur and sulfuric acid) whose sale will reduce the net cost and perhaps even lead to profits. There has even been speculation in business publications that SO_x removal processes might spell the end of the conventional Frasch process for mining sulfur. The catalytic oxidation process converts SO_x in the gas stream into weak (75 to 80%) sulfuric acid by passing it through a vanadium pentoxide (V_2O_5) catalyst and a series of condensers. The char sorption or Reinluft process uses a special activated char (a porous substance composed largely of pure carbon) to absorb the sulfur dioxide. The alkalized alumina process uses a dry sodium-aluminate metal oxide to contact and react with the SO_2 and eventually convert it into elemental sulfur. Several private firms are also developing SO_x removal processes about which information is limited. It is likely to be several years before any of the processes have improved to the point of being commercially attractive. A recent review may be found in Reference 18.

CARBON MONOXIDE. Carbon monoxide emissions from stationary combustion sources can be reduced by modifications of the combustion process or the use of afterburners [19]. In addition, they can be reduced or eliminated

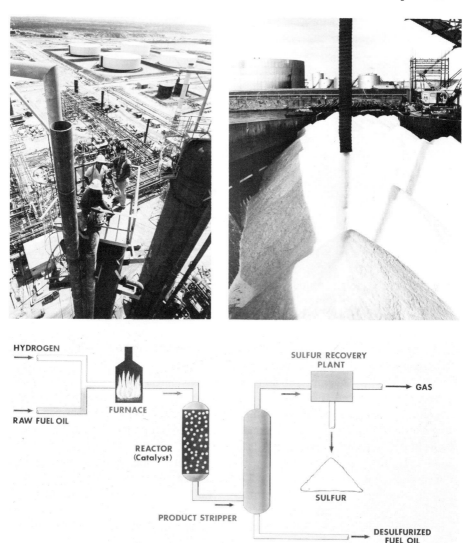

FIG. 5-4. (a) A view of part of the fuel oil desulfurization plant built by Creole Petroleum Company at its Amuay (Venezuela) refinery. (b) Sulfur produced at the sulfur recovery plant. (c) Schematic drawing showing the desulfurization process. A combination of heat, hydrogen, and a special catalyst reduces the sulfur content of the fuel oil to as low as 0.3%. [*Courtesy of Standard Oil Company (New Jersey).*]

by changes in fuel or energy source; for example, natural gas combustion produces less CO than coal or oil combustion and nuclear power eliminates CO emissions altogether.

The amount of CO emitted in a combustion process depends markedly on certain parameters that should be properly adjusted. The air supply must be adequate or else incomplete combustion of the fuel results, with more CO and less CO_2 produced. Longer residence times in the combustion chamber also promote more complete combustion, as does proper mixing. High temperatures are generally desirable to promote combustion but above 1550°C there is noticeable dissociation of CO_2 into CO and oxygen, and flame temperatures above 1650°C are conducive to formation of nitrogen oxides.

The combustion of CO can sometimes be accomplished by the use of afterburners, as in some segments of the iron and steel industry. Since this combustion is a source of heat, the CO is sometimes burned as fuel, as in blast furnaces and petroleum catalytic cracking units.

HYDROCARBONS. Hydrocarbon emissions can sometimes be reduced by operational and process changes or substitution of materials but several control methods are also available [20].

Evaporation of hydrocarbons is an important problem in the petroleum refining industry and numerous systems have been devised to reduce losses by evaporation—floating roof tanks, pressure tanks, vapor recovery systems, etc. Control of vapors emitted during the loading of gasoline tank trucks, their unloading, and subsequent storage has also been emphasized by the industry. Evaporation is also a concern of dry cleaning establishments, municipal sewage treatment plants, and other industries.

Although organic emissions from paints and solvents cannot satisfactorily be controlled, it is possible to make use of materials that are less volatile or less reactive photochemically. Los Angeles County has instituted regulations restricting the emissions of reactive compounds such as olefins, substituted aromatics, and aldehydes.

Hydrocarbon emissions can be controlled by incineration, adsorption, absorption, and condensation. Incineration can be accomplished by direct-flame afterburners, with temperatures of 650 to 750°C and residence times of 0.3 to 0.6 s, or catalytic afterburners, whose efficiencies are lower. Adsorption can be accomplished by the use of activated carbon and subsequent regeneration of the carbon; high removal efficiencies are possible. When hydrocarbon concentrations are high, it is possible to use wet scrubbers to absorb the vapors by water or mineral oil or another suitable liquid. Condensers that collect hydrocarbons by lowering the temperature of the gaseous stream to the condensation point can also be used for concentrated gases; this is common in the petroleum and chemical industries, where the condensers are really production equipment rather than control equipment.

NITROGEN OXIDES. Nitrogen oxide emissions are best controlled by modifications of the combustion process, although other techniques are being tried or developed [21, 22]. Lowering peak combustion temperatures reduces NO_x emissions since NO production is negligible below 550°C but quite significant above 1650° (see Figure 5-5). It is also desirable to avoid strongly

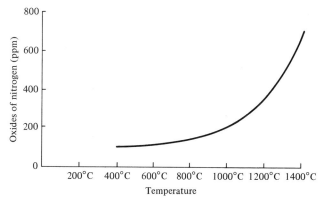

FIG. 5-5. Temperature dependence of nitrogen oxides emissions from coal combustion [21].

oxidizing atmospheres and to avoid too rapid cooling of the exhaust gases—both of which increase NO_x emissions.

Catalytic methods are available for reducing NO_2 to NO ("decolorizing") and several methods have been tried, without marked success, to use catalysts to reduce NO_x to nitrogen and oxygen gases. Scrubbing techniques using solutions of sodium or calcium hydroxide that lead to formation of nitrates and nitrites have also been developed. A number of speculative control techniques are discussed in References 21 and 22.

REFERENCES

1. *Great Speeches of Col. R. G. Ingersoll.* Chicago: Rhodes and McClure, 1885. p. 160.
2. Marsh, Arnold, *Smoke.* London: Faber & Faber, Ltd., 1948.
3. *Clean Air Year Book 1969–70.* London: National Society for Clean Air, 1969.
4. Evelyn, John, *Fumifugium* (1661). London: National Society for Clean Air, 1961.
5. Cannon, L. H., *Smoke Abatement.* St. Louis, Mo.: St. Louis Public Library, 1924.
6. U.S. Bureau of Mines, *Ringelmann Smoke Chart.* Washington, D.C.: U.S. Dept. of the Interior, May 1967. Bureau of Mines Information Circular 8333.
7. Wallin, J. E. W., *Psychological Aspects of the Problem of Atmospheric Smoke Pollution.* Pittsburgh, Pa.: University of Pittsburgh, 1913.
8. *Air Pollution and the Regulated Electric Power and Natural Gas Industries.* Washington, D.C.: U.S. Federal Power Commission, 1968.
9. Stern, A. C. (ed.), *Air Pollution. Volume III. Sources of Air Pollution and Their Control.* 2nd ed. New York: Academic Press, 1968.
10. *Control Techniques for Particulate Air Pollutants.* Washington, D.C.: U.S. Dept. of Health, Education, and Welfare, 1969. N.A.P.C.A. Publication No. AP-51.

11. Elkin, Harold F., "Petroleum Refinery Emissions" in Reference 9, pp. 97–121.
12. Adams, Donald F., "Pulp and Paper Industry" in Reference 9, pp. 243–268.
13. Heller, A. N., et al., "Inorganic Chemical Industry" in Reference 9, pp. 191–242.
14. Bird, David, "Leaking Tear Gas Has Area Crying." *N.Y. Times,* January 1, 1971, p. 13.
15. Faith, W. L., "Food and Feed Industries" in Reference 9, pp. 269–288.
16. *Control Techniques for Sulfur Oxide Air Pollutants.* Washington, D.C.: U.S. Dept. of Health, Education, and Welfare, 1969. N.A.P.C.A. Publication No. AP-52.
17. Heller, Austin, and Edward Ferrand, "Low-Sulfur Fuels for New York City." In Harte, John, and Robert Socolow (eds.), *Patient Earth.* New York: Holt, Rinehart and Winston, Inc., 1971.
18. Malin, H. Martin, Jr., "SO$_2$ Removal Technology Enters Growth Phase." *Env. Sci. Tech.* **6:**688–691 (1972).
19. *Control Techniques for Carbon Monoxide Emissions from Stationary Sources.* Washington, D.C.: U.S. Dept. of Health, Education, and Welfare, 1970. N.A.P.C.A. Publication No. AP-65.
20. *Control Techniques for Hydrocarbon and Organic Solvent Emissions from Stationary Sources.* Washington, D.C.: U.S. Dept. of Health, Education, and Welfare, 1970. N.A.P.C.A. Publication No. AP-68.
21. *Control Techniques for Nitrogen Oxide Emissions from Stationary Sources.* Washington, D.C.: U.S. Dept. of Health, Education, and Welfare, 1970. N.A.P.C.A. Publication No. AP-67.
22. Hall, Homer J., and William Bartok, "NO$_x$ Control from Stationary Sources." *Environ. Sci. Tech.* **5:**320–326 (1971).

6

AIR POLLUTION: MOTOR VEHICLE EMISSIONS AND PHOTOCHEMICAL SMOG

The Chicago contest . . . showed that the smell of gasoline and imperfectly consumed gases was such as to be offensive, not only to the occupants of the vehicle, but to persons in the street. . . . Horseless carriages operated by gas or gasoline motors are not a success at the present time. —*New York Times,* April 19, 1896, p. 28.

This report on the new horseless carriages of the late 19th century commented on the disagreeable odors and noise they produced and came to the conclusion that electric automobiles were more promising if only a suitable storage battery could be developed—a goal that is just as elusive today as it was three-quarters of a century ago.

Motor vehicle emissions occupy a special place in discussion of air pollution because they are the major sources of two of the "imperfectly consumed gases" mentioned in the quotation above—carbon monoxide and hydrocarbons—and because they play an important role in the production of the unique Los Angeles-type smog so irritating to the eyes. In the United States in 1968 (see Table 3-2) gasoline and diesel motor vehicles produced:

59% of the carbon monoxide.

49% of the hydrocarbons.

35% of the nitrogen oxides.

3% of the particulate matter.

1% of the sulfur oxides.

This chapter will begin with a review of some of the scientific detective work undertaken in the early 1950's by A. J. Haagen-Smit, the Stanford Research Institute, and other investigators to learn the causes of the Los Angeles-type smog that was first noticed during World War II. Succeeding sections will discuss the source and control of automotive emissions from the modern internal combustion engine, alternatives to the internal combustion engine, and the special problem of lead about which there has been a great deal of controversy in recent years.

89

PHOTOCHEMICAL SMOG

The first major Los Angeles "smog" to attract public concern occurred in 1943. Recurrences became more and more common in the next few years. Since it appeared to be a new phenomenon, it was at first blamed on the newest industry in the area—a government-financed butadiene plant built to help meet the need for synthetic rubber during the war years. Its closing did not alleviate the problem, however, and attention was then focused on the oil refineries in the area and on the numerous backyard incinerators of the region [1].

As steps were taken to combat these polluters—without noticeable improvement in the air quality—a number of investigators began scientific investigations of the effects, nature and sources of the smog.

Early attention focused on injury to plants in the Los Angeles Basin [2, 3]. Leaf injury to plants, especially leafy vegetable crops, was first noted in 1944 and increased in severity each year. Badly damaged plants included endive, spinach, romaine lettuce, garden and sugar beets, Swiss chard, and oats, while injuries were also noted to alfalfa, barley, eggplant, tomatoes, and head lettuce. The type of plant injury produced was quite different from that caused by sulfur dioxide, a common air pollutant known to have acute and chronic injurious effects on plants [4], and lists of plants in order of susceptibility were quite different for sulfur dioxide and Los Angeles smog [2]. A common injury was silverleaf—a shiny, oily effect on the leaf's lower surface that developed into a glaze and then turned silver. The horticultural experience suggested that Los Angeles smog was rather different in nature from the ordinary ("classical") smog of London and other urban areas, in which high concentrations of sulfur dioxide and particulate matter produced most of the effects.

Since 1940 Los Angeles had also been experiencing a reduction in visibility but no climatic changes had occurred that might be responsible for this reduction. Investigation showed that small (typically 3 μm in diameter), mostly hygroscopic (i.e., easily retaining moisture), acidic particles were largely responsible, although part of the effect was due to carbon, tars, minerals, pollen, and other materials. Salt (NaCl) particles, which provide condensation nuclei for most of the fogs occurring throughout the world, turned out to be very rare in the Los Angeles atmosphere, even during bad smog episodes [3].

The most distinctive feature of the Los Angeles air, especially during smog episodes, was the high level of oxidant (oxidizing material) in contrast to classical smog, which is chemically reducing because of its SO_2 and particulate matter. Investigations showed the oxidant to be largely ozone (O_3), which is normally present in the atmosphere at sea level at concentrations of 0.01 to 0.03 ppm but which has been detected in Los Angeles at up to 1 ppm.

Irritation of the eyes due to oxidant is noticeable at 0.15 ppm and objectionable at 0.25 ppm.

Investigations of smog damage to plants put much of the blame on the ozone. Plants grown in greenhouses did not exhibit smog damage when their air was cleaned by filters and activated carbon (which adsorbs many gases) but the damage resulting when ordinary outside air was used was identical with that noted in the outdoor fields. The occurrence of damage was highly correlated with the ozone level and with smog irritation (as judged subjectively) but not with SO_2 level. Since there was no geographical pattern to the crop damage, such as might result if a specific source was responsible for the damage, the source was apparently very widespread [3].

A number of experiments were carried out to discover the source of the irritating effects of smog and of the crop damage [2, 3]. The only combination of gases that seemed to produce all the typical plant injuries was a mixture of ozone and automobile exhaust gases. In order to test the effects on human beings and make sure that important smog constituents were not being overlooked, a couple of "artificial smogs" were developed by the Stanford Research Institute and given to a panel for their opinion [3]. These smogs contained carefully measured amounts of known gases and particulate matter approximating the natural composition of smog, as far as it was known. One artificial smog contained known amounts of aldehydes, while in another the aldehydes were allowed to form through the reaction of ozone with gasoline vapors; the latter smog had the following ingredients:

0.5 mg/m^3 HNO$_3$ 0.01 mg/m^3 NaCl
1.4 mg/m^3 SO$_2$ 0.75 mg/m^3 O$_3$
0.2 mg/m^3 SO$_3$ 10 mg/m^3 gasoline
0.5 mg/m^3 oil (40% diesel, 60% crankcase oil)
1.0 mg/m^3 particulate matter (lampblack)

The conclusions from the panel's testing were as follows [3]:

1. Atmospheres containing all of the major constituents of Los Angeles smog at the maximum concentrations at which these constituents have been found in smog produced definite irritation of the eyes.
2. Removal of no single constituent of artificial smog eliminated all of the irritating action.
3. Removal of all of the particulate constituents had little effect on the eye-irritating action of the artificial smog.
4. Removal of all the gaseous constituents eliminated the eye-irritating action of the artificial smog.
5. Both formulas of artificial smog produced much more eye irritation than did the reaction products of ozone and gasoline alone at concentrations which one might expect to find in intense Los Angeles smog.

6. The artificial smog in which the aldehydes were produced by the reaction of gasoline and ozone seemed to have nearly the odor of natural smog.

An important finding of these investigations is that a mixture of the products of pollutant emissions from petroleum combustion is not capable of producing the effects of smog unless exposed to sunlight. This is due to the fact that the ozone and other irritants are not produced except through photochemical reactions involving solar radiation. Nitrogen oxides play an important role in this set of reactions. Nitrogen dioxide (NO_2) is an efficient absorber of ultraviolet portions of the solar spectrum (0.3 to 0.4 μm), which cause the NO_2 molecule to photolyze into NO (nitric oxide) and O (atomic oxygen). The atomic oxygen can react with molecular oxygen (in the presence of a collisional molecule to conserve energy) to form ozone. The ozone in turn can react with NO to form NO_2 and O_2 again, completing the atmospheric nitrogen dioxide photolytic cycle [5]. These reactions may be summarized as follows, where M represents a collisional molecule:

$$NO_2 + \text{ultraviolet light} \rightarrow NO + O$$
$$O + O_2 + M \rightarrow O_3 + M$$
$$O_3 + NO \rightarrow NO_2 + O_2$$

TABLE 6-1. Some important reactions involved in the production of photochemical smog [6]. "Hc" refers to a hydrocarbon and an asterisk (*) is used to denote a free radical. R denotes an alkyl group such as methyl (CH_3—) or ethyl (CH_3CH_2—).

$$O + Hc \rightarrow HcO^*$$
$$HcO^* + O_2 \rightarrow HcO_3^*$$
$$HcO_3^* + Hc \rightarrow \text{aldehydes (RCHO), ketones (}R_1R_2CO\text{), etc.}$$
$$HcO_3^* + NO \rightarrow HcO_2^* + NO_2$$
$$HcO_3^* + O_2 \rightarrow HcO_2^* + O_3$$
$$HcO_x^* + NO_2 \rightarrow \text{peroxyacyl nitrates (RCOOONO}_2\text{)}$$

At nighttime, the NO and O_3 levels will be low and the NO_2 level will generally be a little higher. During the day nitrogen oxides are being produced by automobiles and the concentrations of NO and NO_2 rise; the NO_2 photolyzes rapidly to NO + O, which then leads to a rise in ozone concentration. In the presence of hydrocarbons, a number of other reactions occur, some of which are shown in Table 6-1. The partially oxidized hydrocarbons and free radicals react with NO to form more NO_2, upsetting the balance of O_3 consumption by NO. As a consequence the NO level decreases and the NO_2 and O_3 levels rise, with the ozone level peaking about midday. Figure 6-1 shows typical variations of these gases during the day.

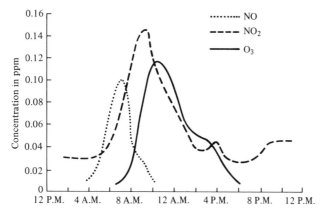

FIG. 6-1. Typical variations of nitric oxide, nitrogen dioxide, and ozone (recorded at Los Angeles, July 19, 1965) [5].

In the 1964 to 1967 period, Los Angeles had peak hourly average levels of oxidant (ozone or the ozone equivalent of other highly oxidizing gases) over 0.15 ppm 30% of the time and over 0.10 ppm 48.5% of the time; the maximum hourly average recorded in that period was 0.58 ppm [5]. Many other U.S. cities (such as Denver, St. Louis, Philadelphia, and Washington, D.C.) exceed 0.10 ppm at least 10% of the time. The highest oxidant peaks tend to occur during the summer and early fall months (June through October in Los Angeles) because of the importance of the solar radiation.

Once ozone is present in the atmosphere, other chemical reactions leading to other irritants occur; these include aldehydes, ketones, and peroxyacyl nitrates (PAN). Figure 6-2 shows the chemical structure of some known eye irritants. The eye irritants known to occur in photochemical smog are not sufficient to account for all the eye irritation that occurs, however, and other compounds apparently remain to be accounted for.

Eye irritation results when instantaneous oxidant concentrations reach 0.10 ppm; increased frequencies of asthma attacks have occurred when oxidant concentrations reached 0.13 ppm. The threshold for nasal and throat irritation appears to be exposure to a concentration of 0.3 ppm for 8 h, and exposure to concentrations of 0.6 to 10 ppm for 1 to 2 h may impair pulmonary function in several ways [5]. In Los Angeles, monitoring stations have recorded short-term oxidant peaks as high as 0.67 ppm and 1-h average concentrations as high as 0.58 ppm [5].

The main effect of photochemical smog as currently produced in urban areas is the production of eye irritation. Photochemical smog has not led to smog episodes as serious as those that have resulted from classical smog. Increased mortality has been noticed in Los Angeles during photochemical smog incidents that occurred during heat waves but the increase has been comparable to that which normally occurs during heat waves.

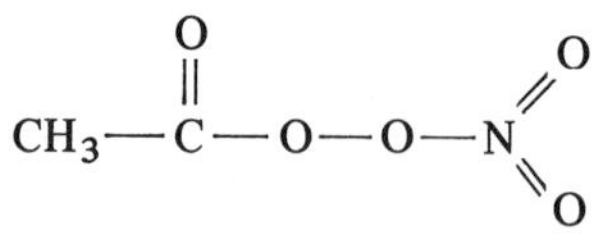

$$CH_3-\overset{\overset{\textstyle O}{\|}}{C}-O-O-N\overset{\nearrow O}{\underset{\searrow O}{}}$$

Peroxyacetyl nitrate ("PAN")

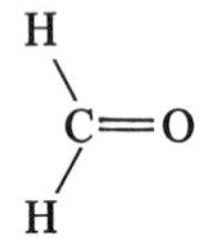

$$\overset{H}{\underset{H}{}}\!\!\diagdown C\!\!=\!\!O$$

$$\overset{H}{\underset{H}{}}\!\!\diagdown C\!\!=\!\!\overset{H}{\overset{|}{C}}\!\!-\!\!\overset{H}{\overset{|}{C}}\!\!=\!\!O$$

Formaldehyde 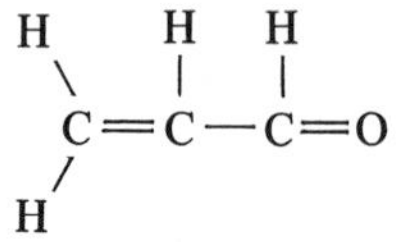   Acrolein

FIG. 6-2. Some eye irritants of photo-chemical smog (see also Fig. 1-1).

Ozone and other photochemical oxidants also affect vegetation adversely, as discussed earlier in this section, and ozone attacks and cracks stretched rubber, with noticeable effects resulting from 1-h exposures to 0.02 ppm [5].

At present, toxicological and epidemiological data relating to photochemical smog and its oxidants are rather scanty and further research is clearly needed.

Although carbon monoxide is the most important motor vehicle pollutant in terms of mass, it has not been proved to play an important role in photochemical smog. There is some indication [7] that it does lead to increased mortality in Los Angeles County, perhaps by as much as 11 deaths on days of high average CO concentration (about 20 ppm).

SOURCES AND CONTROLS OF MOTOR VEHICLE EMISSIONS
FROM THE INTERNAL COMBUSTION ENGINE

The estimated annual production of pollution by an average new 1963 U.S. car with no emission controls was 770 kg of carbon monoxide, 240 kg of hydrocarbons, and 40 kg of nitrogen oxides [8]. The uncertainties in these numbers are quite large and more recent information suggests that these totals may underestimate emissions by up to 50%. Since the average vehicle is driven about 10,000 mi (16,000 km), the emissions listed above correspond to 77 g/mi of CO, 24 g/mi of hydrocarbons, and 4 g/mi of nitrogen oxides (the mixed metric and nonmetric units of grams per mile are those used by the federal government). Practically all the carbon monoxide and nitrogen oxides came from the exhaust, while of the 240 kg of hydrocarbons emitted about 140 kg came from the exhaust; 60 kg were crankcase blow-by gases (gases escaping from the engine cylinders between the piston sealing surfaces and the cylinder wall into the crankcase, from which they are vented into the atmosphere); and 40 kg were evaporated from the fuel tank or the carburetor (see Figure 6-3). The exhaust itself consisted of air and combustion products (largely CO_2 and H_2O), with 3.5% CO, 900 ppm hydrocarbons, and 1500 ppm nitrogen oxides (by volume). Although the CO emissions were high, General Motors Corporation claims that they were even higher in the past—3.75% in 1950, 4% in 1940, and 7% in 1927 [8].

Almost all the vehicles on the road today are powered by an internal combustion engine in which air and gasoline fuel are mixed, compressed, and ignited to produce large volumes of CO_2 and water vapor, which expand and exert forces on the moving parts to turn the wheels. Diesel engines, which constitute less than 1% of U.S. motor vehicles and yet consume 3 to 5% of the fuel consumed because they are common on large trucks [6], are internal combustion engines in which the air alone is compressed and thereby heated to a temperature sufficiently high to ignite the fuel, which is then sprayed into the cylinders and burned. Diesel fuel is generally heavier and less volatile than gasoline (and less expensive). Diesel engines exhaust perhaps one-tenth as much CO as the gasoline internal combustion engine for com-

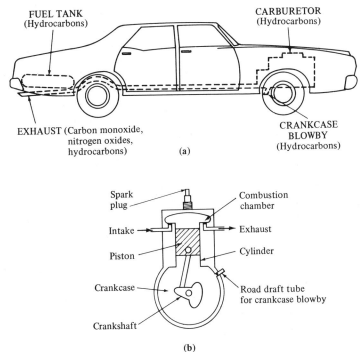

FIG. 6-3. (a) Sources of uncontrolled automotive pollutant emissions. (b) Schematic diagram of four-stroke gasoline internal combustion engine. Road draft tubes were used before the advent of positive crankcase ventilation.

parable amounts of fuel consumed but the amounts of hydrocarbons and nitrogen oxides are roughly the same [8].

There are several possible ways to reduce motor vehicle emissions:

1. Instituting controls on the gasoline internal combustion engine—the route (described below) that has thus far been taken in the United States.

2. Developing new fuels—propane, natural gas, and alcohol have all been considered—to use with the internal combustion engine.

3. Developing new types of engines—steam or other Rankine cycle engines, Wankel engines, gas turbines, etc.—using gasoline or other fuels.

4. Reducing motor vehicle use through educational programs (to encourage car pools, for instance), development of better mass transportation systems for intracity and intercity travel, or governmental controls (such as extra fuel taxes).

Efforts in the U.S. have largely been devoted to reducing emissions on the gasoline internal combustion engine, although some efforts have been directed at developing new fuels or new types of engines.

California, in which air pollution from motor vehicles first became a serious problem in the 1940's, took the lead in requiring pollution control equipment. Crankcase controls to reduce hydrocarbon emissions were required by legislation on all new cars sold in California in 1963; federal legislation later required them on all new cars sold nationwide beginning in 1968. The automobile manufacturers voluntarily installed them on all cars sold in California beginning with the 1961 models and on all cars sold nationwide beginning with the 1963 models [9].

The method chosen to control emissions of hydrocarbons from the crankcase was a technique known for half a century but neglected by the automobile manufacturers, who now refer to it as "positive crankcase ventilation" (PCV for short) [10]. In PCV the gases escaping from the crankcase are recycled back to the engine intake instead of being vented into the atmosphere. The acidic nature of the crankcase blow-by gases can lead to sludging of poor engine oils and plugging of the PCV valve but this will not occur with engine oils containing polymeric dispersants and metallic detergents [6].

Legislation requiring the control of evaporative emissions of unburned hydrocarbons from the carburetor and fuel system applied to all new cars sold in California beginning with 1970 model cars and all new cars sold nationwide beginning with 1971 model cars. These emissions are being controlled by collection of gasoline vapors from the carburetor and the fuel tank either in the crankcase or in activated-carbon canisters, from which they are sent to the engine once it has been started again [9, 11].

Exhaust emission controls for carbon monoxide and hydrocarbons (but not nitrogen oxides) were required on all new cars sold in California beginning with the 1966 model cars and all new cars sold nationwide beginning with the 1968 model cars, i.e., after the institution of crankcase controls but before the requirement of evaporative controls [9]. Both California and the federal government are requiring these controls to meet more and more stringent standards in the early 1970's, including nitrogen oxide emission standards. The numerical values of these exhaust emission standards are discussed later in this section.

Pollutant emissions in the exhaust are influenced by a variety of factors: air-fuel ratio (see Figure 6-4), ignition timing and quality, intake-manifold vacuum, engine compression ratio, engine speed and load, fuel distribution within and between cylinders, coolant temperature, and combustion chamber configuration and deposits [9].

In order to reduce carbon monoxide and hydrocarbon emissions from the exhaust, General Motors, Ford, and American Motors originally used an air injection system in which air was injected into the exhaust manifold near the exhaust valves (where the exhaust gases were hottest and thus most easily ignited) to complete the oxidation into CO_2 and H_2O, without the extraction of usable power. Chrysler instead developed a "Clean Air Package" in which emissions were reduced through engine adjustment and operational modification, particularly the use of leaner and more precisely controlled carburetor mixture ratios, and the other motor vehicle manufacturers are presently making similar modifications. Unfortunately, these adjustments have led to an increase in nitrogen oxide emissions [8].

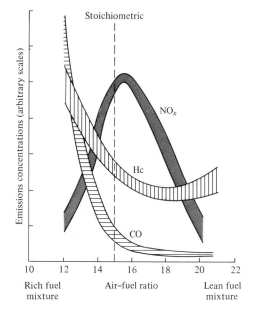

FIG. 6-4. Effect of air-fuel ratio on exhaust emissions of carbon monoxide, hydrocarbons, and nitrogen oxides [12]. The stoichiometric ratio of 15 provides precisely the right amount of oxygen for complete combustion. (*Courtesy of Battelle Columbus Laboratories.*)

The controls have not proved wholly satisfactory for a number of reasons. Only prototype motor vehicles have been submitted for federal certification and there are questions as to how well production models are controlling emissions [10]. In addition, the effectiveness of the controls has proved to have a shorter life than the vehicles. Finally, inadequate testing procedures have permitted cars to be erroneously certified.

In order to meet the stricter standards of the future, better control equipment for exhaust emissions will be required. One promising control method is the "exhaust manifold thermal reactor," which is designed to take the place of the usual exhaust manifold. In it, the exhaust gases are allowed enough reaction time at high temperatures with sufficient air to permit more complete oxidation of the hydrocarbons and carbon monoxide. Figure 6-5 shows the DuPont thermal reactor in cutaway and its operation.

The thermal reactor may require expensive metals capable of withstanding high temperatures and may cause some decrease in fuel economy. These problems may not occur if catalytic reactors operating at lower temperatures are used. These reactors would be located in the usual muffler position and would contain a catalyst (typically platinum or one of the platinum-group metals or various mixtures of oxides of chromium, iron, copper, etc.) promoting the oxidation of hydrocarbons and carbon monoxide. Catalytic reactors were thoroughly researched in the period from 1957 to 1964 and are now making a "comeback" since they may the most economical solution to the exhaust emission problem. These devices are promising since they can reduce emissions to very low levels but there are also disadvantages, particularly the fact that lead compounds resulting from the tetraethyl lead antiknock additive in most gasolines will poison the catalyst very quickly and reduce its effectiveness. Other problems include the expensiveness of the catalyst, the necessity for materials capable of operating at fairly high temperatures for long periods of time, the shortness of the time in which to complete the combustion since the exhaust gases are traveling at high speeds, and the low concentrations of the gases that must be oxidized. Apparently feasible catalytic reactors have been built by a number of companies, such as Universal Oil Products and Oxy-Catalyst, but they have yet to be used commercially.

The control of nitrogen oxides is likely to be difficult. These oxides result from oxidation of atmospheric nitrogen gas during the combustion process, and the engine adjustments and modifications made to complete oxidation of hydrocarbons and carbon monoxide have generally also increased nitrogen oxide emissions. One method that is being tried to reduce NO_x emissions is exhaust gas recirculation—recycling part of the exhaust gas back into the engine to reduce the peak combustion temperature and thus the amount of nitric oxide formed [9]. Several catalytic reactors, involving copper oxide-cobalt oxide-alumina catalysts or nickel-copper catalysts, for example, have been developed to reduce NO_x emissions but their value is not known at the present time. No method has been found to promote the simultaneous chemical reduction of nitrogen oxides and chemical oxidation of

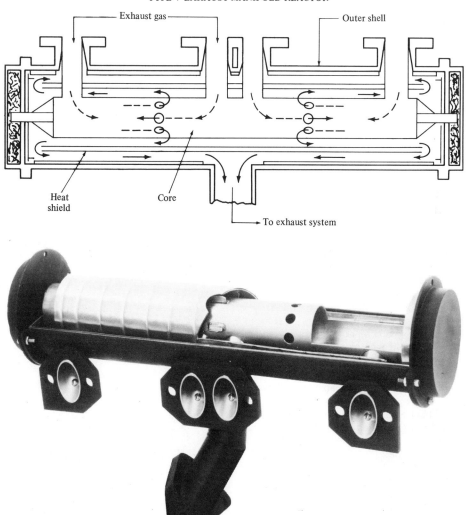

FIG. 6-5. The DuPont exhaust manifold thermal reactor. (a) Cutaway view. The reactor is 10 cm in diameter and 87 cm long. (b) Schematic view. An air pump delivers air to the exhaust ports; the exhaust gases and air are mixed and oxidation initiated in the core; the reacting gases then pass out of the core, around the shield, and to the exhaust system. (*Courtesy of E. I. Du Pont de Nemours & Company, Inc.*)

hydrocarbons and carbon monoxide in the exhaust, although the idea is an appealing one.

Table 6-2 lists automobile exhaust emissions in grams per mile before exhaust controls, as specified by California and federal legislation, and as reported attained for various engines and fuels with or without controls.

TABLE 6-2. Automobile exhaust emission (in grams per mile) as set by law or as attained by various operational vehicles and engines [8, 9, 11, 13]. All data refer to new light-duty passenger automobiles only. Because of differences in test cycles and measurement methods the uncertainties in these numbers are rather large and the numbers should not be considered accurate to within less than 50%.

	HC	CO	NO_x
Actual uncontrolled emissions on pre-1966 cars (based on 1972 test procedures)	17.0	124.0	5.4
1966 California standards	3.4	34	—
1968 California "Low Emission Vehicle" (see text)	0.5	11	0.75
1970 U.S. standards (not met as scheduled)	2.2	23	—
Feasible in 1975 according to Morse Report [8]	0.6	12	1
U.S. standards for 1975–1976	0.41	3.4	0.4
Sunoco test vehicle	0.7	12	0.6
DuPont thermal reactor	0.2	12	1.2
Chrysler-Esso engines:			
Manifold reactor	1.5	20	1.3
Catalytic reactor	1.7	12	1.0
Synchrothermal reactor	0.25	7	0.6
Ethyl Corporation "lean reactor" car	0.7	10.4	2.5
Wankel engine (with controls)	1.8	23	2.2
Natural gas fueled internal combustion engine	1.5	6	1.5
Gas turbine	0.5–1.2	3.0–7.0	1.3–5.2
Rankine cycle engine	0.2–0.7	1.0–4.0	0.15–0.4
Stirling-electric hybrid (GM)	0.006	0.3	2.2
Stratified-charge or "proco" engine (Ford)	0.37	0.93	0.33

The California Pure Air Acts have set progressively stricter standards for successive model years and the state has been granted congressional permission to set standards stricter than the U.S. standards. California's 1968 Low Emission Vehicle Act defined the emission levels (given in the table) for what was defined to be a "low emission motor vehicle"; the state was required to purchase such vehicles whenever they became available at a cost no more than twice that of conventional vehicles purchased the previous year.

The exhaust emissions in grams per mile are carefully defined in terms of a test cycle involving accelerations, cruisings, decelerations, and idling [9]. The vehicles certified as meeting the 1970 U.S. standards (this occurred well before the start of the 1970 model year, of course) actually failed to do so because the test procedures were later recognized to be faulty. Under the old test procedures the uncontrolled emissions on pre-1966 cars were estimated as 11 g/mi hydrocarbons, 80 g/mi carbon monoxide, and 4 g/mi nitrogen oxides—all considerably lower than the values given in Table 6-2.

In 1970 the Clean Air Act (see Chapter 20) specified that 1975 model cars must satisfy standards no more than 10% of those emissions allowable for 1970 model cars in the case of carbon monoxide and hydrocarbons and that 1976 model cars must have nitrogen oxide emissions no more than 10% of those actually measured on 1971 model cars. These are strict standards, well below what the 1967 Morse Report [8] considered to be feasible in 1975, and the automobile manufacturers naturally expressed great displeasure at the legislation. The administrator of the Environmental Protection Agency was empowered to permit a suspension of the federal standards for one year if necessary; in May 1972 he denied the auto manufacturers' requests for such a delay, but the final decision may be made by the courts.

Some of the entries in Table 6-2 suggest that the standards are not unreasonable. Prototype models of Ford's "proco" (for "programmed combustion") engine—a stratified charge engine in which the fuel is mixed with air in the combustion chamber instead of the carburetor—have looked especially promising to the Environmental Protection Agency, which believes the engine could be mass-produced by 1977.

The Clean Air Act amendments of 1970 also specified that automobiles must meet the emission standards not only at the time of manufacture but also for their whole useful life, defined as 5 years or until the vehicle has been driven 50,000 mi, whichever comes first.

Internal combustion engines run with light hydrocarbons as fuel tend to produce fewer pollutant emissions. A number of government units have converted some vehicles to natural gas. California, for example, converted several hundred automobiles at a cost of $400 per vehicle to permit operation on compressed natural gas for 50 to 160 km per day; the natural gas is used for urban driving and the gasoline for steady highway operation. Any large-scale use of natural gas for transportation purposes is currently out of the question because of the present scarcity of natural gas (discussed in Chapter 2). Another possibility is the use of propane ($CH_3CH_2CH_3$), which can be produced by cracking of heavier hydrocarbons. Any of these alternative fuels presently involves problems of distribution and storage of the fuel and high costs for installation and engine modification, but these are problems that could be readily solved should the United States decide to encourage the use of these fuels for transportation.

ALTERNATIVE ENGINES AND POWER SOURCES

The various possible power sources for motor vehicles currently in use or under investigation are listed in Table 6-3. The standard internal combustion engine is by far the most common and the diesel engine also finds some use, especially on trucks. The other heat engines are very uncommon. Some automobiles have been marketed with rotary internal combustion (Wankel) engines or with gas turbines, and in the early days of automobiling there were a number of steam engines in use.

TABLE 6-3. Possible power sources for motor vehicles. The thermodynamic cycles characterizing the different types of heat engines are discussed in engineering thermodynamics texts [14, 15]. A full discussion of the different types of internal combustion engines can be found in Reference 16.

Heat Engines

A. Intermittent-combustion, internal combustion engines
   1. Otto cycle with gasoline fuel
      (a) Standard internal combustion engine (4-stroke, spark-ignited, reciprocating-piston)
      (b) NSU-Wankel internal combustion engine (spark-ignited, rotary-piston)
      (c) Miscellaneous internal combustion engines (two-stroke, stratified-charge, etc.)
   2. Diesel cycle with diesel fuel: Diesel engine
B. Continuous combustion engines
   1. Brayton cycle with various possible fuels: gas turbine engine
   2. Rankine cycle with various fuels: external combustion engines using steam, Freon, or other working substances
   3. Stirling cycle with diesel fuel: Stirling external combustion engine with air as the working substance

---

Electrochemical Devices

A. Batteries: lead-acid, nickel-cadmium, nickel-zinc, silver-zinc, zinc-air, sodium-sulfur, lithium-chlorine, etc.
B. Fuel cells: hydrogen-oxygen, hydrogen-air, hydrazine, methanol, etc.

Hybrids

A. Heat engine-battery hybrid engines: gas turbine-battery, Stirling-electric hybrid, etc.
B. Battery-battery hybrid engines
C. Fuel cell-battery hybrid engines

The Wankel engine is simple and operates smoothly with minimum vibration, and it has high power-to-weight and power-to-volume ratios. Its emission characteristics are not well known, but General Motors has secured rights to the engine and is presumably studying the possibility of using it to meet future emission regulations.

The diesel engine, as mentioned earlier, emits less carbon monoxide than standard internal combustion engines but comparable amounts of hydrocarbons and nitrogen oxides. It also has special smoke and odor problems that are not well understood. Barium additives in the diesel fuel can reduce smoke emissions by about 50%; three-quarters of the barium is exhausted as barium sulfate, which is insoluble in water and apparently harmless to humans [6], although the possibility of unrecognized dangers should not be overlooked.

Gas turbines are continuous combustion engines, the fuel being continuously sprayed into a combustion chamber supplied with air compressed to several atmospheres pressure, the gaseous combustion products then expanding through the turbine to atmospheric pressure. The concentrations of pollutants in gas turbine exhausts are very low because of the large volumes of exhausts but the number of grams per mile produced is comparable to ordinary engines. Nevertheless, some engineers believe that these engines can be designed to produce much less pollution than the standard internal combustion engine [9]. One apparent disadvantage is that they are likely to be rather expensive.

In all the engines discussed thus far the products of the combustion of air and fuel act directly as the working fluid providing the power. In external combustion engines, however, the combustion products transfer their heat to another fluid that acts as the working substance. Steam engines, which operate on the Rankine thermodynamic cycle, are the most common type of external combustion engine, finding their widest application in the electric power industry. A number of different Rankine cycle engines are being developed for possible use in motor vehicles. Their advantages include low pollutant emissions, good safety and reliability, and durability, and they are probably not too expensive (although those currently on the market are naturally rather expensive). Their disadvantages include bulkiness and slow starting (the famous Stanley Steamers of the early 20th century took over an hour to start). Use of water as the working substance would make operation impossible at subfreezing temperatures but other liquids with lower melting points would be suitable for use in such an engine.

Stirling engines, which are being studied by General Motors and other corporations, are claimed to show promise as very low emission engines (see Table 6-2) and their thermal efficiencies are high, but they are bulky, heavy, and expensive [9].

Currently there is also active research and development under way to provide acceptable electric cars, at least for limited performance objectives, such as intown driving. No major inventions or scientific developments are necessary as far as electrical control and transmission systems are concerned since they have been studied and developed for the many other uses of electricity in modern society. It has proved difficult, however, to develop at reasonable cost a source of electrical energy that provides both ample power to move the vehicle and ample energy to give it sufficient range. Electric cars were marketed in the early 20th century (see Figure 6-6) but their maximum speeds and maximum ranges were very small by modern standards.

In order to understand the importance of power and energy in motor vehicle transportation, it is necessary to analyze motor vehicle motion in a little more detail [8]. The heat engines currently used for automotive power

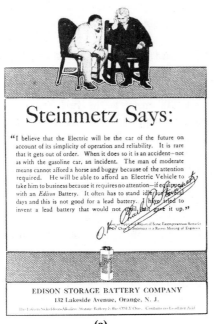

Steinmetz Says:

"I believe that the Electric will be the car of the future on account of its simplicity of operation and reliability. It is rare that it gets out of order. When it does so it is an accident—not as with the gasoline car, an incident. The man of moderate means cannot afford a horse and buggy because of the attention required. He will be able to afford an Electric Vehicle to take him to business because it requires no attention—if equipped with an *Edison* Battery. It often has to stand idle for several days and this is not good for a lead battery. I have tried to invent a lead battery that would not spoil, but gave it up."

EDISON STORAGE BATTERY COMPANY
132 Lakeside Avenue, Orange, N. J.

(a)

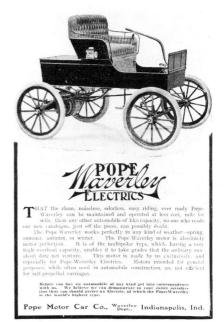

Pope Waverley ELECTRICS

THAT the clean, noiseless, odorless, easy riding, ever ready Pope-Waverley can be maintained and operated at less cost, mile for mile, than any other automobile of like capacity, no one who reads our new catalogue, just off the press, can possibly doubt.

The Pope-Waverley works perfectly in any kind of weather—spring, summer, autumn, or winter. The Pope-Waverley motor is absolutely motor perfection. It is of the multipolar type, which, having a very high overload capacity, enables it to take grades that the ordinary run about dare not venture. This motor is made by us exclusively, and especially for Pope-Waverley Electrics. Motors intended for general purposes, while often used in automobile construction, are not efficient for self-propelled carriages.

Before you buy an automobile of any kind get into correspondence with us. We believe we can demonstrate to your entire satisfaction that you should prefer an Electric, of which the Pope-Waverley is the world's highest type.

Pope Motor Car Co., Waverley Dept., **Indianapolis, Ind.**

(b)

GROUT STEAM CARS

We Use Steam, as it Has Proven Reliable, Simple in Operation, No Noise, No Vibration, No Odor, No Cranks to Turn, No Danger, Boiler Can't Burn. Railroads Use Steam, Ocean Liners Use Steam, Manufacturers Use Steam

WHY?
Because it is Reliable

Records....

Hill Climbing, Nelson Hill, First Prize, 81 competitors.

Long Island Consumption Test, least fuel used, First Prize

New York-Southport 100-mile run, no stop, First Prize

Philadelphia Show, Construction Bronze Medal, First Prize

Records....

500-mile Run, New York to Boston and Return, Three First-class Certificates and Gold Medal, First Prize

England, Dashwood Hill, a 6½ H. P. Grout beat every car on the hill with a rear tire badly punctured

Welbeck Speed Trials, the Grout did a kilometer in 1 minute 20.4 seconds

Grand Tonneau

1903 CATALOGUE READY
100 MILES WITHOUT A STOP

GROUT BROS., ❦ ❦ *Orange, Mass., U. S. A.*

.... Special built work to order, such as Trucks, Busses, Light Deliveries

Frenche Runnerbout

Drop Front

(c)

FIG. 6-6. Early 20th century ads. **(a)** Charles Steinmetz called the electric car "the car of the future" in 1914. **(b)** This 1904 car was powered by 32 cells of 1.25 volts emf each, top speed was 24 km/h (15 mi/h) and range was 64 km (40 mi) on one charge. **(c)** This 1903 steam car developed 5 kW (6.5 hp) and could travel at 24 to 32 km/h (15 to 20 mi/h).

typically have thermal efficiencies of 30 to 40% at best and the same is true of those under development; thus much of the fuel energy can never be used to provide thrust to the car.

The thrust F_t of a car does not all go into acceleration or climbing (against gravitational forces) but some of it counteracts the frictional forces against the rolling tires and the wind resistance. The thrust can be written [8] as

$$F_t = F_r + F_w + Mg \sin \phi + Ma$$

where F_t = thrust
F_r = rolling friction
F_w = wind resistance
M = mass of car
$g = 9.8$ m/s² = acceleration due to gravity
ϕ = climbing angle
a = acceleration of the car

The rolling friction can be approximated by 0.0175 Mg and the wind resistance F_w by $\frac{1}{2}\rho C_D A_f v^2$ where ρ is the mass density of the air (1.3 kg/m³), C_D is a drag coefficient (approximately 0.6 for a small compact car), A_f is the frontal area of the car, and v is the speed of the car (or more properly, the relative speed of the car and the wind). The power that is delivered to the car is vF_t, the product of the speed of the car and the thrust; this is only an approximate expression, especially regarding the rolling friction.

Table 6-4 shows the calculation of the thrust and power necessary to get a small compact car moving at 50 km/h to accelerate 1 m/s² while climbing a 4% grade. The thrust required is 1614 N (363 lb) and the power required 23.8 kW (32 hp). Steady level driving at the same speed, with no provision for acceleration to that speed, would require only 4.5 kW (6.7 hp)— the amount necessary to counteract rolling friction and wind resistance. The thrust and power needed increases sharply with increasing speed, and steady level driving at 100 km/h (62 mi/h) would require 21.6 kW (29 hp) since doubling the speed increases vF_r by a factor of two and vF_w by a factor of eight.

Since the total mass of the car cannot be in the motive power source, the power that must be developed per unit mass of the source must be calculated assuming that some particular fraction of the car's mass is the power source. Suppose in our example that 25% of the car's mass (or weight) constitutes the motive power source, which thus has a mass of 250 kg. A power source capable of furnishing the 23.8 kW of the example in Table 6-4 would then have to have a "specific power" of 23.8 kW/250 kg or 95 W/kg. Not all power sources are this dense and therefore not all power sources can serve in even a small compact car.

Even if a power source has a sufficiently large specific power to be suitable, it may not be able to furnish energy for very long. If the car of Table

6-4 were driven 200 km (124 mi) at 50 km/h under level driving conditions, it would require 18 kWh of energy (4.5 kW for 4 h), and if it were driven 200 km at 100 km/h, it would require 43.2 kWh of energy (21.6 kw for 2 h). A motive power source capable of taking the car 200 km at 50 km/h would then have to have a "specific energy" of at least 18 kWh/250 kg or 72 Wh/kg. Not all power sources have this high a specific energy.

TABLE 6-4. Calculation of thrust and power delivered to a small compact car. See text for explanation.

Assumptions:	$M = 1000$ kg (weight $= 2205$ lb)
	$A_f = 2.0$ m^2 (21.5 ft^2)
	$v = 13.9$ m/s $= 50$ km/h (31 mi/h)
	$a = 1.0$ m/s^2 (3.28 ft/s^2 or 2.24 mi/h/s)
	$\sin \phi = 0.04$ ("4% grade")
Thrust:	$F_r = 172$ N
	$F_w = 150$ N
	$Mg \sin \phi = 392$ N
	$Ma = 1000$ N
	Total $= F_t = 1614$ N (363 lb)
Power:	$vF_r = 2.4$ kW
	$vF_w = 2.1$ kW
	$vMg \sin \phi = 5.4$ kW
	$vMa = 13.9$ kW
	Total $= vF_t = 23.8$ kW (32 hp)

The suitability of any particular motive power source—whether internal combustion engine, lead-acid battery, fuel cell, or other type—must be judged on the bases of both specific power and specific energy. Figure 6-7 shows the ranges of specific powers and specific energies for a number of different power sources, together with lines to represent the specific power requirements for different vehicle speeds and curves to represent the specific energy requirements for different combinations of vehicle ranges and speeds. It should be recognized that specific power and energy requirements increase rapidly with increasing speed and range, respectively.

The superiority of the internal combustion engine (and gas turbines and external combustion engines) is clearly evident. Battery systems are generally capable of supplying sufficient power but are energy-limited and cannot provide sufficient range. Fuel cells are quite capable of supplying sufficient energy to move the car long distances without refueling but their power is limited and incapable of providing anything except very low speeds.

In time, technological developments may lead to electrochemical devices of sufficient specific power and specific energy to be useful in motor vehicles. At present there are some limitations to the systems available. The silver-zinc battery, for example, is limited by the expensiveness of silver, while the sodium-sulfur (Ford Motor Company) and lithium-chlorine (General

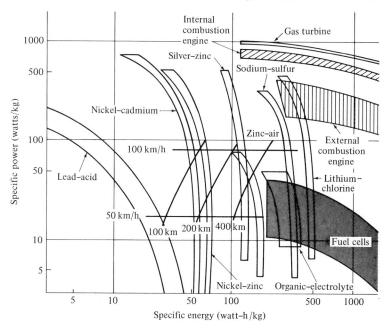

FIG. 6-7. Summary chart of specific power versus specific energy for various motive power sources. Power and energy are taken at the output of the conversion device. A 1000-kg vehicle with a 250-kg motive power source and steady level driving have been assumed in constructing the specific power requirements for different vehicle speeds and the specific energy requirements for different vehicle ranges [8].

Motors) batteries operate with very high temperatures and are not yet feasible. In fact, expensiveness is a limitation to most of these devices, including fuel cells, but this limitation may disappear with the passage of time.

TABLE 6-5. Example of a hybrid power system: half battery and half fuel cell.

	Battery	Fuel Cell	Hybrid
Mass (kilograms)	250	250	250
Maximum specific power (watts per kilogram)	400	40	220
Specific energy (watt-hours per kilogram)	40	400	220
Maximum power (kilowatts)	100	10	55
Energy (kilowatt-hours)	10	100	55

Hybrid power sources can be constructed combining an energy source characterized by high specific power with one characterized by a high specific energy (as shown in Table 6-5) to obtain an energy source characterized by satisfactory intermediate values. At present, hybrids are too expensive to consider for actual automotive use.

Although pollutant emissions of all types would be low for motor vehicles with electric drives powered by electrochemical devices or hybrid engines, there remains the possibility of indirect production of air pollution. If electricity is generated at an electric power plant and then stored in a storage battery that is periodically recharged, there will be air pollution produced at the electric power plant. At present, with most electric power plants being fossil-fuel plants, these emissions are quite large (see Chapter 16). Whether an electric car produces more air pollution at the power plant than an internal combustion engine produces on the motor vehicle itself depends on how one weighs the harmfulness of the different pollutants and what emission controls are required at power plants and on vehicles. At present it is difficult to see that electric vehicles could reduce overall air pollution in the U.S., even though important reductions could occur in urban areas where they are most needed. There may also be unrecognized problems associated with some of the newer types of batteries being developed; it is necessary to be very alert for the environmental effects of any device that may someday be present in the hundreds of millions.

AIRCRAFT EMISSIONS

Pollutant emissions from aircraft are a small but noticeable component of the total air pollution problem in the U.S. As seen in Table 3-2, aircraft are responsible for about 2.5% of the carbon monoxide emissions in the U.S. and about 1% of the hydrocarbon emissions, but only negligible amounts of the other major air pollutants. These emissions, unlike most air pollutant emissions, do not occur mainly in urban areas. The amounts of pollutants emitted by jet aircraft per kilogram of fuel consumed or per passenger-mile are considerably lower than the corresponding values for motor vehicles.

The visibility of smoke from jet aircraft has proved a frequent source of complaints. This smoke is composed largely of fine carbon particles approximately 0.5 μm in diameter, which were not burned properly. Since these small particles scatter light quite well, they reduce visibility and are thus quite conspicuous. Modern turbofan engines have combustion chambers ("smoke burner cans") for smoke reduction that provide a leaner (higher air-to-fuel ratio) mixture and reduce carbon formation. These avoid the visible smoke but the other emissions, which were more serious to begin with, still remain.

Although they do not constitute an important pollution source from the point of view of public health, space rocket engines produce many dangerous exhaust products. One study [17] found that the engine of the Apollo Lunar Module, which took two astronauts to the surface of the moon and brought them back to the command ship circling the moon, produced ammonia, water, carbon monoxide, nitrous oxide, oxygen, carbon dioxide, and nitric oxide as major exhaust products and a wide variety of different minor exhaust products. These constitute a possible source of contamination of the lunar surface and of the lunar samples returned to earth.

LEAD

Lead is a natural constituent of air, water, and the biosphere, and human beings ingest a certain amount in food, water, and air. It is difficult to know what the natural levels would be in the absence of man's activities but one study [18] based on geochemical relationships and materials balances suggests that while primitive man might have absorbed daily amounts of 20 μg from food, 0.5 μg from water, and 0.01 μg from air, modern man absorbs daily amounts of 20 μg from food, 1 μg from water, and up to 10 μg from urban air. Natural conditions are estimated to have consisted of 0.01 ppm Pb in food, 0.0005 ppm in water, and 5×10^{-4} μg/m^3 in air, the latter arising almost completely from silicate dusts, with smaller contributions from volcanic gases and other sources. By contrast, large U.S. urban areas have annual average lead concentrations of 1 to 3 μg/m^3 in the air, with short-term mean concentrations up to 44 μg/m^3 occurring in heavy traffic and vehicular tunnels [6]. The increase of lead aerosols in the air over prehistoric levels is due to lead compounds (mainly lead halides) exhausted from motor vehicles and to lead emissions from lead smelters.

One interesting study [19] reported that analyses of lead in annual ice layers from northern Greenland and the Antarctic continent showed an increase in lead concentrations from less than 0.001 μg/kg at 800 B.C. to 0.01 μg/kg about 1753, 0.03 μg/kg about 1815, to over 0.20 μg/kg today, with the sharpest rise occurring after 1940, i.e., coincident with the sharp rise in the use of tetraethyl lead in gasoline as an antiknock additive. Lead emissions from smelters were estimated to be twice as important as those from motor vehicles in the 1930's, while today the latter are perhaps 50 times as great as the former (relatively very small amounts of lead are also emitted from coal combustion) [19].

Blood lead concentrations in Americans average about 0.25 ppm (levels of 0.80 ppm and above are associated with chronic lead poisoning) but it is not known if they are increasing or even how much greater they are than those of persons in underdeveloped countries whose exposure to atmospheric lead should be much less [20]. The body burdens of Americans may be about 200 mg (for a 70-kg adult) on the average, with great variation [18]. The lead in human bodies in the U.S. is probably in substantial equilibrium, with daily absorption approximately equal to daily elimination [20].

The health effects of the existing blood levels and body burdens is not known, and many questions concerning lead are quite controversial in the scientific literature [21]. The clear-cut cases of lead poisoning that do occur tend to be due to ingestion of lead paints—notably by ghetto children living in substandard housing [21] and by zoo animals [22]. The long-term effects of exposures to lead are not known, although it has been argued that the ancient Romans as well as Europeans in the last century may have been harmed on a large scale by the use of lead containers for storage and drinking [18].

Efforts to "get the lead out" of gasolines, to which on the average about 2 or 3 g used to be added per gallon (3.8 liters), are thus based mainly on concern with the growing amounts of lead aerosols in urban atmospheres and their possible effects over long periods of time. Gasoline without tetraethyl lead additive is lower in octane rating and therefore less suitable for modern high-compression internal combustion engines. Gasoline of sufficiently high octane rating without lead can be produced and has been on the market for many years (the "Amoco" brand of the American Oil Company). Large-scale use of unleaded gasoline, however, will require additional refinery capital investments of $1.5 to 4.25 billion (according to different estimates) and the unleaded gasoline will cost a few cents more per gallon.

A change from leaded to lead-free gasoline is likely to involve a number of effects. Cars burning unleaded gasoline are said to have lower maintenance costs because of such things as longer spark plug and muffler life. On the other hand, the lead in leaded gasoline acts as a lubricant around the engine valves and its removal will increase valve seat wear unless substitute lubricants are used.

An important advantage of unleaded gasolines is that they will not interfere with the operation of catalytic reactors oxidizing unburned hydrocarbons and carbon monoxide, as mentioned earlier in this chapter.

Since unleaded gasolines will differ in fuel composition from leaded gasolines, the possibility remains that there may be an effect on exhaust emissions. In fact, since the octane ratings are generally raised through increased amounts of aromatic and other higher-octane hydrocarbons that are photochemically more reactive, emissions from unleaded gasoline may increase the potential to produce photochemical smog [23].

The switch to unleaded gasolines has already begun, with major oil companies investing in new refining equipment and the automotive manufacturers using lower-compression engines capable of using lower-octane fuel. It will be necessary, however, to guard against the appearance of new pollution problems or the aggravation of old ones as this trend continues. Nonleaded gasolines may also be less efficient and thereby increase the rate of depletion of petroleum resources, and this effect will also have to be considered carefully by society.

REFERENCES

1. Carr, Donald E., *The Breath of Life*. New York: W. W. Norton & Co., Inc., 1965.
2. Haagen-Smit, A. J., et al., "Investigation on Injury to Plants from Air Pollution in the Los Angeles Area." *Plant Physiology* **27**:18–34 (1952).
3. Stanford Research Institute, *The Smog Problem in Los Angeles*. Los Angeles: Western Oil and Gas Assn., 1954.
4. Hindawi, I. J., *Air Pollution Injury to Vegetation*. Raleigh, N.C.: U.S. Dept. of Health, Education, and Welfare, 1970. N.A.P.C.A. Publication No. AP-71.

5. *Air Quality Criteria for Photochemical Oxidants.* Washington, D.C.: U.S. Dept. of Health, Education, and Welfare, 1970. N.A.P.C.A. Publication No. AP-63.

6. *Cleaning Our Environment: The Chemical Basis for Action.* Washington, D.C.: American Chemical Society, 1969.

7. Hexler, Alfred C., and John R. Goldsmith, "Carbon Monoxide: Association of Community Air Pollution with Mortality." *Science* **172**:265–267 (1971).

8. Panel on Electrically Powered Vehicles, Richard S. Morse, Chairman, *The Automobile and Air Pollution: A Program for Progress,* Parts I and II. Washington, D.C.: U.S. Dept. of Commerce, 1967.

9. *Control Techniques for Carbon Monoxide, Nitrogen Oxide, and Hydrocarbon Emissions from Mobile Sources.* Washington, D.C.: U.S. Dept. of Health, Education, and Welfare, 1970. N.A.P.C.A. Publication No. AP-66.

10. Esposito, John C. (Project Director), *Vanishing Air. The Ralph Nader Study Group Report on Air Pollution.* New York: Grossman Pub., 1970.

11. Bowen, D. H. M., "The Drive to Control Auto Emissions." *Environ. Sci. Tech.* **5**:492–495 (1971).

12. Trayser, D. A., and F. A. Creswick, "Effect of Induction-System Design on Automotive-Engine Emissions." *Mechanical Engineering* **92**(2):69 (February 1970). American Society of Mechanical Engineers Preprint 69-WA/APC-7, 1969.

13. "Post-1974 Auto Emissions: A Report from California." *Environ. Sci. Tech.* **4**:288–294 (1970).

14. Crawford, F. H., and W. D. Van Vorst, *Thermodynamics for Engineers.* New York: Harcourt, Brace & World, Inc., 1968.

15. Van Wylen, Gordon J., and Richard F. Sonntage, *Fundamentals of Classical Thermodynamics.* New York: John Wiley & Sons, Inc., 1965.

16. Obert, Edward F., *Internal Combustion Engines.* 3rd ed. Scranton, Pa.: International Textbook Co., 1968.

17. Simoneit, B. R., et al., "Apollo Lunar Module Engine Exhaust Products." *Science* **166**:733–738 (1969).

18. Patterson, C. C., "Contaminated and Natural Lead Environments of Man." *Arch. Environ. Health* **11**:344–360 (1965).

19. Murozumi, M., et al., "Chemical Concentrations of Pollutant Lead Aerosols, Terrestrial Dusts and Sea Salts in Greenland and Antarctic Snow Strata." *Geochimica et Cosmochimica Acta* **33**:1247–1294 (1969).

20. Kehoe, R. A., Letter to the Editor. *Arch. Environ. Health* **11**:736–739 (1965).

21. Engel, R. E., et al., *Environmental Lead and Public Health.* Research Triangle Park, N.C.: Environmental Protection Agency, March 1971. Air Pollution Control Office Publication No. AP-90.

22. Bazell, Robert J., "Lead Poisoning: Zoo Animals May be the First Victims." *Science* **173**:130–131 (1971).

23. Eccleston, B. H., and R. W. Hurn, *Comparative Emissions From Some Leaded and Prototype Lead-Free Automobile Fuels.* Washington, D.C.: U.S. Dept. of the Interior, 1970. Bureau of Mines Report of Investigations No. 7390.

7
NOISE

It would be difficult to select one night out of three hundred and sixty-five, during which the entire population of New-York are permitted to rest in peace. . . . Surely a city kept in a fever of excitement through the day ought to be permitted to rest in tranquility at night.
—Editorial, *New York Times,* September 13, 1859, p. 4.

The editorial writer in the *New York Times* had probably been awakened once too often by the fire department's tolling of bells for some petty fire in the city. Although that practice has long since disappeared, today's New York Fire Department sends its vehicles out sounding the fierce oscillating note of the "Grover T" siren. Noise has come to be regarded as an important type of urban pollution, capable of causing annoyance and hearing loss, and perhaps even adverse physiological and psychological effects. This chapter will discuss the nature and effects of noise.

SOUND AND HEARING

Noise is unwanted sound. Sound is due to the presence of mechanical waves in matter—gaseous, liquid, or solid. These waves are longitudinal; i.e., the atoms and molecules transmitting the wave oscillate in the direction in which the wave is traveling. Sound waves in a material are characterized by alternate regions of compression and rarefaction of matter, as shown in Figure 7-1(a). The accompanying pressure wave, shown in Figure 7-1(b), varies sinusoidally along the wave at any instant in time, with regions of compression corresponding to regions in which the pressure exceeds atmospheric pressure, and regions of rarefaction corresponding to regions in which the pressure is less than atmospheric pressure. The amplitude of the pressure wave is generally denoted P. The intensity of the sound wave is the energy flowing each second through a unit area perpendicular to the direction of propagation of the wave and is measured in watts per square meter; Figure 7-1(c) shows the variation in intensity. When the density of the air is ρ and the speed of propagation of the sound is v, the intensity is related to the pressure amplitude by $I = P^2/2\rho v$ [1].

The human ear is capable of receiving sound waves and transmitting signals to the brain to create the sensation of hearing. The ear's response is

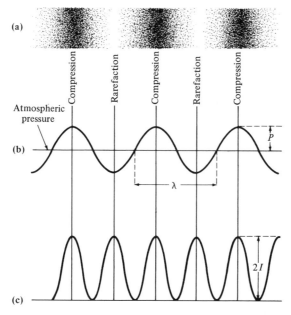

FIG. 7-1. Sound waves. All parts of the figure show the spatial variation along the wave at a particular instant of time. (a) Regions of compression and rarefaction in the air. (b) Pressure wave; $P =$ pressure amplitude ($P_{rms} = P/\sqrt{2}$) and $\lambda =$ wavelength. (c) Intensity wave; the average intensity is denoted I.

not directly proportional to the intensity or the pressure of the sound wave but is more nearly proportional to the logarithm of the intensity or the pressure; i.e., the ear subjectively judges the "loudness" of two sounds by the ratio of their intensities or pressures. Consequently, sounds are measured on a logarithmic scale, either the *intensity level*

$$\text{IL} = 10 \log_{10} (I/I_0)$$

where the reference intensity $I_0 = 10^{-12}$ watts/m², or the *sound pressure level*

$$\text{SPL} = 20 \log_{10} (P_{rms}/P_0)$$

in terms of the rms (root-mean-square) pressure $P_{rms} = P/\sqrt{2}$, where the reference pressure level $P_0 = 2 \times 10^{-5}$ N/m². These are two independent definitions but they do not differ significantly since P_0 is approximately the rms pressure corresponding to an intensity I_0. The sound pressure level is most often used since it is easier to construct instruments to measure pressure than to construct those to measure intensity. These levels are expressed in decibels (dB). Thus 0 dB corresponds to an intensity I_0 or an rms pressure P_0, 20 dB corresponds to an intensity of 100 I_0 or an rms pressure $10P_0$, etc. A sound 1 dB higher than another sound is characterized by an intensity $\sqrt[10]{10} = 1.259$ times as great or a pressure $\sqrt[20]{10} = 1.122$ times as great.

The "threshold of hearing" is the decibel level at which a sound can just barely be perceived. For a young person with good hearing, the maximum sensitivity occurs at a frequency of about 3000 Hz (hertz) (3000 cps), and the threshold of hearing is about 0 dB over a flat range extending from about 1000 to 5000 Hz. At higher and lower frequencies, the threshold of hearing is higher; the ear is unable to perceive sounds of frequencies lower than about 20 Hz or higher than about 20,000 Hz, the actual cutoff frequencies varying quite a bit from person to person.

The decibel level of a sound does not correspond to the subjective loudness of the sound except at a given frequency. For this reason a number of other loudness scales have been devised, including some instrumental scales. In the scale based on *phons,* the loudness level in phons of a sound is taken to be numerically equal to the decibel level of a 1000-Hz tone that is judged by the average observer to be equally loud [2]. The *perceived noise decibel* (PNdB) scale is based in a complicated way on the frequency distribution of the noise [3] and is often encountered in discussions of aircraft noise. Sound level meters for noise measurement weight sounds at different frequencies in such a way as to approximate the total loudness level. A, B, and C weighting networks (whose results are expressed as dB(A), dB(B), and dB(C)) are intended for use in the regions below 55 dB, between 55 dB and 85 dB, and above 85 dB, respectively.

Figure 7-2 shows the approximate threshold of hearing and loudness levels of different frequencies [1]. The figure also shows the "threshold of pain" or level at which a sound begins to feel uncomfortable due to tickling sensations; sounds 5 or 10 dB above the threshold of pain as plotted correspond to rather severe pain and discomfort.

Table 7-1 lists the sound pressure levels in ordinary decibels for various locations near different noise sources. In general, sound intensities decrease at greater distances from the source, the intensity being proportional to the inverse square of the distance (so that the pressure is inversely proportional to the distance) if there is no attenuation of the sound. In practice some attenuation always exists from the conversion of sound energy into heat in the air and by plants, buildings, and other substances so that sound intensities will in fact fall off with distance even more rapidly.

OCCUPATIONAL NOISE EXPOSURES

Many of the louder noises listed in Table 7-1 will be seen to be associated with particular occupations, especially construction and machinery work. Prolonged exposures to noise are known to lead to a gradual deterioration of the inner ear and to subsequent deafness. This occurs in addition to the normal loss of hearing that accompanies aging, which is referred to as pres-

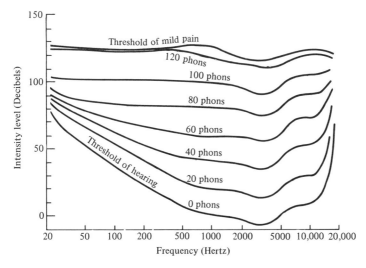

FIG. 7-2. The auditory area between the thresholds of hearing and of mild pain, as a function of frequency of sound. (Adapted from *Vibration and Sound* by Philip M. Morse [1]. Copyright 1948 by McGraw-Hill Book Co. *Used with permission of McGraw-Hill Book Company.*)

bycusis. Either presbycusis or noise-induced deafness will generally begin at higher frequencies; i.e., the threshold of hearing rises more at high frequencies than at low frequencies. Hearing loss thus first makes it difficult to distinguish among the different fricative consonants (*f, v, s, z, th, ch, sh*, etc.). As hearing loss continues, lower frequencies also become more and more difficult to hear.

Hearing-loss problems are especially severe in certain industries: iron and steel manufacture, motor vehicle production, metal products fabrication, printing and publishing, heavy construction, lumbering and wood products, mechanized farming, and textile manufacturing (rock music may have to be added to this list in a few years). Medical authorities are generally of the opinion that constant occupational exposure to levels of 90 dB in the range of human hearing are dangerous, while levels of 80 dB are not; these values are based on many studies of the effects of long-term exposure to occupational noise. As a result, an 8-h daily exposure to 85 dB is felt to be the limit that should be tolerated. The 1970 Occupational Safety and Health Act requires administrative or engineering controls to be applied whenever noise exposure exceeds 90 dB for 8 h a day, 95 dB for 4 h a day, or 100 dB for 2 h a day.

Awards for hearing loss to workers have been increasing in recent years. A government study estimates that perhaps 6 to 16 million workers in the United States are presently being exposed to noise levels capable of damaging their hearing and that there may be 4.5 million workers who could claim hearing loss awards at the present time [4]. At the 1966 average claim

TABLE 7-1. Noise levels in decibels.

Large rocket engine (nearby)	180	
	170	
	160	
Jet takeoff (nearby)	150	
Carrier deck jet operation	140	
		Threshold of pain
Hydraulic press (1 m)	130	
Jet takeoff (60 m)	120	
Automobile horn (1 m)		Maximum vocal effort possible
Construction noise (3 m)	110	
Jet takeoff (600 m)		
Shout (15 cm)	100	
Subway station or train		Very annoying
Heavy truck (15 m)	90	Constant exposure endangers hearing
Inside car in city traffic		Limit for industrial exposures
Noisy office with machines	80	Annoying
Freight train (15 m)		
Freeway traffic (15 m)	70	Telephone use difficult
Conversation (1 m)		Intrusive
Accounting office	60	
Light traffic (15 m)		
Private business office	50	Quiet
Living room in home		
Bedroom in home	40	
Library		
Soft whisper (5 m)	30	Very quiet
Broadcast studio	20	
Rustling leaves in breeze	10	Barely audible
	0	Threshold of hearing

of $2000, this would cost some $9 billion. At present, the Veteran's Administration is spending $65 million annually in rehabilitation programs for about 90,000 veterans with service-connected hearing disabilities.

Occupational noise exposures can be decreased by reducing the noise at the source, by shielding workers from the noise (which may be impossible if the workers must operate near noisy equipment), or by protecting workers with devices to shield the ear [5, 6]. Simple dry cotton plugs will provide 5 to 15 dB attenuation (more at higher frequencies than at the lower frequencies), while rubber, plastic or wax protectors inserted into the ears can provide up to 25 to 40 dB attenuation. Further improvement will result from the use of ear muffs.

PUBLIC NOISE EXPOSURES

EFFECTS ON HUMANS. Even short exposures to noise will produce temporary hearing losses, generally manifested as a shift in the threshold of hearing. This "temporary threshold shift" can be 20 dB or more following exposure for several minutes to a pure note with a sound pressure level of 100 dB or more. The shift typically varies with frequency and is a maximum for a frequency somewhat higher than the frequency of the pure note [3].

Longer exposures to noise lead to persistent or permanent hearing losses. These may be difficult to separate from presbycusis. It is known that presbycusis curves vary among different peoples [7]. The Maabans, an isolated primitive tribe in the Sudan, live in a very quiet environment—typically 35 to 40 dB—and do not use noisy weapons or musical instruments; their hearing loss into what we consider "old age" (into the 70's) appears to be only 5 to 15 dB at 2000 to 4000 Hz compared to perhaps 50 to 70 dB for Americans [7]. Although this study shows a correlation between noise exposure and hearing loss, there are probably many other factors that are important—genetic factors, nutrition, climate, stress, etc.—and some of these may be working to the advantage of the Maabans.

Noise also clearly produces annoyance in human beings and can interfere with proper rest and sleep. In general, annoyance seems to increase with the loudness of the sound (although it is not limited to loud sounds) and with higher frequencies, is worse for discrete tones than for those extending over a range of frequencies, and is generally worse for variable or irregular sounds. Some persons are annoyed by certain types of noise (rattling of metal garbage cans, sounds from transistor radios, power lawn mowers, piped music in public places), perhaps as much for psychological reasons as any other, although this does not make the annoyance any less valid. Many court cases have been based on annoyance. In one early example, the 1908 case of *Le Blanc vs Orleans Ice Company,* the plaintiff argued that noises and vibrations produced by operation of the defendant's ice plant in New Orleans kept his family constantly annoyed and deprived them of their sleep but the court decided the noise was unavoidable and that neighbors had to submit to it for the public good (see also the quote at the beginning of Chapter 20).

Noise can interfere with speech and work. Traffic noise that might almost pass unnoticed suddenly becomes annoying when one attempts to make a telephone call from a roadside booth. Conversations in the presence of noise require louder talking. Noise of up to 60 dB does not interfere with speech, higher levels cause unintentional raising of voices, still higher levels cause intentional raising of voices, levels of 80 to 85 dB make understanding barely possible with loud shouts, and levels of 90 dB and over make understanding impossible [8]. Noise also produces inefficiencies in work, especially in tasks requiring a high degree of concentration, such as learning in schools. Experiments have clearly established that noise makes it harder to perform simple vigilance tasks, such as watching for the appearance of three successive odd digits presented in sequence on a screen [8].

Noise also produces physiological effects in the human body. One important effect is the vasoconstriction reflex, in which the small blood vessels of the body constrict and reduce the flow of blood. Vasoconstriction occurs even for short noises and persists for several minutes after cessation of the noise [7]. Other physiological effects include dilation of the pupils, paling of the skin, tensing of the voluntary and involuntary muscles, diminution of gastric secretion, increase in diastolic blood pressure, and the sudden injection of adrenalin into the blood stream, which increases neuromuscular tension, nervousness, irritability, and anxiety [7]. Recently there have been reports from research physicians that noise has similar effects on unborn children, who are not as well protected from the outside environment as formerly thought. The physiological effects on humans are of unknown importance but noise may be contributing adversely to the health of Americans, especially those with heart disease, high blood pressure, and emotional problems who need to be protected from additional stresses.

Impulsive noises are generally thought to be even worse than continuous noise and capable of causing sudden damage to hearing if loud enough. Sudden noises also startle persons and lead to accidents.

SOURCES OF NOISE. Community noises that affect the public include construction noises (especially in large cities in which there is always some construction or repairing taking place); transportation noises from cars, trucks, motorcycles, and airplanes; and noises from appliances, power lawn mowers, power saws, etc.

In New York it has been estimated that there are 80,000 street repair jobs and 10,000 construction and demolition projects each year. Very loud noises—often 110 to 120 dB in the immediate vicinity—are produced by pneumatic riveters and chippers, air compressors, hammers, and other heavy equipment. It is possible to construct the equipment with mufflers and other noise suppression devices, and citizens' groups (such as the Citizens for a Quieter City in New York) have urged the adoption of less noisy equipment such as some European cities require. One U.S. firm manufactures an air compressor producing a noise level of 82 dB at a cost only 10% higher than that of an unmuffled compressor producing 110 dB.

Motor vehicle noises are of concern to occupants of the vehicle and to persons living or working near highways. Part of the noise arises from the rolling of the tires, partly due to the roughness of the road and partly due to the shape of the tires and their tread design; this noise is especially important at high speeds. At lower speeds, engine noises predominate. Noises are also produced by brakes, cooling fans, heating and air conditioning equipment, the transmission, etc. Exhaust noises due to nonlinearity and turbulence in the flow of the exhaust gases are partly controlled by the automobile muffler with its baffles and tubes; the importance of the muffler is readily recognized when it wears out. Finally there is the automobile horn, which is too often used in nonemergency situations.

Loud noises are also produced by the steel wheels of subway trains moving on steel rails, especially at high speeds.

Aircraft noises have been troublesome in the vicinity of airports for many years, and their existence represents a real cost to society that has not been fairly assessed against the air transportation industry, just as numerous other pollution costs have not been assessed against polluters. The situation is being aggravated by the continued growth of the airline industry. Areas near airports are subjected to noise from landing approaches, from takeoffs, and from the "sideline" noise of jet engines on the ground. Subjective annoyance increases with increasing peak noise level and it increases with increasing number of aircraft using the airport; a "noise and number index" has been devised to measure the annoyance [3]. Annoyance begins with noise levels in excess of 80 dB, and quadrupling the number of aircraft per day has been shown to correspond to keeping the number unchanged but adding about 9 dB to the noise level. Most of the noise from jet aircraft comes from the jet engine. Part of it, "jet noise," arises from turbulence, where the high-velocity gas stream leaving the exhaust nozzle mixes with the outside air; this noise extends over broad frequency ranges and is generated outside the engine. "Fan noise" is generated in the turbofan engine and propagated forward out of the air inlet and backward out of the fan ducts; it contains broad band noise from the flow of turbulent air through the fan, but it also contains some annoying discrete tones arising from the chopping of the blades.

Rocket engines used in the space program also produce noises of concern near launch or test sites; these noises tend to be very low in frequency.

Noise basically represents the transfer of energy into the form of acoustic energy, but the power required to produce loud sounds is actually quite small. Typical acoustic power outputs are [6]:

10^4 watts by jet engine.

1 watt by pneumatic chipping hammer.

10^{-1} watt by automobile traveling 45 mi/h.

10^{-5} watt by ordinary speech.

10^{-8} watt by small electric clock.

The huge Saturn V engine radiates about 200 million watts of acoustic power during launch, producing a roar of 110 dB at a distance of 16 km, but only 0.5% of the total power developed by the rocket engine is in the form of acoustic power [9].

Sound pressure levels can be determined from acoustic power outputs or vice versa. If the acoustic power $\mathcal{P}$ is radiated isotropically (uniformly in all directions), the intensity of the sound wave at a distance r from the source of the sound is $I = \mathcal{P}/4\pi r^2$ in the absence of attenuation processes. Thus the intensity level 60 m away from a 10^4-watt jet engine would be

$$\text{IL} = 10 \log_{10} I/I_0 = 10 \log_{10} \frac{10^4/4\pi(60)^2}{10^{-12}}$$

$$= 10 \log_{10}(2.2 \times 10^{11}) = 113 \text{ dB}$$

NOISE CONTROLS. Noise control can generally be accomplished by one of the following methods: (1) reduction of noise at the source, (2) substitution of a different machine or operation, and (3) reduction of noise at the listener.

Proper building construction can reduce noises from outside and inside the building [4, 5, 6]. Many older homes are much quieter than newer homes because of their more massive construction, their large rooms, their heavy doors, and the use of heavy, sound-absorbing furnishings. Modern dwellings seem to be much noisier, as demonstrated by the frequent complaints of apartment residents. The reasons for this include greater mechanization (noisy domestic appliances; noisy high-pressure systems for heating, cooling, and plumbing, etc.), poor acoustical design (such as lack of separation of noisy areas from those requiring quiet and privacy), light-frame construction (thin wall and floor constructions, hollow-core doors, etc.), poor workmanship (such as carelessness in sealing holes, cracks, and noise leaks), and the avoidance of the costs of sound insulation construction (which can run 2 to 10% of the cost of the building). The trend to high-rise apartment buildings with their greater concentration of families and the general lack of mandatory acoustic criteria and enforcement procedures have also contributed to the problem.

TABLE 7-2. Sound transmission losses through various substances [6].

25 dB	Cinder block, 10 cm thick
30 dB	Glass, ⅝ cm thick
42 dB	Cinder block, 10 cm thick, plastered on one side
45 dB	Cinder block, 10 cm thick, plastered on both sides; Brick, 10 cm thick
55 dB	Two cinder blocks, each 7.5 cm thick separated by equal amount of air, with plaster on both outside surfaces

The sound transmission loss through various materials is given in Table 7-2. In general, heavier materials cause greater losses and the use of intervening air spaces and resilient suspensions helps to increase the loss.

Noise from motor vehicles can be reduced by controlling it at the source or by improving the design, construction, and location of highways. Tire tread pattern is important in noise production and automobile tire patterns are designed to avoid strongly tonal sounds. There is much that can be done to reduce engine and exhaust noises and noise controls can be specified by purchasers of motor vehicles. In recent years, New York has replaced its noisy old escalator garbage collection trucks—the source of numerous complaints over the years—by quieter compactor trucks.

The problem of aircraft noise can be alleviated by operational procedures, such as flying over unpopulated or sparsely populated areas wherever possible. In addition, specific takeoff and landing procedures have been developed to reduce noise exposure of persons near airport runways. In general it is desirable to increase altitude as rapidly as possible and to cut back power as soon as possible after takeoff to reduce noise.

The question of zoning around airports is often a crucial one because of the noise levels. Some extraordinary steps have been taken in some places. When Los Angeles International Airport recently opened a new runway near several residential neighborhoods, the protests against the noise were so vociferous that the airport decided to spend $80 million to buy and demolish or move 1994 homes (at prices up to $115,000), a school, and several commercial buildings [10]. The Frankfurt (West Germany) airport has built a concrete barrier 14 m high to reduce sideline noise to nearby homes.

Modifications of engine and airframe design have also helped. Lining the engine nacelle with sound-absorbing fibers has reduced fan noise levels in turbofan engines by about 10 dB. The use of suppressor nozzles to reduce the jet noise has not proved as successful, however. Newer turbofan engines with high bypass ratios (i.e., with a greater fraction of the air taken in being expelled through the fan nozzle, bypassing the burner and turbine sections) have proved to be much quieter than previous engines but there are still many older, noisier engines in use.

The Aircraft Noise Abatement Act of 1968 authorized the Federal Aviation Administration to set standards for the measurement and control of aircraft noise (including sonic booms) and prescribe regulations for noise abatement and certification of aircraft.

Quiet is not always desired; some persons are able to study or concentrate better with some background noise present. Some devices have even been developed to provide extra background noise in the hopes of masking other sounds [11]. Sears, Roebuck, and Co. has marketed a swoosh-maker called "Sleep-Mate" and other companies have placed "Sleep Sound" and the three-sound "Sleepatone" ($120 price tag) on the market. (Some persons simply turn on an electric fan to accomplish the same purpose.) One firm provides background noise systems (such as waterfall sounds broadcast from loudspeakers) to offices for up to $25,000 per building. There are some sounds that hearers prefer to be loud, such as rock and roll music, but hearing losses result from them just as they do from unwanted noise [12].

SONIC BOOMS

A new type of noise in recent years is the sonic boom, which is produced by objects traveling faster than the speed of sound, whose value in air at sea level is about 330 m/s (740 mi/h) [13, 14]. Sonic booms are produced constantly, not just when the object first reaches a speed greater than that of sound. The object sets up shock waves, and at a distance from it there will generally be two strong shock waves associated with the front and the back of the object, as shown in Figure 7-3(a). The pressure in the region between the shock waves (the "region of overpressure") deviates from atmospheric pressure. The bow wave is associated with higher-than-atmospheric pressures, and the tail wave with lower-than-atmospheric pressures. As the object speeds

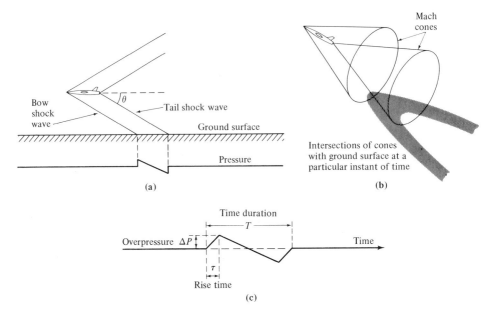

FIG. 7-3. Sonic booms. (a) Cross-sectional representation of bow and tail shock waves and pressure at ground level. (b) Three-dimensional view of supersonic plane, bow and tail Mach cones and region of overpressure between the intersections of the cones with the surface of the ground. (c) "N signature" of the sonic boom experienced at a given point on the ground.

along, the bow and tail waves have approximately the shape of cones ("Mach cones") of half-angle θ where $\sin \theta = v_s/v_0$; v_s is the speed of sound and v_0 is the speed of the object. (The pressure and temperature variations in the air lead to variations in v_s so that the shock waves do not exactly have the shape of cones.) The region of overpressure between the cones intersects the ground over an area shown shaded in Figure 7-3(b); attenuation will take place in the atmosphere, of course, and the overpressures are generally detectable only near the object's track along the ground.

The pressure disturbance actually heard by a person on the ground will have the shape shown in Figure 7-3(c). This "N signature" has the general shape of the letter N: A sharp rise over a rise time τ to a maximum overpressure ΔP, a slow decrease to a lower-than-atmospheric pressure, and a sharp rise back to normal atmospheric pressure. The total time duration depends on the size and shape of the supersonic object, being about 400 ms for a 56.4 m-long Concorde supersonic transport [15], but being so short for a bullet that only a single loud crack is heard, not the typical double boom of a supersonic plane. The rise time may be only 3 ms or so. A noticeable sonic boom is usually heard over a width of perhaps 80 to 130 km for supersonic planes flying at an altitude of about 20 km.

The resulting overpressures are small compared to atmospheric pressure, typically 100 N/m² (about 2 lb/ft²) or 0.1% of atmospheric pressure.

They are large enough to be very loud however; a sound wave with rms pressure of 100 N/m^2 would correspond to 137 dB, but a sonic boom is not sinusoidal and the apparent loudness depends on the shortness of the rise time (shorter rise times sound louder) as well as the amount of overpressure [13]. U.S. Air Force studies have shown that sonic booms of about 100 N/m^2 correspond to about 120 PNdB [14]. These pressures are also capable of rattling dishes and shattering glass in addition to startling people. The studies that have been carried out in Oklahoma City and other communities to determine what sorts of complaints result from sonic booms and what damage claims are filed suggest that roughly $600 in damage claims can be expected for each million man-booms experienced and that there might be 1000 claims totaling about $500,000 annually for each supersonic plane [16].

Students of the sonic boom believe that even after a slow buildup of sonic boom frequencies as supersonic transports come into general use, some 25 to 50% of the U.S. population will not be able to adapt to the boom, and extensive social, political, and legal reactions will occur against supersonic flights [14, 16]. There remains the possibility of banning such flights over populated areas (which might very well obviate some of the economic benefits of SST's) but even supersonic flights over oceans would expose some people to the booms. Opponents of development of supersonic planes have felt that once the investment in development has been made, it might be very hard to institute restrictions.

There remains the small possibility of avoiding the sonic boom altogether by proper design. Although several shapes are known to be capable of supersonic speeds without production of a sonic boom (i.e., the configuration is such that there is no shock wave with a pressure discontinuity), these shapes are incapable of providing lift. A design that provides no lift would be useless for aircraft, although it could conceivably be useful in ballistic propulsion.

The supersonic transports being developed will also cause great engine noise problems. At one time, the U.S. aircraft industry wanted to strive for a limit of 120 PNdB 3 mi away from the SST during the start of the takeoff roll and felt that the limit of 105 PNdB proposed by the federal government was too stringent. At the time that the U.S. Congress decided to stop federal funding of the SST, in 1971, the industry had decided that the lower limit was probably attainable after all. A noise reduction of 15 dB is quite significant, of course, since it corresponds to a reduction in sound intensity by a factor of 32.

REFERENCES

1. Morse, Philip M., *Vibration and Sound.* New York: McGraw-Hill Book Co., 1948.
2. Kinsler, L. E., and A. R. Frey, *Fundamentals of Acoustics.* New York: John Wiley & Sons, Inc., 1962.

3. Burns, William, *Noise and Man*. London: John Murray, Ltd., 1968.

4. Federal Council for Science and Technology, Committee on Environmental Quality, *Noise—Sound Without Value*. Washington, D.C.: U.S. Govt. Printing Office, 1968.

5. Beranek, Leo L. (ed.), *Noise Reduction*. New York: McGraw-Hill Book Co., 1960.

6. Harris, Cyril M. (ed.), *Handbook of Noise Control*. New York: McGraw-Hill Book Co., 1957.

7. Rosen, Samuel, "Noise, Hearing and Cardiovascular Function." In Welch, Bruce L., and Annemarie S. Welch (eds.), *Physiological Effects of Noise*. New York: Plenum Press, 1970.

8. Carpenter, A., "Effects of Noise on Performance and Productivity." In *The Control of Noise*. London: Her Majesty's Stationery Office, 1962. pp. 297–306.

9. Ribner, H. S., "Jets and Noise." In *Aerodynamic Noise*. Toronto: University of Toronto Press, 1969.

10. Lindsey, Robert, "Jet Noise Dooming Homes Near Los Angeles Airport." *New York Times,* July 21, 1971, p. 1.

11. Leger, Richard R., "If You Can't Shut Out the Racket, Drown It With Your Own Noise." *Wall Street Journal,* January 28, 1970, p. 1.

12. Smitley, E. K., and W. F. Rintelmann, "Continuous Versus Intermittent Exposure to Rock and Roll Music." *Arch. Environ. Health* **22**:413–420 (1971).

13. Hubbard, Harvey H., "Sonic Booms." *Physics Today* **21**(2):31–37 (February 1968).

14. Kryter, Karl D., "Sonic Booms from Supersonic Transport." *Science* **163:** 359–367 (1969).

15. *Sonic Booms and Other Aircraft Noise in Studios*. London: B.B.C., April 1968. B.B.C. Engineering Monograph No. 73.

16. Shurcliff, W. A., *S/S/T and Sonic Boom Handbook*. New York: Ballantine Books, 1970.

8
WATER POLLUTION: INTRODUCTION

They that use filthy, standing, ill-coloured, thick, muddy water, must needs have muddy, ill-coloured, impure, and infirm bodies. And because the body works upon the mind, they shall have grosser understandings, dull, foggy, melancholy spirits, and be really subject to all manner of infirmities. —Robert Burton, *The Anatomy of Melancholy* (1621 A.D.).

Many of the wastes of human society are disposed of in bodies of water—rivers, lakes, seas—and some of those disposed of in the air or on land eventually wind up in waterways. When the human population is concentrated too much, the waterways are unable to handle the quantities of material that are turned into them and they become polluted—with unhappy results such as those Robert Burton mentioned.

In this chapter we shall consider some quantitative data about the earth's water balance and then discuss the various types of water pollutants, their sources, and their costs to society.

THE EARTH'S WATER BALANCE

The total amount of water on the earth is about 1.35 billion km³ (3.5×10^{20} gal). Over 97% of this is found in the earth's oceans, and the earth's fresh water totals only about 37 million km³, of which four-fifths occurs in the polar ice caps and glaciers. Table 8-1 shows the locations of the world's waters as estimated by the U.S. Geological Survey [1]. Surface water in streams amounts to a very tiny portion of the total, despite its importance to mankind. Much more water is found below the earth's surface.

Generally, water will percolate down into the ground. Some moisture is present in the soil, near enough to the surface to be reached by plant roots. Additional water, beyond the roots, is called *vadose* water; it is water that is descending through an unsaturated region under the influence of gravity. Below a level known as the *water table*, the ground is saturated with water. This "groundwater" is capable of flowing in a fashion similar to the flow of surface water in streams. When the water table intersects the land surface, free water will be present above ground as river water, lakes, or marshes. The presence of impermeable (nonporous to water) rock sets a lower limit to

TABLE 8-1. Locations of the world's waters [1].

Location		Volume (km³)	Fraction (%)
Surface water:	Freshwater lakes	120,000	0.009
	Saline lakes, inland seas	100,000	0.008
	Stream channels (average)	1,200	0.0001
Subsurface water:	Soil and vadose water	65,000	0.005
	Groundwater (to 800 m)	4,000,000	0.3
	Groundwater (deep-lying)	4,000,000	0.3
Other water:	Ice caps and glaciers	29,000,000	2.1
	Atmosphere	13,000	0.001
	Oceans	1,315,000,000	97.3
Total (rounded)		1,350,000,000	100.0

the depth of groundwater, although there may be more groundwater at even greater depths in other permeable regions. Water wells are drilled into the groundwater region, from which water may be pumped to the surface. If the aquifer (groundwater supply) is subject to pressure, it may be possible to drill "artesian" wells that require no pumping. Figure 8-1 illustrates some of these points.

The earth's water is in constant circulation through the hydrological cycle involving evaporation, precipitation, percolation, and runoff. The total amounts of precipitation and evaporation are in balance on a worldwide basis (see Table 8-2), with the oceans evaporating more water than they receive in precipitation and the continents receiving more in precipitation than they evaporate. Since the oceans' water level is not changing noticeably, the difference is made up by river and groundwater flow into the oceans.

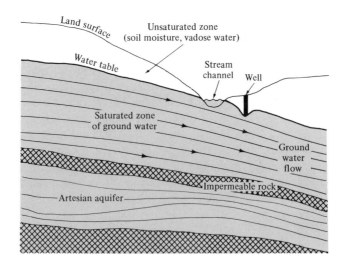

FIG. 8-1.   Schematic diagram of subsurface water supplies.

TABLE 8-2. The water balance of the earth [2].

	Oceans	Continents	Whole earth
Precipitation (km³/yr)	324,000	99,000	423,000
Evaporation (km³/yr)	361,000	62,000	423,000
Gain by inflow (km³/yr)	+37,000	−37,000	0
Precipitation (cm/yr)	90	67	83
Evaporation (cm/yr)	100	42	83
Gain by inflow (cm/yr)	+10	−25	0

The annual water balance of the United States as estimated in 1962 is shown in Table 8-3. Of the total annual precipitation of nearly 6000 km³ (which amounts to about 65 cm/year), 71% is evaporated or evapotranspirated by vegetation before ever getting into streams, and the other 29% becomes streamflow. Some of the streamflow (especially that used for irrigation purposes) eventually evaporates too, so only 27% of the original precipitation flows into the oceans and 73% is evaporated into the atmosphere. Groundwater is another important source of public water supplies and withdrawal has increased from about 28 km³ in 1935 to 64 km³ in 1955 to about 90 km³ in 1970. In many places, especially in the arid regions of the West and Southwest, groundwater is being withdrawn faster than it is being re-

TABLE 8-3. Water balance in the United States in 1962 [3]. Data are for the 48 coterminous states. Streamflow withdrawals for irrigation, industry, and municipalities could, in principle, add up to more than the total streamflow since much of the water is eventually returned to streams and is available again to downstream users.

	Amount (km³/yr)	Percent
Evaporation and evapotranspiration	**4140**	**71**
Croplands and pasture	1350	23
Forests and browse vegetation	940	16
Noneconomic vegetation	1880	32
Streamflow	**1730**	**29**
Not withdrawn	1305	22
Withdrawn for irrigation	195	3.3
Withdrawn for industry	195	3.3
Withdrawn by municipalities	35	0.6
Total annual precipitation	**5870**	**100**
Flow into oceans	**1580**	**27**
Streamflow not withdrawn	1280	22
From irrigation water	80	1.4
From industrial water	190	3.2
From municipal water	30	0.5
Total evaporation into atmosphere	**4290**	**73**
Total annual precipitation	**5870**	**100**

charged, and the water table is dropping. As a result, the practice of artificially recharging groundwater reservoirs has been instituted upon occasion.

The average yearly rainfall on the United States is about 5870 km³ (1.5 × 10¹⁵ gal), of which about 1650 km³ are available as surface runoff. Only about 900 to 1100 km³ are economically available for human use, however. Table 8-4 shows the actual usage in 1965—a little over 370 km³—and projections for 1980 and 2000 by the Water Resources Council [4]. The fastest-growing use is for cooling water in steam-electric plants, which will be using 60% of the total water used by 2000 if the projections are accurate (see also the discussion of thermal pollution in Chapter 14). It should be noted that much of the water used is eventually returned to streams or reused and is therefore not used consumptively; very little of the industrial water is used consumptively, for example, and most cooling water is returned to streams or lakes unchanged except for a temperature rise. Table 8-4 includes the figure for consumptive use.

TABLE 8-4. U.S. water supply and projected water use in cubic kilometers per year [4].

| | Annual usage in cubic kilometers | | |
	1965	1980	2000
Rural domestic	3.3	3.4	4.0
Municipal (public supplied)	32.7	46.3	70.0
Industrial (self-supplied)	64	104	176
Steam-electric power	117	267	650
Agriculture	155	191	212
Total water usage	**372**	**611**	**1110**
Total consumptive usage	118	144	177

Although these figures suggest that the United States has ample water resources, these resources are not distributed evenly and some portions of the U.S. are quite short of water. The Mississippi River carries 40% of the total U.S. streamflow into the oceans. The Pacific Northwest has abundant water resources. The arid southwestern regions of the U.S., however, are quite concerned about ensuring ample water supplies in the decades to come.

There have been suggestions that the western states purchase water from Canada (and statements by U.S. politicians that it would be immoral of Canadians to refuse to sell the water) and some Canadians are fearful that this would not only impede Canadian development but also invite U.S. interference in Canadian domestic affairs should Canada ever decide to halt sales of water. Engineers have already proposed a $100 billion North American Water and Power Alliance (NAWAPA) that would block off parts of north-flowing Canadian and Alaskan rivers and pump the water up 300 m through huge pipelines into the Rocky Mountain Trench (an 800-km long natural gorge containing the Columbia, Fraser, and Kootenay Rivers), from which the water could flow across the Canadian prairies to the Great Lakes and southward across the U.S. drylands to Mexico [5].

It is not unusual for nations to eye water supplies across their borders, as the U.S. is doing, and international water disputes are quite common and often quite troublesome. Disputes of current importance in the world involve the Colorado River and the Rio Grande between the U.S. and Mexico, the Indus River between India and Pakistan, and the Jordan River between Israel and the Arab world.

Water is immensely important to urban life. According to Wolman [3] the typical input into a city for each million inhabitants each day is:

565,000 metric tons of water.

1800 metric tons of food.

2700 metric tons of coal.

2500 metric tons of oil.

2400 metric tons of natural gas.

900 metric tons of motor fuel.

The typical output is:

450,000 metric tons of sewage (waste water with perhaps 110 metric tons of suspended solids).

1800 metric tons of refuse.

860 metric tons of air pollutants.

Water consumption in U. S. cities ranges between about 0.34 and 0.75 m³ (90 to 200 gal) per person per day including both domestic and local industrial usage. A survey of domestic usage carried out in 1964 by the U.S. Geological Survey showed the following breakdown [6]:

41% for flushing toilets.

37% for washing and bathing.

6% for kitchen use.

5% for drinking water.

4% for washing clothes.

3% for general household cleaning.

3% for garden watering.

1% for car washing.

Domestic use exhibits interesting cyclic variations—seasonal (highest in the hot summer months because of lawn watering, air conditioning, increased frequency of bathing, etc., but with secondary peaks in winter months when water is left running to prevent pipes from freezing and bursting), daily (low on Sundays, high on Mondays), hourly (peaks in the early morning and early evening). Short, sharp peaks occur during television commercials, as on the hour and half hour, and are also reflected in peaks in sewage flow.

In the past, our national water policy has focused on such problems as the control of disease, the control of floods, soil conservation, rural electrification, and regional economic development. Our concerns have changed from time to time, however, and proper water resources planning will involve making choices that satisfy current objectives in such a way as not to exclude the possibility of meeting other objectives in the future. Today there is growing concern about pollution of our lakes and rivers, and our water policies will have to take this into account.

WATER POLLUTANTS

In 1960 a Senate committee [7] classified water pollutants into eight categories in a manner that has become standard. In this section these eight groups will be discussed one by one and some common parameters of water pollution will be defined.

1. SEWAGE AND OTHER OXYGEN-DEMANDING WASTES. These are largely carbonaceous organic material that can be oxidized by microorganisms to carbon dioxide and water. A common measurement of this type of pollution involves the amount of molecular oxygen required to decompose the material through aerobic biochemical action. The standard test is the 5-day BOD (biochemical oxygen demand) test, in which the amount of dissolved oxygen required for oxidation over a 5-day period is measured, the results being expressed in milligrams of oxygen per liter (mg/l). When the amount of sewage discharged is relatively minor, a river will not become badly polluted since biological degradation will soon remove most of the wastes; however, strong sewage or other oxygen-demanding wastes from industry or agriculture can lead to the depletion of the dissolved oxygen in the water. "Septic" conditions are said to be present when the dissolved oxygen level is very low. Fish and other aquatic life require dissolved oxygen for survival and the DO (dissolved oxygen) level should normally be at least 5 mg/l (higher in cold water, especially in spawning areas, which require at least 7 mg/l) [8]. The DO level of water saturated with oxygen is 9.2 mg/l at 20°C (68°F) (see Table 14-4 for values at other temperatures). If there is insufficient dissolved oxygen for degradation of organic materials, oxygen may be obtained from dissolved nitrates and sulfates, with the accompanying production of disagreeable-smelling gases such as H_2S. If there is ample dissolved oxygen, microorganisms can oxidize nitrogen compounds and certain inorganic compounds such as ferrous salts, sulfides, and sulfites. COD (chemical oxygen

demand) tests make use of strong oxidizing compounds, such as potassium permanganate, to oxidize even some materials that are not biologically degradable; COD values will be larger than BOD values.

Sewage and other oxygen-demanding wastes are classified as water pollutants because their degradation leads to oxygen depletion, which affects (and even kills) fish and other aquatic life; because they produce annoying odors; because they impair domestic and livestock water supplies by affecting taste, odors, and colors; and because they may lead to scum and solids that render water unfit for recreational use.

2. INFECTIOUS AGENTS. Waste water from municipalities, sanitoria, tanning and slaughtering plants, and boats may be sources of bacteria or other microorganisms capable of producing disease in men and animals, including livestock. Any fair-sized community at any given time is likely to have *some* persons who are diseased so that disease microorganisms are almost always present in sewage.

There are several types of human infections, not all of which are transmissible through water (see Table 8-5). Many of the diseases whose epidemics

---

TABLE 8-5. Examples of human infections (adapted from Reference 9, © Copyright by The Macmillan Company, 1969).

A. Animal infections that are of public health importance because they are transmissible to man.
   1. Tetanus from horses and cattle transmitted by inoculation or contact with animal feces.
   2. Bubonic plague from wild rodents by insect (flea) bite.
   3. Anthrax from herbivorous animals by direct contact.
   4. Rabies from dogs, bats, etc., by bites.
   5. Bovine tuberculosis from cattle through ingestion or airborne transmission.
   6. Jungle yellow fever from monkeys through mosquito bites.
   7. Several types of encephalitis from birds and fowl through mosquito bites.
   8. Trichinosis from swine through ingestion.
B. Primarily human infections in which the infective agent has a certain period of extrahuman residence before transmission.
   1. Schistosomiasis ("snail fever") from water from snails.
   2. Urban yellow fever from mosquitoes.
   3. Hookworm from soil by skin penetration.
   4. Malaria (also a mosquito infection) from mosquitoes.
   5. Typhus from lice.
C. Infections that persist or multiply in the external environment and are transmissible from man to man.
   1. Cholera, typhoid fever, bacillary dysentery, poliomyelitis, and infectious hepatitis from water and food through ingestion.
   2. Staphylococcal and streptococcal diseases from food, air, and the proximate environment through contact and inhalation.
   3. Smallpox from air, dust, and the proximate environment through inhalation.
   4. Coxsackie and ECHO virus diseases from water through ingestion.

---

recurrently decimate human populations are transmitted by water, however, cholera and typhoid being important examples. The work of Dr. John Snow in London in the late 1840's and early 1850's led him to the conclusion that cholera epidemics were the result of some microorganism present in the feces of cholera victims and that the microorganism was transmitted through water supplies [10]. In 1854, when a severe cholera epidemic struck London, he was able to point to the Broad Street pump as the main source, and the authorities' removal of the pump handle at his suggestion immediately stayed the epidemic. In the late 19th and early 20th centuries (see the historical discussion in Chapter 9) public health officials in many countries were able to secure ample and pure water supplies, with accompanying decreases in death rates from waterborne diseases. Typhoid death rates annually per 100,000 persons in the United States, for example, decreased from 31.3 in 1900 to 22.5 in 1910 to 7.6 in 1920 to 4.8 in 1930 to 1.1 in 1940 to essentially zero in 1950. Waterborne epidemics still occur today in some countries, as the cholera epidemic in Pakistan in 1971 demonstrates.

The identification of pathogens in water requires very large samples and many sophisticated techniques and is too time-consuming and expensive for routine pollution tests. The standard method involves determination of the most probable number (MPN) of coliform organisms in the water sample. Coliform bacteria like *Escherichia coli* are normal inhabitants of human and animal intestines, and the daily per capita excretion in human feces may number from 125 to 400 billion [11]. These organisms will be reduced in number in the water by death in the nonnormal environment and by their removal and destruction in waste-water and drinking-water treatment processes. Although coliform organisms are not pathogens and are not affected by the water environment in exactly the same manner as pathogens, their existence and density has proved to be a fairly reliable indicator of the adequacy of treatment for reduction in pathogens, and coliform tests are therefore widely used [11]. Various standards have been suggested for various water uses [8] but typical maximum levels for fecal coliform counts are about 2000/100 ml for general recreational use, 200/100 ml for primary contact use (swimming), and 1/100 ml for drinking water.

3. PLANT NUTRIENTS. Plant nutrients such as nitrogen and phosphorus can stimulate the growth of aquatic plants, which interfere with water uses and later decay to produce disagreeable odors and add to the BOD of the water. Excess algal growths have been of particular concern since they lead to low dissolved oxygen levels and create treatment problems for municipalities and industries, in addition to interfering with recreational uses. Plant nutrient concentrations are generally expressed in ppm by weight or milligrams per liter (mg/l), which are essentially equal in water. In general it is the concentration of soluble inorganic nitrogen and phosphorus that is of importance and care must be taken in expressing the levels to distinguish

between elements and compounds, such as between phosphorus and phosphates.

The enrichment of waters by nutrients is referred to as *eutrophication*. Over periods of many millennia, the aging of lakes and slow-moving waters through eutrophication leads to their conversion into swamps and marshes. Man-made eutrophication hastens this process, as it has with Lake Erie, the lower Potomac River, the southwestern part of Lake Michigan, the Madison Lakes in Wisconsin, and Lake Washington in Washington, as well as waterways in foreign countries. Many streams and lakes that had clear, fresh water before the arrival of European settlers have become clogged with water hyacinth and other aquatic weeds, whose "rampant growth . . . has come to be one of the symptoms of our failure to manage our resources" [12].

Algal growth requires many different nutrients—carbon dioxide, nitrogen, phosphorus, iron, manganese, boron, cobalt, vitamins, hormones, etc. Presumably elimination of any one of the essential nutrients would prevent algal growth but there is an active scientific controversy about which nutrients should be controlled [13]. It has generally been felt that control of nitrogen and phosphorus would be best but there are scientists who think that the availability of carbon is a much more important limitation, feeling that the large amounts of phosphorus present in the sediments in lakes and rivers provide a vast reservoir available for the growth of algae. More research is needed to establish, for example, whether or not CO_2 from air and rain is sufficient to stimulate algal growth or whether CO_2 production by bacterial decomposition of organic matter is necessary [14]. The most likely answer is probably that phosphorus is the limiting element in some bodies of water but not in others where it is present in large concentrations. The deterioration of Lake Washington near Seattle was promptly slowed down when sewage effluent formerly dumped into the lake was diverted and phosphorus levels decreased to about one-fourth the previous levels; abundance of algae seems to be more closely related to phosphorus concentrations than to nitrogen or carbon dioxide [15].

There is not much information available about comparative natural and man-made sources of nutrient in waterways. The estimates made by Ferguson [14] are reproduced in Table 8-6. Using the total streamflow in the U.S. and taking the minimum phosphorus tonnage from the table, he estimates the average concentration of phosphorus in U.S. waterways to be 0.26 ppm, of which 0.08 ppm come from phosphates in detergents. Phosphate removal from detergents might alleviate algal growth problems in some areas but not in all, and the same is true of phosphate removal from waste water by advanced treatment methods.

4. EXOTIC ORGANIC CHEMICALS. The exotic organic chemicals include surfactants in detergents (see Chapter 9), pesticides (see Chapter 12), various industrial products, and the decomposition products of other organic com-

TABLE 8-6. Natural and man-generated sources of nitrogen and phosphorus in the United States [14]. Data are in millions of metric tons per year.

	Nitrogen	*Phosphorus*
Natural	**935–3800**	**222–643**
Man-generated	**3610**	**620–920**
Domestic sewage	1200	350–405[a]
Runoff from urban land	180	17
Runoff from cultivated land	1850	100–345
Runoff from land on which animals are kept	380	153
Total	**4545–7410**	**842–1563**

[a] The contribution from detergents is estimated to be 254 million metric tons.

pounds. Analyses of polluted waters show the presence of a wide variety of these compounds and many others are probably not being detected. Concentrations are generally expressed in ppm by weight (equal to mg/l). Some of these compounds are known to be toxic to fish at very low concentrations, such as 1 ppm phenol. Many are not biologically degradable, or are degraded only very slowly. Since many new chemical compounds are introduced each year without much knowledge of their effects on natural ecosystems, there is always a possibility that irreversible damage might be caused before scientists realize it.

5. INORGANIC MINERALS AND CHEMICAL COMPOUNDS. Inorganic chemicals of many types find their way into water from municipal and industrial waste waters and urban runoff. They are also measured in ppm by weight or mg/l. These pollutants can kill or injure fish and other aquatic life and they can interfere with the suitability of water for drinking or industrial use.

A prominent example is the occurrence of mercury in water [16]. A number of industrial processes make use of mercury, some of which is eventually disposed of in waste-water effluents. Metallic mercury was once believed to sink into the sediments and remain there in a chemically inert form. It is now known [17], however, that anaerobic bacteria in bottom muds can convert inorganic mercury into methyl mercury (CH_3Hg^+), which can be concentrated in living things and lead to mercury poisoning.

One potential pollutant arises in petroleum drilling, where brine is discharged along with crude oil when the latter is pumped to the surface. In mid-continent and Gulf Coast oil fields, the volume of brine produced is typically three times the volume of crude oil recovered. The pollution that could result is avoided by reinjecting the brine into underground strata. In some places, notably Michigan and California, the brines have proved valuable sources of important minerals and elements, such as bromine, iodine, and magnesium.

Another very important problem, especially in the Ohio River Basin, has been acid mine drainage. On exposed coal mine surfaces, minerals containing sulfur (most notably iron pyrite, FeS_2) come into contact with air and water, forming sulfuric acid that is carried into streams by waters draining from the mines. This occurs from abandoned mines as well as operating mines and has been most pronounced in bituminous coal mines east of the Mississippi River. In the late 1930's it was estimated [18] that about 2.45 million metric tons of the acid (as 100% H_2SO_4) were being produced annually, about 45% each from active mines and from abandoned mines (in which acid formation could be prevented by sealing techniques to cut off the air circulation) and 10% from idle or marginal mines. According to the Environmental Protection Agency, some 4 million metric tons of acid are being contributed to 17,000 km of streams and other receiving waters today.

6. SEDIMENTS. Sediments are soil and mineral particles washed from the land by storms and floodwaters, from croplands, unprotected forest soils, overgrazed pastures, strip mines, roads, and bulldozed urban areas [4]. Sediments fill stream channels and reservoirs; erode power turbines and pumping equipment; reduce the amount of sunlight available to green aquatic plants; plug water filters; and blanket fish nests, spawn, and food supplies, thus reducing the fish and shellfish populations.

Rough estimates suggest that the amount of suspended solids from surface runoff in U.S. streams may be 700 times the amount from sewage discharges. The total mass may be about 4 billion metric tons annually. Development of land for agricultural purposes may increase erosion rates to four to nine times the rates from the same land undeveloped [4]. On a global scale it has been estimated [19] that the mass of material moved annually by rivers to the ocean was 9.3 billion metric tons before man's intervention and is now 24 billion metric tons, which would mean that the continents are now being lowered at the rate of 5.8 cm (2.3 in.) every 1000 years. (Man also moves small amounts of sediments from the sea to the land, as in dredging.) The problem of the disappearance of the continents is not as immediate as the problem of the loss of valuable topsoil from agriculturally productive land, of course.

7. RADIOACTIVE SUBSTANCES. Harmful radiation may result in water environments from the wastes of uranium and thorium mining and refining, from nuclear power plants; and from industrial, medical, and scientific utilization of radioactive materials. Radiation is discussed in Chapter 15.

8. HEAT. Vast amounts of water are used for cooling purposes by steam-electric power plants (and other industries to a lesser extent), as indicated in Table 8-4. Cooling water is discharged at a raised temperature, and some rivers (such as the Mahoning River at Youngstown, Ohio) may have their temperatures so high (even up to 40°C) that fish life is completely

eliminated and the river becomes useless for assimilation of pollution or further cooling.

Increased temperatures have a number of effects on water [20]. The density and viscosity are decreased, permitting suspended solids to settle at a faster rate—about 2.5 times as fast at 35°C (95°F) as at 0°C (32°F). The evaporation rate increases very fast, being about five times greater at 32°C than at 15.5°C. The rate at which chemical reactions occur increases at increased temperatures, generally by about a factor of two for each 10°C rise. This leads to faster assimilation of waste and therefore faster depletion of oxygen. The amount of dissolved oxygen present at saturation in water decreases with increasing temperature, being only 70% as much at 35°C as at 15°C. Fish are also greatly affected by temperature and subtle behavior changes in fish are known to result from temperature changes too small to affect survival; actual fish kills from thermal pollution are not very common, however.

SOURCES OF WATER POLLUTION

The major sources of water pollution can be classified as domestic, industrial, agricultural, and shipping waste waters.

DOMESTIC WATER POLLUTION. This consists of waste water from homes and commercial establishments. Domestic waste water arises from many small sources spread over a fairly wide area but not so highly scattered as to make it impossible to collect the wastes and transmit them by sewers to a municipal waste treatment plant. Only about two-thirds of the U.S. population is presently served by sewerage systems, with the rest relying on disposal with septic tanks (see Figure 8-2) and cesspools, which are capable of producing localized water pollution problems. Municipal waste waters in the United States average by volume about 55% domestic and 45% industrial waste water.

INDUSTRIAL WATER POLLUTION. This tends to occur in large amounts in specific locations, making collection and treatment fairly simple to accomplish. There are over 300,000 water-using factories in the United States, discharging wastes with a total BOD load about three to four times as large as the load from the sewered population [4]. Only about 7 or 8% of industrial waste waters are disposed of in municipal sewer systems but, as mentioned above, they constitute about half the total municipal load. Industrial pollution is discussed further in Chapter 10.

AGRICULTURAL WATER POLLUTION. This includes sediments, fertilizers, and farm animal wastes. These pollutants can all enter waterways as runoff from agricultural lands but farm animal wastes are an especially large prob-

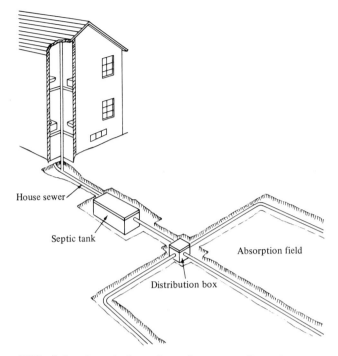

FIG. 8-2. A typical septic tank sewage disposal system
with a subsurface disposal system [21]. The liquid contents of the house sewer are discharged into the septic tank,
where some solids settle out and some anaerobic decomposition of sewage takes place. Clarified effluent flows to
the distribution box and thence through a tile system into
the subsurface absorption field. The septic tank must be
cleaned out occasionally or else sludge will clog the disposal field.

lem near the large feedlots on which thousands of animals are concentrated.
Agricultural pollution is discussed in Chapter 11.

SHIPPING WATER POLLUTION. This includes both human sewage and
other wastes, the most important of which is oil. There are about 8 million
watercraft on navigable waters of the U.S. and their combined waste discharges are equivalent to a city with a population of 500,000 [4]. Oil pollution, an oxygen-demanding waste, is of concern not only from sensational
major spills from ships and offshore drilling rigs but also from small spills and
cleaning operations. Shipping pollution is discussed in Chapter 10.

Not all water pollutants arise from waste waters. Some water pollution
can arise from solid wastes that are not prevented from contaminating surface and groundwater, and some can arise from the settling of air pollutants.

THE COSTS OF WATER POLLUTION

The costs of water pollution are probably even more difficult to estimate than the costs of air pollution. Who can measure the cost of not being able to swim in some river or not being able to catch fish in it? Who can measure the aesthetic cost of a polluted lake or waterway? Is it even valid to try to express these costs in monetary terms?

It is clear that for a very modest price people in the U.S. are able to enjoy a plentiful and generally pure supply of drinking water. The cost is certainly small in comparison with the cost that would be incurred by society by repetitions of the waterborne epidemics of the recent past. The investment in municipal water treatment facilities and sewers in the United States is probably about $25 billion, although the current replacement cost might be three times as much. Much remains to be done, however. Some communities have no sewer systems and many that are sewered have inadequate sewage treatment facilities. Many communities have a combined sewer system for storm water and domestic sewage without provisions for holding storm water for treatment, and large capital investments would be required to separate the sewer systems or to provide some sort of treatment of storm runoff.

In 1939 the National Resources Committee [18] estimated that the United States needed to invest a total of over $2 billion during the next 10 to 20 years for municipal waterworks and industrial waste treatment plants where practical methods existed. More recently, Sigurd Grava [6] has estimated that $100 to $110 billion will be required by the year 2000 to obtain adequate secondary treatment plants for domestic and industrial waste waters —implying average annual expenditures of nearly $4 billion.

The Council on Environmental Quality has recently estimated [22] that the total annualized costs (i.e., costs of capital investment, operation, and maintenance) to meet federal water quality standards needs to be $5.8 billion annually by 1975, compared to actual costs of $3.1 billion in 1970. These include both municipal and industrial costs. The 1975 costs would amount to about $25 per person.

Although these figures are large, the total annual costs amount to much less than 1% of the gross national product of the United States, just as in the case of air pollution abatement costs.

REFERENCES

1. U.S. Geological Survey, *Water of the World.* Washington, D.C.: U.S. Dept. of the Interior, 1968.
2. Defant, Albert, *Physical Oceanography,* Vol. I. New York: The Macmillan Co., 1961.
3. Wolman, Abel, *Water, Health and Society: Selected Papers by Abel Wolman*

edited by Gilbert F. White. Bloomington, Ind.: Indiana University Press, 1969.

4. *Environmental Quality. The First Annual Report of the Council on Environmental Quality.* Washington, D.C.: U.S. Govt. Printing Office, August 1970.

5. Sewell, W. R., et al., "NAWAPA: A Continental Water System." *Bulletin of the Atomic Scientists* **23**(7):8–27 (September 1967).

6. Grava, Sigurd, *Urban Planning Aspects of Water Pollution Control.* New York: Columbia University Press, 1969.

7. U.S. Senate Select Committee on National Water Resources, *Water Resource Activities in the United States. Pollution Abatement.* Washington, D.C.: U.S. Govt. Printing Office, 1960.

8. *Water Quality Criteria. Report of the National Technical Advisory Committee to the Secretary of the Interior.* Washington, D.C.: U.S. Dept. of the Interior, April 1, 1968.

9. Kilbourne, E. D., and W. G. Smillie (eds.), *Human Ecology and Public Health,* 4th ed. New York: The Macmillan Co., 1969.

10. Snow, John, *On the Mode of Communication of Cholera,* 2nd ed. London: John Churchill, 1855. Reprinted in *Snow on Cholera.* New York: The Commonwealth Fund, 1936.

11. Clark, John W., and Warren Viessman, Jr., *Water Supply and Pollution Control.* Scranton, Pa.: International Textbook Co., 1965.

12. Holm, L. G., et al., "Aquatic Weeds." *Science* **166**:699–708 (1969).

13. Bowen, D. H. M., "The Great Phosphorus Controversy." *Environ. Sci. Tech.* **4**:725–726 (1970).

14. Ferguson, F. Alan, "A Nonmyopic Approach to the Problem of Excess Algal Growths." *Environ. Sci. Tech.* **2**:188–193 (1968).

15. Edmondson, W. T., "Phosphorus, Nitrogen, and Algae in Lake Washington after Diversion of Sewage." *Science* **169**:690–691 (1970).

16. Goldwater, Leonard J., "Mercury in the Environment." *Scientific American* **224**(5):15–21 (May 1971).

17. Jernelov, A., "Conversion of Mercury Compounds." In Miller, M. W., and G. C. Berg (eds.), *Chemical Fallout.* Springfield, Ill.: C. C. Thomas, Pub., 1969. pp. 68–74.

18. National Resources Committee, *Water Pollution in the United States.* Washington, D.C.: U.S. Govt. Printing Office, February 16, 1939.

19. Judson, Sheldon, "Erosion of the Land, or What's Happening to Our Continents?" *American Scientist* **56**:356–374 (1968).

20. Pacific Northwest Water Laboratory, Federal Water Pollution Control Administration, *Industrial Waste Guide on Thermal Pollution.* Corvallis, Ore., F.W.P.C.A., 1968.

21. U.S. Dept. of Health, Education, and Welfare, Public Health Service, *Manual of Septic-Tank Practice.* Washington, D.C.: U.S. Govt. Printing Office, 1957. Public Health Service Publication No. 526.

22. *Environmental Quality. The Second Annual Report of the Council on Environmental Quality.* Washington, D.C.: U.S. Govt. Printing Office, August 1971.

9

WATER POLLUTION: MUNICIPAL

When a natural watercourse traverses a town, and its banks become built upon, the easiest way of getting rid of filth and house wastes is to throw them into the stream. Every man's instinctive impulse is to get rid of what annoys him, and not to mind how his neighbor will be affected. After a while, when the watercourse has become sufficiently nasty, the people come to a realizing sense of what they have brought upon themselves, and then they try to devise a remedy.
—The Sanitary Engineer 8:559 (1883).

Great cities and small need a water supply that is adequate quantitatively and qualitatively, and they need a means of disposing of their water wastes. As towns have grown, they have had to build sewer systems and later sewage treatment plants but the remedies have too often lagged behind the need for them. Nevertheless, the success that cities have had in the 19th and 20th centuries in bringing clean water into homes and getting the dirty water out must rank as one of the most successful pollution control efforts of society. In this chapter, the history of municipal water supplies and sewerage systems will be reviewed and then the various types of sewage treatment methods in use or in development will be discussed in turn. The chapter concludes with a discussion of detergents because of their importance to water pollution in the past two decades.

MUNICIPAL WATER SUPPLIES

Ancient Rome was supplied with water from nine or more aqueducts, which were really open streams gently sloping from the hills to a pool in the city, from which most of the citizens had to collect the water in pitchers, although the wealthy had plumbing, baths, pools, and taps. A description of the water supply was written about 97 A.D. by the curator of the aqueducts, Sextus Julius Frontinus, whose figures seem to indicate a supply (before correcting for leakage and diversion) of close to 1 m^3/day for each inhabitant [1]. This is an ample amount of water and even if it overestimates the per capita share by a factor of five, as is sometimes supposed, the Romans did not lack for water.

The ancient Persians also had an ingenious system of underground channels (the "qanats") to convey water from groundwater aquifers in the highlands to the surface at lower levels by gravity, and this 3000-year-old system is still in use today, supplying 75% of all the water used in modern Iran [2].

The Roman aqueducts fell into disuse during the dark ages but interest in them revived during the Elizabethan period in Great Britain. In 1591 Drake's Leat, authorized by Parliament in 1585, was completed to carry water 18 mi from a weir on the River Mewe (Meavy) to the town of Plymouth, and it served as that town's main source of water for 3 centuries. In 1613, after 4½ years of work, the "New River" was completed, bringing water 39 mi from springs in Hertfordshire to London; although shortening and straightening have reduced its length to 24 mi, it is still in use today [3].

As early as the 13th century, London was using a 3-mi conduit and by about 1600 it had nine major conduits—pipes of earthenware or lead—to conduct water by gravity flow through the city. In 1582 the London Bridge Waterworks built by Morris were opened; they raised water from the Thames with plunger force pumps run by a waterwheel turned by the tidal races on the Thames at London Bridge and were London's chief water supply for over 40 years until the completion of the New River. Today London obtains two-thirds of its water from the Thames and the rest from the River Lee and from chalk wells [3].

Water supplies in other old cities came from other sources as well. Venice, which was founded in the fifth century A.D. on 100 islands, had to depend for 1300 years on rainwater caught and stored in cisterns, supplemented occasionally by water brought on boats from the mainland. Most of the cisterns were equipped with sand filters to remove impurities from the water, which drained through sand into a storage cistern of greater depth [4].

In the late 18th century, U.S. cities drew most of their water from internal supplies—springs, wells, and cisterns. Since the largest city in the 1790 census, Philadelphia, had only 42,250 persons, the quantity and quality of the water was usually reasonably adequate. The supply generally proved to be inadequate to fight fires and to maintain a clean urban environment during major outbreaks of disease, however. A fire in New York in 1776 destroyed 493 houses—one-quarter of those in the city. Epidemics of smallpox, influenza, typhoid fever, diphtheria, scarlet fever, and yellow fever recurrently caused health crises in U.S. cities during the 17th and 18th centuries.

Philadelphia was especially hard hit. In 1699 a yellow fever epidemic affected one-third of the population of 4000 and killed 220; another in 1793 killed over 4000 persons (10% of the population) and caused one-third of the residents to flee the city; another in 1798 killed 3500 persons and forced 40,000 (75% of the population) to flee, and 2000 died in New York in the same year. Medical opinion was divided on the cause of the epidemic. Dr. William Currie and others believed the yellow fever to be a disease of the tropics brought in from the West Indies on boats and urged strict quarantines. Dr. Benjamin Rush said it was due to domestic filth and urged a campaign to clean up the city, including the use of water to flush out the streets at regular intervals. (It was observed that mosquitoes were prevalent during epidemics but it was another century before Dr. Walter Reed proved that they were actually the transmitters of the disease.)

Fortunately, the municipal authorities in Philadelphia chose to follow the advice of both groups of doctors. In 1798 the engineer Benjamin Henry Latrobe urged the city to use water from the Schuylkill River to supply the city, pumping it up with steam engines. His plan was accepted and the works were completed in 1801. Although the number of paying customers increased slowly (water at the conduits emptying into the streets was free) and the steam engines and wooden pipes were frequently troublesome, Philadelphia had taken a major step forward in considering an adequate water supply to be the responsibility of the local government. Expansion of the waterworks continued over the years; by 1837 there were nearly 20,000 customers receiving over 11,000 m³ (3 million gal) of water daily from four reservoirs with 83,000-m³ (22 million-gal) capacity. The water supply in the meantime had led to a great reduction in the problems of fire and disease [5].

While Philadelphia was providing water publicly, many other cities were letting private water companies furnish the water supply, usually with very unsatisfactory results. Through the efforts of Aaron Burr, New York chartered the Manhattan Company in 1799 to provide the city with water but Burr was really interested in the lucrative banking profession and the charter permitted the company to engage in banking with its surplus capital. (Burr had been stymied in attempts to obtain a legitimate bank charter.) As a result, the company concentrated on banking with great success but provided only an inadequate water supply from wells to a portion of the city— the wealthier portion. The Bank of the Manhattan Company continued in existence until 1955, when it merged with the Chase National Bank to form the Chase Manhattan Bank. New York did not have an adequate water supply until 1842 when the city-owned Croton Aqueduct was completed to carry water from a reservoir behind a dam on the Croton River, 64 km away [5]. Although New York has secured other supplies since then, the Croton Reservoir is still in use but only in areas less than 12 m above sea level. It was the source of supply for a Harlem neighborhood around East 137th Street that discovered "wormlike organisms" in the water commencing September 17, 1969; health officials determined that tiny insect larvae (Chironomidae) were responsible and posed no danger but complaints from residents continued until December 4, when the supply was changed to the Delaware-Catskill system [6].

By the year 1800 there were actually 17 water utilities in the United States; the number rose to 100 by 1850, 1000 by 1885, 3000 by 1895, and 9000 by 1924. In 1963 there were 140 million persons being served in 22,000 communities by 18,500 systems [7]. Almost all are public systems; the only two major investor-owned water companies, American Water Works and Hackensack Water Company, serve a total of about 2 million customers.

As time passed, it became necessary for communities to protect their citizens by purification of the water supply before use. Some purification methods had been known for centuries. A body of Indian medical lore dating back to about 2000 B.C., the *Sus'ruta Samhita*, contained the injunction, "Impure water should be purified by being boiled over a fire, or being heated in the sun, or by dipping a heated iron into it, or it may be purified by filtration through sand and coarse gravel and then allowed to cool" [4]. The Roman aqueducts contained settling reservoirs and had pebble catchers. Francis Bacon's *Sylva Sylvarum* (1627) contained 10 experiments dealing with water purification. Venice used filter cisterns for many centuries.

In 1791 James Peacock was awarded a British patent for a remarkable sand filter in which the water was filtered while flowing upward under pressure, the filter being cleaned when necessary by reverse flow (see Figure 9-1). In the early 1800's, filtration plants to supply whole cities were put into use at Paisley and Glasgow in Scotland and in Paris, France [4]. The 19th century was a period of great progress in the design and construction of these plants. For many years filtration was regarded merely as a straining process to clarify the water but, by the mid-19th century, research indicated that it also removed organic matter; by the end of the 19th century it was realized that it was very effective in removing bacteria and was thus an important public health measure.

Other purification measures that date back many centuries are sedimentation (settling by repose) and coagulation [generally using alum (aluminum sulfate) but sometimes using almonds, beans, nuts, or Indian meal]. Omaha, Neb., used alum coagulation as early as 1889, many years before it used rapid filtration. In the early 1900's many U.S. cities adopted the practice of coagulation followed by rapid filtration.

Recognition of the role played by pathogens in waterborne diseases led to disinfection practices. Disinfection actually dates back to the use of boiling for purification (or boiling water to infuse tea, as in China). Today the preferred disinfection method is chlorination, although in some circumstances ozonation is used. A patent on chlorination of water was granted in 1888 to Dr. Albert R. Leeds, Professor of Chemistry at Stevens Institute of Technology (Hoboken, N.J.) and the first chlorination of a public water supply was at Adrian, Mich. in 1899. The real impetus to chlorination of U.S. water supplies came after its large-scale introduction in 1908 at the

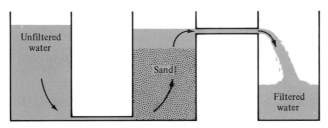

FIG. 9-1. James Peacock's 1791 upward-flow filter with reverse-flow wash using three tanks [4].

Booton Reservoir of the Jersey City, N.J., water works. By World War II, chlorination was securely established in the U.S. despite some competition from ozonation, treatment with ultraviolet rays, and even treatment with silver compounds (used at one time in swimming pools, including that at the Congressional Country Club in Washington, D.C. in 1935) [4].

Aeration of the water supply was used in the U.S. as early as 1860 in Elmira, N.Y. Although it was originally intended to help reduce the amount of organic matter (and is so used today in sewage treatment processes), it is now used mainly to reduce odors and to remove iron and manganese through oxidation.

An important problem encountered by many communities in the 19th century was the presence of intolerable odors and tastes from algal growths in surface water supplies, such as open reservoirs. The only known solution was to cover a reservoir (when it was small enough to do so) to shut out the sunlight and prevent the algal growths. In 1904, however, George T. Moore and Karl F. Kellerman proved that copper sulfate was an efficient algicide that could be sprayed on the surface or distributed from bags immersed in the water and carried by boats [4].

In the late 19th and early 20th centuries, water-softening methods employing lime and soda ash and the zeolite method (ion-exchange resins) were introduced to reduce the hardness of the water and permit soap to be used without precipitation of scum.

During the years 1923 to 1933, iodization of water supplies was tried in an effort to reduce the incidence of goiter but it was never widely adopted and after World War II iodized salt became generally available for those who wished to use it. After World War II, also, many communities began adding 1 ppm fluoride to the water supply to reduce the incidence of dental caries. Long-term studies indicate that the public health benefits from fluoridation outweigh the risks, but the practice remains a matter of great scientific and political controversy.

A typical city water supply and the chemistry involved in purification is detailed in Table 9-1.

MUNICIPAL SEWERS AND SEWAGE

The problem of getting a water supply into a city is matched by the problem of getting it out again. The disposal of waste water is a problem of long standing. Sanitation in London was originally accomplished by draining the waste water into cesspools, which was reasonably adequate as long as the amounts of waste water were not too large and the population was not too concentrated. The city streets were also provided with a series of drains that were really just open troughs for rainwater.

The amounts of waste water were greatly increased by the use of the water closet. This amenity of life was apparently invented in its present form

TABLE 9-1. Typical steps involved in securing and purifying a water supply for a town (Ames, Iowa). Water is delivered to homes at a cost of about \$0.18/m³ (\$0.65/1000 gal).

Source of water: Groundwater in extensive glacial sand and gravel deposits beneath the downtown area, pumped to the surface from wells 30 to 40 m deep.

Undesirable qualities of raw water
1. Dissolved iron in ferrous (Fe^{2+}) form, readily oxidized to ferric (Fe^{3+}) form, which precipitates out, leaving rust stains.
2. Hardness due to about 400 ppm of dissolved minerals (mainly Ca and Mg) that increase the amount of soap needed in using the water for washing.
3. Dissolved gases, such as CO_2 and H_2S, producing undesirable tastes and odors.

Purification steps
1. Aeration by means of a "waterfall" to remove gases and convert iron from the soluble ferrous form to the insoluble ferric form.
2. Removal of much of the hardness by addition of lime [$Ca(OH)_2$] and soda ash (Na_2CO_3). Calcium and magnesium salts, such as the bicarbonates and sulfates, are thereby removed by conversion to the insoluble forms $CaCO_3$ and $Mg(OH)_2$:
   (a) $Ca(HCO_3)_2 + Ca(OH)_2 \rightarrow 2CaCO_3 \downarrow + 2H_2O$
   (b) $Mg(HCO_3)_2 + 2Ca(OH)_2 \rightarrow Mg(OH)_2 \downarrow + 2CaCO_3 \downarrow + 2H_2O$
   (c) $CaSO_4 + Na_2CO_3 \rightarrow CaCO_3 \downarrow + Na_2SO_4$
   (d) $MgSO_4 + Ca(OH)_2 + Na_2CO_3 \rightarrow Mg(OH)_2 \downarrow + CaCO_3 \downarrow + Na_2SO_4$
   Total hardness is reduced to about 100 ppm (this is still fairly hard and many households have water softeners using ion-exchange resins). Each cubic meter (264 gal) of water requires the use of about 250 g of lime and 50 g of soda ash.
3. After about 20 to 30 min of mixing, the water flows (for 2 or 3 h) through a quiet settling tank, precipitating the $CaCO_3$ and $Mg(OH)_2$.
4. After the softening process is completed, generally after carbonation (bubbling of CO_2) or another mixing-settling stage, the water is filtered through very fine sand to remove suspended particles that impart a cloudy appearance to the water.
5. Chlorine is added to ensure bacteriological safety and fluoride is added to help reduce dental caries (tooth decay) before the water is sent to the tap. Total usage amounts to about 1.4 g/m³ of chlorine and 1.3 g/m³ of fluoride.

by the Elizabethan court poet, Sir John Harington, who wrote a humorous social document, *The Metamorphosis of Ajax* ("a jakes") (1596), describing in detail the construction and virtues of his "shyting place." The real increase in the popularity of the water closet came in the early 19th century and the serious overflow problems in cesspools led to connections with the open street drains. This naturally led to increased filth and smelliness and to epidemics of waterborne diseases such as cholera, typhoid, dysentery, and diarrhea. Eventually underground drainage systems came into being. In the mid-19th century, Sir Joseph Bazalgette built two drainage systems in London, one north and one south of the Thames, using 9-ft sewers to the Beckton (northern) and Crossness (southern) Outfalls [3].

FIG. 9-2. Nineteenth-century water closet advertisements.

At first sewers were simply means of conveying domestic sewage directly to a river or other waterway, with no treatment of any kind. The immediate result of London's sewerage system was to make the Thames lifeless near the city. Human wastes that had been removed as "night soil" and transported to fields now went into the river, creating conditions that made fish life impossible, especially after the Beckton and Crossness Outfalls came into use in 1864 and 1865, respectively [8].

As the cities grew, it became necessary to build sewage treatment plants to avoid these problems. London adopted sewage treatment in 1889 to 1891 and enlarged the main sewers, and fish returned to the London part of the Thames during the 1890's. (Improvements continued only up until World War I and aquatic life disappeared again during the 1930's, but the fish have been reappearing in the London Thames in the late 1960's with the rebuilding of the major sewage outfall works [8].)

Sewer systems have grown in absolute and relative terms during the 20th century (see Figure 9-3). At the beginning of the century only about 30% of the U.S. population was served by sewers (as opposed to septic tanks and cesspools) but the fraction rose to 50% by 1930 and is now about 70%.

Treatment facilities have also expanded in the U.S. during the 20th century (see Figure 9-3) and today of the two-thirds of the population served by sewers, perhaps 60% are also served by secondary sewage treatment facilities, 30% by only primary treatment, and about 10% by no treatment at all [9]. Even today, about 25% of New York's raw sewage goes directly into the Hudson and East Rivers, and in some large towns there is only the most rudimentary treatment. Vancouver, British Columbia, which obtains pure, unpolluted water from high in the Canadian Rockies, dumped its raw sewage into the Fraser River, Burrard Inlet, False Creek, and English Bay until the early 1960's, when pollution of the beaches in the area had become intolerable.

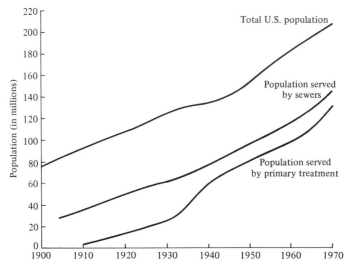

FIG. 9-3. Growth of U.S. population, population served by sewers, and population served by primary treatment or better during the 20th century [7].

The purification of water supplies and the improvements in the handling and treatment of sewage and other wastes has been very important in controlling the incidence and spread of various infections, especially waterborne diseases, and thereby in lowering the death rate. Death rates for individual cities generally showed sharp declines after the introduction of filtration of water supplies and again after the introduction of chlorination [7]. The death rate from typhoid fever is now essentially zero in the U.S. but in 1900 it was 31.3 per 100,000 population, which would correspond to about 65,000 deaths annually today. Nevertheless there are occasional reports in the U.S. and other developed countries of outbreaks of waterborne diseases, such as a hepatitis outbreak in Riverside, Calif., in 1965.

Reductions in diseases and death rates have been due primarily to public health measures rather than preventive or therapeutic measures by individual physicians, and water quality has been only part of the story. Much of the improvement has also been due to the control of mosquitoes and other insects, to massive immunization programs, to the proper sanitary control of milk and food, and to similar measures.

COMPOSITION AND TREATMENT OF DOMESTIC SEWAGE

The composition of municipal sewage varies greatly according to the wealth and habits of the population and the amount and character of industrial waste waters that are added to the domestic sewage. Table 9-2 shows the typical

TABLE 9-2. Average composition of domestic sewage in milligrams per liter (equivalent to ppm) [10]. The 5-day BOD refers to the milligrams per liter of oxygen required in the first 5 days for decomposition by microorganisms.

State of solids	Mineral	Organic	Total	5-Day BOD
Suspended	65	170	235	110
Settleable	40	100	140	50
Nonsettleable	25	70	95	60
Dissolved	210	210	420	30
Total	275	380	655	140

composition of the suspended and dissolved solids in domestic sewage [10]. It should be recognized that these pollutants in domestic sewage are rather dilute, rarely totaling as much as 0.1% of the total mass. Dissolved solids are generally those with a size of less than 1 nanometer (10^{-9} m or 10 A), while the settleable suspended solids are generally over 100 μm in size, the intermediate region consisting of colloidal and supracolloidal particles.

Sewage treatment processes are designed to remove the BOD, solids, and bacteria of the waste water, although in recent years removal of some new substances (such as phosphorus and nitrogen) has become important.

TABLE 9-3. Typical primary and secondary treatment processes in a moderate-sized (40,000 population) town (Ames, Iowa). This water pollution control plant is shown in Figure 9-4.

1. The raw sewage passes through comminutors (grinders) to chop the solids so that they will not later clog or damage equipment.
2. The flow is monitored and the sewage lifted by centrifugal pumps in the control building.
3. The water flows for about 30 min through aeration-grit removal tanks that add air to the waste water and permit grit and sand to settle out.
4. The water then flows for about 2 h through primary settling tanks to permit most of the suspended solids to settle to the bottom.
5. The settled waste water is sprayed evenly by revolving sprinkler arms over the surface of several coarse rock bacterial contact beds some 40 m across; these beds are called "trickling filters" although it is not the filtering action that is important. The water trickles down through these beds and organic matter is removed by aerobic action of bacterial and fungal growth on the rocks. Thick slime layers eventually form, slough off, and are removed.
6. The waste water from the trickling filters flows for about 2 h through final settling tanks to remove the small amounts of solids still carried.
7. The effluent is discharged to a nearby river through a large pipe; it can be chlorinated if necessary (such as in an epidemic).
8. Sludge from settling tanks and sloughed trickling-filter slimes are digested for about a month at 35°C in a digestion tank from which methane is recovered and used for fuel. Digested sludge is dried in the open for several weeks and then given away as a soil conditioner.

FIG. 9-4. An aerial view of the Ames (Iowa) water pollution control plant described in Table 9-3. The three round
trickling filters are in the center of the picture. Visible in
an arc from left to right around the filters are the sludge
drying beds, the two round sludge digestion tanks, the
control building, the tanks for aeration-grit removal and
primary setting, and at the far right the three round final
settling tanks and the rectangular chlorination contact
tank. Effluent from the final settling tanks is discharged
into the Skunk River, marked by the line of trees in the
upper part of the photograph. (*Courtesy of the City of
Ames, Iowa, Water Pollution Control Plant.*)

Treatment processes are generally divided into three groups:

Primary: Generally mechanical processes.

Secondary: Biological processes.

Tertiary: Advanced biological, chemical, and physical processes.

These are discussed in turn in the succeeding sections. Table 9-3 lists the
steps involved in a typical town's primary and secondary treatment, and the
plant itself is shown in Figure 9-4.

PRIMARY TREATMENT

Primary treatment processes involve grit removal, screening, grinding, flocculation, sedimentation, and skimming. Grit removal and screening systems are
designed to remove the larger suspended or floating materials—paper, rags,
twigs, glass, wires, roots, etc. These materials often interfere with equipment
in the rest of the treatment plant if not removed. Coarse screens often have
metal bars or heavy wires spaced 25 to 50 mm apart, while finer screens may

range down to 0.8-mm openings. All screens must be cleaned and the fine screens are usually being cleaned continually with automatic equipment. The materials that are removed by the screening may be incinerated, buried, or digested later with other solid materials.

Grinding is accomplished using comminutors or cutting screens, such as rotating screens with cutting teeth that can chop solids down to 6 mm in size. Comminutors usually have a bypass with a coarse screen so that their value is not completely lost during periods of maintenance.

Flocculation is the agitation of waste water by mechanical stirring or air injection to cause small suspended solids to collide and form larger particles (flocs) that can settle out more rapidly.

Sedimentation is the removal of suspended solids by gravitational settling, usually in basins having a continuous flow that are normally about 3 m deep and 30 m across, with detention times of a few hours. Continuous removal of the sludge is often desirable.

Filtration using porous materials (sand, anthracite, diatomaceous earth, etc.) is not generally used in sewage treatment, although it is still used to some extent in the treatment of drinking water, despite high construction costs and the need for large land areas. Filters involving biological treatment are considered to be secondary treatment processes.

Certain pretreatment processes may be necessary before secondary treatment when special wastes from industrial sources are present [11]. These may include removing oil or grease in a skimming tank or separator, neutralizing excessive alkalinity or acidity, and precipitation or ion exchange (see the section on tertiary treatment) for heavy metals.

Primary treatment of municipal wastes generally costs about $0.03 to 0.04/1000 gal (3.8 m^3), although the cost may range higher when industrial wastes are present [9]. This is less than one-tenth the cost of purified drinking water, for which a typical figure in the U.S. is $0.50/1000 gal. The efficiency in removing pollutants is not very great, however, and typically might be about 35% of BOD; 30% of COD; and 60% of suspended solids, including 20% of total nitrogen and 10% of total phosphorus but none of the dissolved minerals [9].

SECONDARY TREATMENT

Secondary treatment involves the use of biological methods, especially trickling filters and activated sludge, that approximate natural degradation processes. Secondary treatment plants sometimes include chlorination to accomplish chemical oxidation and disinfection.

In the activated sludge process, which is slightly more popular than the use of trickling filters in the United States, biologically active growths are continuously circulated through organic waste in the presence of oxygen (generally supplied as fine bubbles of compressed air). In the usual operation, effluent from primary settling tanks is given secondary treatment by microorganisms from some of the settled sludge. The suspended and dissolved organic wastes, well aerated and mixed, undergo adsorption, flocculation, oxidation, and biological degradation for a few hours before being passed into the secondary sedimentation tank, where the sludge, rich in growing organisms, settles out. Part of the sludge is used to seed the next batch of wastes from the primary settling tanks. The activated sludge does two things:

1. It clarifies the water by adsorbing most of the colloidal and suspended solids on the surfaces of the sludge particles.
2. It oxidizes the organic material.

The major capital expense of the activated sludge plant is for land and the major operating expense is for adding the compressed air to the waste water. In the late 1960's a competing method involving the use of pure oxygen instead of air has emerged and may lead to capital investment savings of 16 to 20% and operating savings of up to 50% [12]. The use of pure oxygen permits the support of more bacteria.

Trickling filters are manufactured beds of crushed stone used to percolate the waste water, bringing it into contact with air and with the biological slimes on the filter. It is not filtration that is important here but the adsorption of organic matter at the surfaces of the slimes and their decomposition by the bacteria and fungi in the slime. Near the top of the trickling filter, aerobic growth is important but further down, where less atmospheric oxygen diffuses, anaerobic action becomes predominant. When the slime layers become very thick, they slough off and can be removed.

Handling and disposal of the sludge from secondary treatment plants is an important problem and may account for 25 to 50% of the total capital and operating costs. Chicago's Metropolitan Sanitary District handles 6 million m^3 of waste water daily and produces 900 metric tons of solids (on a dry basis) daily, and the solid disposal costs are just under half the total costs [9]. The actual solids content of wet sludge is 2.5 to 5% from primary sedimentation tanks, 0.5 to 5% from trickling filters, and 0.5 to 1% from the activated sludge process, the remainder in each case being water.

Several methods are used in handling sludge. It can be concentrated by gravity thickening and other methods. It is often digested in digesters 6 to 10 m deep and circular in shape. The sludge is maintained at a temperature of about 35°C (95°F) and the organic materials are eventually reduced to gases —about 70% methane and 30% carbon dioxide, with small amounts of ammonia, hydrogen sulfide, hydrogen, and nitrogen. The methane can be recovered as fuel and burned to keep the digester heated. After a month or two of digestion the sludge is converted into a stable humus material that can be used as a soil conditioner. Some cities actually market their product, such as Milwaukee's milorganite.

Other sludge-handling methods are dewatering by drying on sand beds, using vacuum filtration, centrifuging, or various mechanical processes. Dewatering on sand beds is accomplished by water filtration through the sand and by evaporation of the water from the sludge surface; in a few days the moisture content will be reduced to less than 75%, permitting easy handling. This method is often used to dewater well-digested sludge. In vacuum filtration a horizontal drum covered with filter media (cloth, metal mesh, steel coils, etc.) is rotated so part of the drum is submerged in wet sludge and a vacuum is applied on the inside to draw water out of the sludge and into the drum; at the same time, the vacuum holds a layer of sludge on the drum, from which it can be scraped or lifted off. Centrifuging, which has also been used to dewater paper-mill wastes, packing-house wastes, foundry sludges, and refinery and water-softening sludges, can be used to obtain moisture contents of less than 70%, which is adequate for handling these wastes.

A common method of sludge disposal that can lead to air pollution, unfortunately, is incineration of dried sludge. The sludge itself furnishes sufficient heat to maintain combustion and sometimes even to dry new sludge once the process has begun. The patented Zimmerman or wet oxidation process involves combustion of organic matter through the introduction of compressed air; the system is under pressure, the combustion temperatures are 200 to 374°C, and only about an hour is required to complete the oxidation.

An alternative to incineration is composting, which is being applied to some extent to both sewage sludge and municipal refuse (see Chapter 13). Composting permits recovery of some of the organic matter and plant nutrients of the sludge. For similar reasons, the effluents from sewage treatment plants can sometimes be used for irrigation purposes and would in fact be preferable to conventional irrigation water because of the plant nutrients present [13]. These plant nutrients are thus put to good use and are not given a chance to promote undesirable plant growth in waterways. In 1972 Chicago began experimenting with using liquid sludge to reclaim strip-mined land in Fulton County, Ill.

Since it is a solid waste, sludge is sometimes placed in sanitary landfills along with other refuse (see Chapter 13). Many coastal cities prefer to dump these wastes into the ocean. Los Angeles has a 53-cm pipeline extending 11 km into the ocean to dump 19,000 m³ (5 million gal) daily of digested sludge and plant effluent into water 100 m deep in Santa Monica Bay. New York used to dump wastes off the New Jersey shore but New Jersey won a Supreme Court decision in 1933 forbidding this practice; New York still dumps wastes 20 km out into the Atlantic Ocean (see Chapter 13). As a result of these practices and of the presence of wastes in our rivers, the parts of the ocean near the coasts contain great concentrations (compared to the open ocean) of suspended matter—soot, fly ash, processed cellulose, and combustible organic matter. These wastes tend to move parallel to the shore rather than directly seaward from the mouths of rivers and estuaries [14].

Secondary treatment combined with primary treatment may typically cost $0.05 to 0.10/1000 gal (3.8 m³); the pollutant removal efficiency might be 90% of BOD; 80% of COD; 90% of suspended solids, including 50% of total nitrogen and 30% of total phosphorus, and perhaps 5% of dissolved minerals [9]. Primary and secondary treatment methods have been largely developed to reduce BOD and suspended solids and in the process have accomplished little in the way of reducing dissolved minerals, heavy metals, viruses, drugs, hormones, and other exotic chemicals. In addition, their efficiencies can be strongly affected by temperature; secondary treatment is often inadequate during cold weather.

ADVANCED WASTE TREATMENT METHODS ("TERTIARY TREATMENT")

A number of more advanced ("tertiary") treatment processes, some of them known for many years, have been tested in recent years, often through federal research and development grants. The aim of these processes is not only to improve on primary and secondary treatment or to replace biological methods by physical or chemical methods but actually sometimes to improve the waste water quality to the point at which it can be reused. The increasing costs of water supplies and the increasing emphasis on waste water quality is leading engineers to consider more seriously the possibility of total recycling of contaminated water [15]. The city of Windhoek, South-West Africa, is actually using its purified sewage effluent as part of the community's water supply because of the scarcity of water in the area [16].

A wide variety of methods are used in advanced waste treatment and they may be introduced at any stage of the total treatment process, not necessarily only after conventional primary and secondary treatment. Their purpose may be more complete removal of pollutants largely removed by primary or secondary treatment or removal of other pollutants of importance, such as phosphates and other dissolved inorganic compounds.

CHEMICAL COAGULATION AND FILTRATION. These are used together in water purification and can also be used as a waste water treatment method. There are a number of different compounds, known as chemical coagulants, that react with suspended matter to form flocs—alum [aluminum sulfate, $Al_2(SO_4)_3 \cdot 12H_2O$], copperas (ferrous sulfate, $FeSO_4 \cdot 7H_2O$), ferric sulfate [$Fe_2(SO_4)_3$], ferric chloride ($FeCl_3$), and others [10]. In recent years a number of synthetic, high-molecular-weight, water-soluble polymers have been developed for this purpose, one example being Calgon Corporation's poly-diallyldimethyl ammonium chloride. The filtration step subsequent to coagulation can be accomplished by sand, diatomaceous earth, or any of a number of multimedia filter materials that have been developed or through the use of microscreens. Phosphates can be removed in this way using lime to precipitate them but the cost may run close to $0.05/1000 gal (3.8 m³) [17].

FIG. 9-5. The world's largest micro-
strainer was built in 1970 by the
Crane Company for the North Side
sewage treatment plant (Skokie, Ill.)
of the Metropolitan Sanitary District
of Greater Chicago. The drum is
nearly 4 m in diameter and over 9 m
in length and has a mass of 9 metric
tons. The screening fabric will contain
billions of microscopic holes only
23 μm across. The unit will be used
for advanced treatment of 57,000 m³
(15 million gal) of waste water daily.
(*Courtesy of Environmental Systems
Division, Crane Company.*)

CARBON ADSORPTION. Adsorption of tastes and odors is an old process.
"Activated carbon" is a porous and highly adsorbent form of carbon with a
very large surface area. In granular or powdered form (the latter is more
efficient but also harder to handle) it will adsorb many refractory organic
compounds dissolved in the water. The carbon must eventually (perhaps once
a year) be regenerated by heating to about 925°C in an air-steam atmosphere
to burn off the adsorbed organic material; under proper regeneration conditions the carbon losses are only a few percent. Pilot plants using columns of
granular activated carbon in place of conventional secondary treatment
methods have proved so successful that a plant at Rocky River, Ohio, handling
38,000 m³/day (10 million gal/day) is being built that uses only carbon
adsorption after a coagulation-settling step [12, 18].

CHEMICAL OXIDATION. Waste water treatment can also be accomplished
using strong oxidants such as ozone, hydrogen peroxide (H_2O_2), or the free
hydroxyl radical (OH). Chlorination with chlorine or chlorine dioxide is
also possible.

Dissolved inorganic compounds are a problem since they are more
common in waste water, even after secondary treatment, than they are in the
water supply so that they can easily build up in a water reuse cycle. Several
methods exist for demineralization: distillation, freezing, ion exchange, electrodialysis, and reverse osmosis. Distillation and freezing are apparently not
economical but the other methods are all useful.

ION EXCHANGE. Ion-exchange can be accomplished by the use of
natural materials (such as zeolite) and synthetic materials (such as ion-
exchange resins). Cation-exchange resins exchange their hydrogen ions for
metallic cations in the solution passing through the ion-exchange column,

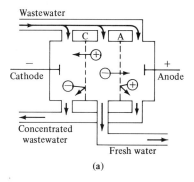

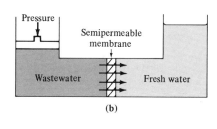

FIG. 9-6. Schematic diagrams of membrane processes for water treatment. **(a)** Electrodialysis; C is a membrane permeable to cations only and A is a membrane permeable to anions only. **(b)** Reverse osmosis.

while anion-exchange resins exchange their hydroxyl ions (OH^-) for chloride and other anions in the solution. The resins can be regenerated by treatment with sulfuric acid (for cationic resins) or sodium hydroxide (for anionic resins). Ion exchange is very effective and produces high quality effluents, and it is possible to mix treated water with untreated water to produce effluent of any desired quality. The cost of this treatment method is fairly high at present. One natural zeolite, clinoptilolite, seems to adsorb both phosphate and ammonium ions and may prove to be of great value [12].

ELECTRODIALYSIS. Placing an electrical potential difference across the waste water produces an electric current, causing the cations to migrate toward the cathode and the anions to migrate toward the anode [see Figure 9-6(a)]. Membranes (really ion-exchange resins in sheet form) permeable to only cations or only anions are used to control the migration of the ions and permit demineralized water to be taken out of the appropriate chambers. Organic molecules are not removed and they can collect on and clog the membranes. Another disadvantage of this method is that it still leaves concentrated waste water to be disposed of in some fashion.

REVERSE OSMOSIS. Another membrane process concentrates the impurities in part of the solution and thereby purifies the other part; it is shown schematically in Figure 9-6(b). Ordinary osmosis involves the movement of water across a semipermeable membrane (i.e., one that is permeable to water but not to the dissolved material) in such a way as to tend to equalize the concentrations. Reverse osmosis is the opposite phenomenon produced by applying pressure to the more concentrated solution and thereby forcing water to the other side of the membrane. Reverse osmosis reduces both the organic and inorganic content of the waste water but some fouling of membranes can still occur. As with electrodialysis there is highly concentrated waste water to be disposed of. An experimental reverse osmosis plant at Pomona, Calif., led to reductions of 88% for total dissolved solids, 84% for COD, 98.2% for phosphate, 82% for ammonia, and 67% for nitrate [9].

AIR STRIPPING. Another process can be used for removing ammonia (NH_3) from waste water. The pH of the water is raised, usually by addition of lime, and the ammonia driven out of the solution by vigorous agitation with air [12].

ADVANCED BIOLOGICAL SYSTEMS. New biological methods are also being considered for waste water treatment. The use of shallow (perhaps a meter deep) oxidation ponds permits water to be purified by the action of aerobic bacteria and algae. These ponds use solar radiation for photosynthesis and the organic material is used for both bacterial and algal growth, greatly reducing BOD (and also coliform organisms, perhaps through the production of antibiotic substances by the algae). The ponds may have disagreeable odors if anaerobic conditions are permitted to exist but under proper conditions there is plenty of oxygen produced and the effluent may even be super-saturated with dissolved oxygen. The ponds eventually have to be cleaned out (perhaps every few years) and weeds must be kept under control (this is part of the reason for keeping the ponds shallow) [10].

Michigan State University is building a set of several lakes to purify and indeed "recycle" waste water from the city of East Lansing, Mich. [19]. Effluent from the secondary treatment (activated sludge) plant will first pass into a lake about 2 m deep containing rooted aquatic plants chosen for their high demand for phosphates and nitrates, which they will then remove, and for their food value for livestock and possibly humans, which will permit harvesting. Succeeding lakes will also contain fish and be used for recreational purposes and some of the water will also be used for spray irrigation. Similar work is being carried out at Iowa State University and other places.

COSTS. The costs of advanced treatment will be quite a bit higher than those of primary and secondary treatment but not very great considering the importance of pure water. The most advanced large-scale tertiary treatment plant is that of the South Tahoe Public Utility District at Lake Tahoe, Calif. It handles close to 30,000 m³ (about 7.5 million gal) of waste water daily and incorporates the usual primary and secondary treatment processes plus flocculation and phosphate removal with lime, ammonia removal by air strip-ping, multimedia filtration aided by synthetic polyelectrolytes, organic removal by adsorption on activated carbon, disinfection by chlorination, and recovery of the lime and activated carbon [9]. An effluent meeting U.S. Public Health standards for drinking water is produced for total capital and operating costs of under $0.40/1000 gal (3.8 m³) [20]. Such treatment of all U.S. municipal waste waters would thus cost about $20 per person per year ($4 billion total per year), which may be compared with the U.S. per capita personal income of approximately $4500 per year.

DETERGENTS

Detergents have attracted special attention in the U.S. and other countries because of a variety of pollution problems involving their constituents. Detergents are substances with notable cleansing properties so that soaps probably qualify as being the oldest detergents. Soap is usually produced by acting on animal fat with alkali and it consists of the sodium (or potassium) salts of fatty acids such as stearic acid $[CH_3(CH_2)_{16}COOH]$. Soap is not suitable for use in acidic solutions or in hard water solutions. Beginning about the time of World War I, chemists developed a number of synthetic compounds with better cleansing properties and it is these that are generally referred to as "detergents."

Detergents generally consist of a "surfactant" or surface-active agent and a number of "builders." The surfactant lowers the surface tension of the liquid in which it is dissolved by concentrating at surfaces and interfaces, and its cleansing properties arise from its ability to replace dirt on surfaces by being preferentially adsorbed at surfaces and by helping the dirt to be carried away as a stabilized emulsion or suspension. The builders sequester calcium and magnesium ions that would interfere with the surfactant; they also maintain a proper level of alkalinity in the solution and help keep dirt in suspension.

Surfactants generally consist of polar or hydrophilic groups (such as COO^-, SO_3^-, or NH_4^+) that are soluble in water and oily or lipophilic groups that are soluble in lipids. They generally fall into three groups, although some surfactants ("amphoteric compounds") can under different conditions act as members of either of the first two groups:

1. Anionic surfactants in which the negative ion is surface-active. These are the most commonly used; soaps can be classified as anionic surfactants. Some examples are shown in Figure 9-7(a). In the early 1960's the most common surfactants were alkyl benzene sulfonates ("A.B.S.") but by the mid-1960's they had been replaced by linear alkyl sulfonates ("L.A.S.") for reasons discussed below.

2. Cationic surfactants in which the positive ion is surface-active. These are usually quaternary ammonium salts of the type shown in Figure 9-7(b).

3. Nonionic surfactants in which the whole molecule is surface-active. A popular type is polyoxyethylene nonylphenol, shown in Figure 9-7(c).

A wide variety of different surfactants are marketed for a wide variety of different solutions, including nonaqueous solutions.

(a)   R—O—SO₃M   R—SO₃M

      Primary alkyl sulfates   Linear alkyl sulfonates ("L.A.S.")

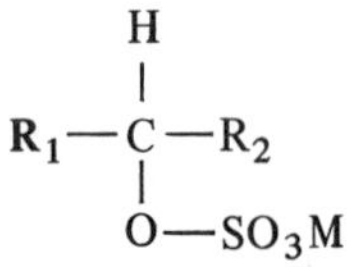

$$R_1-\underset{\underset{O-SO_3M}{|}}{\overset{\overset{H}{|}}{C}}-R_2 \qquad\qquad R-\underset{}{\bigcirc}-SO_3M$$

    Secondary alkyl sulfates 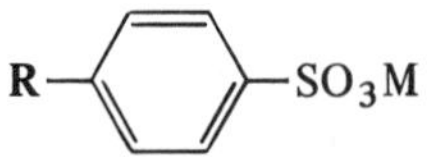   Alkyl benzene sulfonates ("A.B.S.")

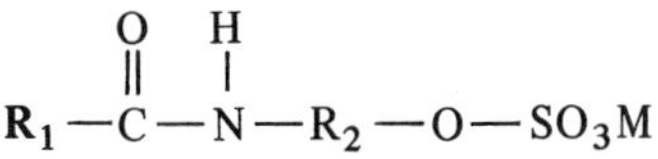

$$R_1-\overset{\overset{O}{\|}}{C}-\overset{\overset{H}{|}}{N}-R_2-O-SO_3M$$

        Fatty amide sulfates

(b)
$$\left[\,R_4-\underset{\underset{R_3}{|}}{\overset{\overset{R_1}{|}}{N}}-R_2\,\right]^+ \qquad X^-$$

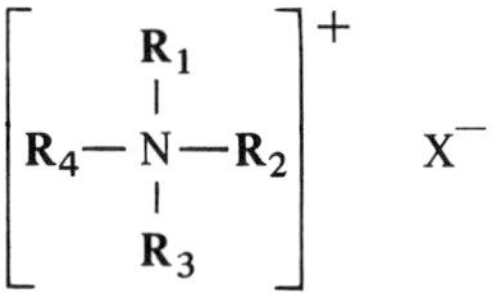

    Quaternary ammonium salt

(c) $CH_3(CH_2)_8$ —$\bigcirc$— $O-(CH_2-CH_2-O)_x\,H$

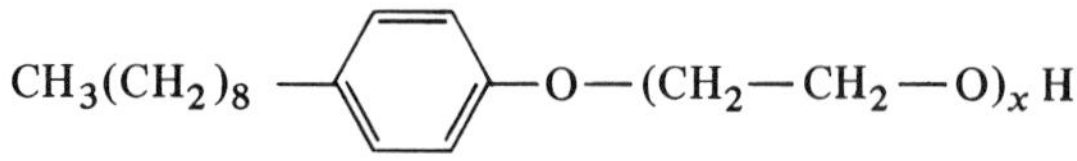

    Polyoxyethylene nonylphenol ($x \approx 10-30$)

FIG. 9-7. Typical surfactants. **(a)** Anionic surfactants; M = cation, **R** = straight hydrocarbon chain. **(b)** Cationic surfactant: quaternary ammonium salt, X⁻ being a halogen or acid group, **R**₁, **R**₂, and **R**₃ being straight hydrocarbon chains, and **R**₄ being an aromatic chain. **(c)** Nonionic surfactant: polyoxyethylene nonylphenol.

The A.B.S. surfactants in common use a decade ago showed remarkable resistance to biological agents; for example, perhaps only 30 to 50% of the A.B.S. might biodegrade over a 180- to 200-km trip down a waterway [21]. They could also inhibit the oxidation of certain molecules (such as phenol), perhaps by enveloping them and thus isolating them from dissolved oxygen in the water. They were also noted for interfering with waste treatment processes by stabilizing small particles in colloidal suspension and reducing the activity of biological filter beds and activated sludge (but not of sludge digestion). Probably the most notorious effect was their tendency to produce stable foam in rivers, sometimes (especially below weirs and at lockgates) several meters high over an extent of several hundred meters. The problems in Europe were so severe that steps were taken to require the use of detergents proved (by specified testing procedures) to be biodegradable. These requirements were promulgated in the early 1960's and U.S. detergent manufacturers changed to biodegradable surfactants (largely L.A.S.) in the mid-1960's.

The builders added to detergent preparations are generally sodium salts of phosphoric, carbonic, sulfuric, silicic, and boric or similar acids. In their absence, unsequestered hardness ions such as Ca^{2+} and Mg^{2+} can poison the surfactant molecules, preventing detergent action.

Most laundry detergents sold at the beginning of the 1970's contained 35 to 50% of the builder sodium tripolyphosphate ($Na_5P_3O_{10}$), which leads to the formation of stable soluble complexes [such as $Ca(PO_3)_3^-$] with hardness ions. Total annual use of detergents in the U.S. was about 2.5 million metric tons, resulting in the release of about 1 million metric tons of phosphates to waste water [22]. These phosphates contribute to eutrophication of waterways, as discussed in Chapter 8, but there is no consensus as to their importance. They might account for perhaps 50% of the phosphates in waste water [22] but a lower fraction of the total phosphates entering waterways from all sources (see Table 8-6). In recent years the phosphate eutrophication problem has been attacked in two ways: by upgrading sewage treatment facilities to permit removal of phosphates (with 80% removal a common goal) and by outlawing the use of phosphates in detergents, as several countries and some local governmental units in the U.S. have done. Neither of these methods will affect certain phosphorus sources such as runoff from fertilized lands.

A search is presently underway for other builders. For several years there was strong interest in nitrilotriacetate ("NTA"), shown in Figure 9-8, but it has exhibited a number of undesirable and possibly dangerous properties: hygroscopicity (i.e., it absorbs moisture from the air, which can cause the detergent to cake); difficulty in degrading in anaerobic systems such as those found in some septic tanks; and possible teratogenic effects (effects that damage the fetus *in utero*) in combination with heavy metals [22]. NTA can also tie up some biologically essential elements and thus harm natural ecosystems. Other builders have shown more difficulties, some being nonbiodegradable. Some products are being marketed with precipitating builders such as carbonates and silicates, which precipitate an insoluble residue with hardness ions but the residue can build up on clothes or washing machines. A search is also being carried out for surfactants that do not require builders.

A completely suitable detergent for household use would have to be inexpensive, safe (especially since 2000 to 3000 children swallow cleaning

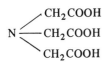

FIG. 9-8. Nitrilotriacetate (NTA).

products annually in the U.S.), and not environmentally troublesome, in addition to doing a good job of washing [22]. Some products advertised in recent years as being low- or no-phosphate have made use of caustic chemicals capable of producing eye and skin irritation or illness among children who accidentally ingest them, and the Surgeon General of the United States has found it necessary to warn against their use. Other products on the market appear satisfactory in all regards, even though they may not provide the "whiteness" so highly advertised by the leading detergent manufacturers.

REFERENCES

1. Winslow, E. M., *A Libation to the Gods: The Story of the Roman Aqueducts.* London: Hodder & Stoughton, Ltd., 1963.
2. Wulff, H. E., "The Qanats of Iran." *Scientific American* **218**(4):94–105 (April 1968).
3. Stephens, J. H., *Water and Waste.* London: Macmillan & Co., Ltd., 1967.
4. Baker, M. N., *The Quest for Pure Water.* New York: American Water Works Assn., 1948.
5. Blake, N. M., *Water for the Cities. A History of the Urban Water Supply Problem in the United States.* Syracuse, N.Y.: Syracuse University Press, 1956.
6. Robinson, Douglas, "City Changes Water Supply in Harlem to Eliminate Wormlike Organisms." *N.Y. Times,* December 7, 1969, p. 49.
7. Wolman, Abel, *Water, Health and Society: Selected Papers by Abel Wolman edited by Gilbert F. White.* Bloomington, Ind.: Indiana University Press, 1969.
8. Wheeler, Alwyne, "Fish Return to the Thames." *Science Journal* **6**(11):28–33 (November 1970).
9. *Cleaning Our Environment: The Chemical Basis for Action.* Washington, D.C.: American Chemical Society, 1969.
10. Fair, G. M., and J. C. Geyer, *Elements of Water Supply and Waste-Water Disposal.* New York: John Wiley & Sons, Inc., 1958.
11. Eckenfelder, W. W., Jr., *Water Quality Engineering for Practicing Engineers.* New York: Barnes & Noble, Inc., 1970.
12. Holcomb, R. W., "Waste-Water Treatment: The Tide is Turning." *Science* **169**:457–459 (1970).
13. Wilson, C. W., and F. E. Beckett (eds.), *Municipal Sewage Effluent for Irrigation.* Ruston, La.: The Louisiana Tech. Alumni Foundation, 1968.
14. Manheim, F. T., et al., "Suspended Matter in Surface Waters of the Atlantic Continental Margin from Cape Cod to the Florida Keys." *Science* **167**:371–376 (1970).
15. Partridge, E. P., and E. G. Paulson, "Water: Its Economic Reuse Via the Closed Cycle." *Chemical Engineering* **74**(21):244–248 (October 9, 1967).
16. Stander, C. J., and L. R. J. Van Vuuren, "The Reclamation of Potable Water from Wastewater." *Jour. Water Poll. Control Fed.* **41**:355–367 (1969).
17. "Phosphate Removal Processes Prove Practical." *Environ. Sci. Tech.* **2**:182–185 (1968).

18. Rizzo, J. L., and R. E. Schade, "Secondary Treatment with Granular Activated Carbon." *Water & Sewage Works* **116:**307–312 (1969).
19. Knapp, Carol E., "Recycling Sewage Biologically." *Environ. Sci. Tech.* **5:**112–113 (1971).
20. South Tahoe Public Utility District, *Advanced Wastewater Treatment as Practiced at South Tahoe.* Washington, D.C.: Environmental Protection Agency, 1971.
21. Prat, J., and A. Giraud, *The Pollution of Water by Detergents.* Paris: Organisation for Economic Co-operation and Development, 1964.
22. Hammond, A. L., "Phosphate Replacements: Problems with the Washday Miracle." *Science* **172:**361–363 (1971).

10

WATER POLLUTION: INDUSTRIAL AND COMMERCIAL

In a sanitary point of view, the purity of our water-supplies is of so great importance that we shall not, as a country, reap the full advantage to be derived from sanitary measures until our rivers have been freed from the abominations poured into them.
—Baldwin Latham in *The Engineer* **21**:261 (April 6, 1866).

The abominations poured into American waters are numerous and have increased over the years in many places. The history of the pollution of the Great Lakes between the United States and Canada is especially illuminating. In 1909 the United States and Great Britain signed a Boundary Waters Treaty applicable to the boundary waters between the United States and Canada that stated, in Article IV, "It is further agreed that the waters herein defined as boundary waters and waters flowing across the boundary shall not be polluted on either side to the injury of health or property on the other" [1]. An International Joint Commission on the Pollution of Boundary Waters was formed to report on the state of affairs and make recommendations; this commission is still in existence today, but without enforcement powers it has been unable to prevent violations of the treaty by both countries.

By 1951, however, after another study extending over several years, the Commission reported that "Industrial wastes, which were of little concern in 1912, are now a major problem . . . . An appreciable volume of harmful pollutants is discharged daily. These include 13,000 pounds of phenols, 8000 pounds of cyanides, 25,000 pounds of ammonium compounds and large quantities of oils and suspended solids of all types . . . . The indus-

The Commission's first report in 1918 was able to conclude that "The Great Lakes beyond their shore waters and their polluted areas at the mouths of the rivers which flow into them are, except so far as they are affected by vessel pollution, in a state of almost absolute purity" [2]. Vessel pollution was distinctly traceable in the customary navigation channels of the time and the Commission urged the remedy of disinfection of vessel sewage. In 1918 the Commission was still able to concern itself with the fact that the Great Lakes and other connecting bodies of water were becoming unsuitable for drinking water. Industrial wastes were mentioned as causing "appreciable injury" at times, especially in the case of sawmill wastes.

trial wastes produce a greater oxygen demand on the receiving streams than the combined total of the domestic wastes of the area . . . . The estimated cost of treatment of industrial wastes in the three areas under reference (the St. Clair River-Lake St. Clair-Detroit River area, the St. Marys River between Lake Superior and Lake Huron, and the Niagara River between Lake Erie and Lake Ontario) is $22,650,000 in the U.S. and $3,450,000 in Canada" [1].

During the 1950's and 1960's industrial and municipal pollution of the Great Lakes became of still greater concern and the Commission has had to deal with such problems as accelerated eutrophication due to phosphates, mercury pollution, and possible environmental dangers from oil and gas drilling. Its most recent study, carried out in the late 1960's, urged immediate action, including the removal of all phosphates from detergents used in the area and the construction of municipal and industrial waste treatment facilities totaling $1.373 billion on the U.S. side and $211 million in Canada [3].

The history of pollution on the Great Lakes is a long and tragic one and some scientists fear the waters have been irreversibly harmed. Similar harm has occurred in other parts of the United States. In this chapter water pollution from shipping and industry will be discussed along with some of the treatment and disposal methods in use.

SHIPPING POLLUTION

The major water pollution problems of watercraft are sewage and oil pollution.

VESSEL SEWAGE. The 8 million watercraft of all types (most of them recreational) that ply U.S. waters have been estimated to contribute sewage and other oxygen-demanding sewage equal to the domestic sewage of 500,000 persons, most of the wastes being dumped raw into the water [4]. There are even floating industries contributing organic wastes, such as Alaskan crab canneries. Although vessel sewage is only a small part of the overall water pollution problem, there are places in which it is the major problem, and most states have enacted legislation regulating or restricting watercraft wastes.

The economies of large-scale municipal waste water treatment plants are impossible to attain on even large vessels. Current sewage disposal systems for vessels include the following [4]:

1. Holding tanks to receive and store sewage until they can be unloaded on the shore. Despite their simplicity they are relatively large and heavy and may require the use of odor control chemicals, and shore facilities for unloading are generally inadequate.

2. Incinerators for incineration of human wastes.

3. Biological treatment systems similar to municipal secondary treat-

ment systems. These are large and heavy and applicable only to large vessels.

4. Maceration-disinfection units that simply macerate and disinfect the wastes, without actually reducing the BOD load.

5. Chemical recirculating flush toilets such as those manufactured by Monogram Industries for use on jet aircraft. Small units are available that can be used 160 times with only 15 liters (4 gal) of water and a single charge of chemicals.

OIL POLLUTION. Oil pollution from accidents involving large tankers and offshore drilling rigs has periodically attracted national and international attention in the past few years. In the best-known recent accident, the *Torrey Canyon*, carrying over 100,000 metric tons of Kuwaiti oil for the Union Oil Company, ran aground off Cornwall, England on March 18, 1967, eventually releasing most of its cargo into the water as the ship broke into two and later three parts [5]. The Royal Air Force spent several days trying to set the oil afire with high explosives, aviation kerosine and napalm, which destroyed perhaps one-third of the cargo. Attempts were then made to disperse the oil with 10,000 m³ (2.5 million gal) of detergents but vast amounts of petroleum eventually washed upon British and French beaches, destroying wildlife, harming marine ecosystems, and temporarily ending recreational uses. The detergents and their aromatic hydrocarbon solvents were even more toxic than the crude oil. In late 1969 the oil and tanker companies agreed to pay $7.2 million in damages for the incident, divided equally between the British and French governments.

Since the *Torrey Canyon* incident, several dozen large tanker accidents have been reported each year, and a great deal of effort has gone into trying to contain the oil spills that result or prevent some of their harmful effects. Three methods have been tried without much success [6]: (1) surrounding the oil slick with some sort of mechanical barrier until it can be removed; (2) collecting the oil by mechanical means such as suction pumping or absorption by straw; (3) dispersing the slick with chemicals. All these methods are slow and expensive.

A more continual problem with oil tankers is that part of each cargo is generally discharged into the ocean. A tanker will carry a cargo of crude oil from the source of the oil to some port, discharge it, and fill the tanks up with seawater to act as ballast on the return trip. Before the tanker docks, the seawater ballast, contaminated with the small residue of oil left in the tanks after discharge, is returned to the ocean. Ordinary vessels that use oil for fuel can also produce oil slicks when waste oil or seawater ballast is dumped from their fuel tanks. This practice is illegal in most countries but the prohibitions are difficult to enforce and the legal alternative—pumping the wastes into storage tanks at port—is inconvenient and expensive.

The number of oil tankers on the world's oceans is about 4000, and the total annual transport is approaching 1 billion metric tons. The amount that is lost to the oceans accidentally or intentionally is probably less than 1%, but this still comes to several million metric tons annually. Crude oil shipping is increasing year by year and tanker sizes have increased dramatically in recent years.

Oil on water spreads rapidly into a thin layer and the lighter-molecular-weight hydrocarbons soon evaporate. The others are degraded biologically but only at a slow rate. The heavier hydrocarbons often persist for long periods of time as tarry lumps. These lumps are widely distributed on the sea surface [7]. The explorer Thor Heyerdahl, who crossed the Atlantic in the papyrus vessel Ra II in 1970, reported that petroleum lumps were a constant companion during the voyage, in contrast to his 1947 voyage on the Kon-Tiki from Peru to Polynesia on parts of the Pacific Ocean not heavily traveled by shipping [8].

The most visible effect of oil pollution is the large-scale killing of seabirds. The oil apparently penetrates the feathers, displacing the air which is normally trapped in the feathers and which provides insulation and buoyancy. The birds become colder and more susceptible to diseases and experience difficulty flying. Perhaps 100,000 birds were killed in the *Torrey Canyon* disaster; only about 100 birds survived out of the 5700 that were caught and cleaned off in an effort to save their lives.

The effects of oil on other ocean life—such as plankton—are not well-known and much more research into the long-term effects on marine and estuarine ecosystems is needed. In one spill of diesel oil—by the tanker *Tampico Mara* in March 1957 at the mouth of a small cove in Mexico—almost the entire population of plants and animals was killed [6]. Some carcinogenic hydrocarbons, such as 3,4-benzpyrene, have been reported in marine environments and they are occasionally concentrated by shellfish.

Oil is occasionally released by natural seepage, as George Vancouver observed off the Santa Barbara Channel, Calif., in 1793, and natural tars were used by California Indians for waterproofing of baskets. Presumably the ecosystems in which natural oil is present have adapted to its existence.

INDUSTRIAL WATER WASTES

The importance of industrial water wastes can be seen from Table 10-1, which lists the estimated volumes, BOD, and suspended solids before treatment for U.S. industries as of 1964 [9]. The table also includes, for comparison, the corresponding figures for the sewered population of 120 million Americans assuming daily per capita production of 0.46 m³ waste water, 75 g BOD, and 91 g suspended solids. A detailed inventory of industrial wastes would, of course, have to include such important quantities as the amounts of toxic metals, refractory organic compounds, and other water pollutants. In 1971 the Environmental Protection Agency began trying to acquire such information from industry using authority granted by the Refuse

TABLE 10-1. Estimated volume of untreated waste water and mass of BOD and suspended solids for U.S. industries and U.S. sewered population (of 120 million persons) as of 1964 [9].

	Waste water (billion m³)	BOD (10^6 kg of O_2)	Suspended solids (10^6 kg)
Primary metals	16.3	220	2130
Chemical	14.0	4400	860
Paper and allied products	7.2	2670	1360
Petroleum and coal	4.9	230	208
Food and kindred products	2.61	1950	3000
Transportation equipment	0.91	54	n.a.[a]
Rubber and plastics	0.61	18	23
Machinery	0.57	27	23
Textile mill products	0.53	400	n.a.[a]
Electrical machinery	0.34	31	9
All other manufacturing	1.70	177	420
Total manufacturing	50	10,000	8200
Domestic (sewered)	20	3300	4000

[a] n.a. = not available.

Act of 1899 (see Chapter 20). Overall, the industrial waste water pollution amounts to several times the domestic waste water pollution and it is certainly increasing at a faster rate.

In some areas with heavy industrial concentration, industrial wastes are many times as great as domestic wastes. In 1960 the Mississippi River in the St. Louis Metropolitan Area (which had a population of 2.2 million) was receiving 5.25 million population equivalents of wastes, of which only 1.25 million were domestic sewage, the other 4 million being industrial. [Population equivalents are generally based on the values typical for untreated domestic sewage—a 5-day BOD of 0.17 lb (77 g) of O_2 and 0.20 lb (91 g) suspended solids per capita per day.] In the mid-1950's the Kalamazoo River between Comstock and Oswego, Mich., was receiving about 95% of its population equivalents from paper-mill wastes.

FOOD AND KINDRED PRODUCTS. Wastes from the food processing industry—meat and dairy products, beet sugar refining, brewing and distilling, canning, etc.—tend to be troublesome mainly because of their high content of putrescible (decomposable) organic matter, which can lead to oxygen depletion and water supply impairment in much the same way as domestic sewage. The meat (cattle, hogs, poultry) processing wastes come from stockyards, slaughterhouses, packing plants, and rendering plants and contain blood, fats, proteins, feathers, and other organic wastes. Until a sewage pretreatment plant was built in the late 1960's, the stockyards and packing plants at Omaha, Neb., dumped untreated wastes into the Missouri River, parts of which were sometimes red from the blood; hair, entrails, and mats of congealed grease floated downstream and collected on islands and along the shoreline. The dairy industry produces organic wastes high in protein, fat, and lactose from milk and cheese processing. Whey from the production of cheese is an important BOD source in parts of states with an established cheese industry [10]. The beet sugar refining industry produces wastes of high BOD content, including sugar and protein. Breweries and distilleries produce organic solids containing nitrogen and fermented starches from grain processing and alcohol distilling. The processing of food to produce canned or frozen products leads to enormous amounts of wet solid wastes. In the early 1960's the 3- to 4-month canning period of California's Central Valley was generating about 500,000 metric tons of wet wastes each year. Tomato wastes amounted to 7 to 10% of the original crop and tree fruit wastes (pits, peels, green and defective fruits) amounted to 12 to 14% [11]. Food processing often involves some danger of infectious disease, although this is rarely a public health problem in the United States.

TEXTILE PRODUCTS. Textile mill wastes, generated by cooking the fibers and desizing the fabrics, have high BOD and are quite alkaline, requiring neutralization and other treatment. The wastes arise from impurities in the fiber and from the chemicals used in processing. The production of 1000 kg of wool leads typically to 1500 kg of impurities (wool fibers, sand, grease, burrs, etc.) and 300 to 600 kg of process chemicals, with a total of 200 to 250 kg BOD [12]. Cotton processing leads to relatively less total BOD but the waste water may have 200 to 600 mg/1 BOD.

PAPER AND ALLIED PRODUCTS. Paper and pulp mills, in addition to being notorious air polluters (see Chapter 5), produce a great amount of water pollution. The Macmillan Bloedel mill at Port Alberni, British Columbia, for example, produces about 1650 metric tons of pulp and paper daily, using 160,000 m³ (42.5 million gal) of fresh water and dumping about the same volume of polluted water [13]. The effluent, like the effluent of any pulp mill, is a complex mixture of the chemicals used in the kraft process (see Chapter 5), stray wood chips, bits of bark, cellulose fibers, and dissolved lignin (woody tissue carbohydrate). About 50% of the wood used as input is eventually discarded as waste waterial. About 25 metric tons of solids are dumped daily into the narrow inlet leading to the mill, most of which sink to the bottom to produce a sludge several feet thick, blanketing fish spawning grounds, destroying bottom-feeding aquatic life, and even harming fish directly by irritating or clogging their gill membranes [13]. Mill effluent tends to be deep brown in color and can thus interfere with aquatic photosynthesis. It also contains compounds toxic to fish such as methyl mercaptan and paper and wood pulp preservatives such as pentachlorophenol and sodium pentachlorophenate. Sulfite liquor is also toxic to shellfish.

CHEMICAL INDUSTRY. A wide variety of water pollutants are produced by chemical plants manufacturing acids, bases, synthetic fabrics, pesticides, detergents, and many other compounds, organic and inorganic [12]. Acid wastes result not only from acid manufacturing plants but also from almost all other chemical plants as well and are customarily neutralized to give a pH of at least 6. Wastes from DDT and rayon manufacturing are especially acidic because a large amount of sulfuric acid is used. The production of the herbicide 2,4-D leads to dichlorophenol in the water, and the effluent may contain 25 ppm dichlorophenol despite treatment that removes 95 to 98% of the compound. The waste waters of the phosphate industry contain elemental phosphorus, fluorine, silica, and large amounts of suspended solids. In recent years mercury wastes from chemical plants (as well as other types of factories) have caused great concern. The list of chemical wastes could probably be extended to many thousands of different compounds and it is quite possible that some particularly dangerous ones are currently being released without suspicion.

PETROLEUM INDUSTRY. Oil drilling wastes include drilling muds, salt-water brines pumped out of the well with the crude oil, and some oil as well. Oil refineries and petrochemical plants produce an astounding number of different pollutants [12, 14]. These include hydrocarbons, acids, alkalis, cyanides, numerous sodium salts, phenolic compounds, numerous inorganic and organic sulfur compounds, and halogenated and nitrogenated hydrocarbons. Many of these compounds cause detectable tastes and odors at concentrations in the ppb range: ethyl mercaptan at 0.00019 mg/l, isoamyl acetate at 0.0006 mg/l, hydrogen cyanide at 0.001 mg/l, etc. Others cause fish flesh to acquire adverse tastes at concentrations of less than 1 ppm: *o*-chlorophenol at 0.015 mg/l, ethyl benzene at 0.25 mg/l, kerosine at 0.1 mg/l, etc. [14].

COAL INDUSTRY. Coal mines lead to acid mine drainage (discussed in Chapter 8), including typically 100 to 6000 ppm sulfuric acid [12]. Coal-preparation wastes from coal washeries contain very large amounts of suspended solids—coal, shale, clay, sandstone, etc.

RUBBER AND PLASTICS. Wastes from rubber production have a high BOD, taste, and odor [12]. Synthetic rubber is made from butadiene and styrene in a soap solution and coagulated with an acid-brine solution, and the wastes contain some of all the materials used. Odors from rubber plants can lead to unpalatable water for several hundred kilometers. Plastics manufacture produces wastes with hydrocarbons and other organic compounds, as well as various reagents.

METAL INDUSTRIES. Steel mills produce water wastes from the coking of coal, the washing of blast furnace flue gases, and the pickling of the steel.

These wastes tend to be acidic and contain cyanogen, phenol, ore, coke, limestone, alkali, oils, mill scale, and fine suspended solids [12]. Larger mills may recover by-products but smaller mills do not find this economical and may simply neutralize the acidity with lime, which leads to a large volume of sludge. Other metal industries have similar problems and their wastes generally contain the metal being produced or plated—chromium, lead, nickel, cadmium, zinc, copper, silver, etc.—as well as acids, alkaline cleaners, grease, and oil.

OTHER INDUSTRIES. Other industries have a wide variety of water pollutants. Leather tannery wastes have high total solids, hardness, salt, sulfides, chromium, alkalinity, lime, and BOD. Polishing of optical glass produces red (from iron) wastes containing detergents and suspended solids that do not settle readily. Laundries have turbid wastes that are alkaline and contain organic solids. Radioactive wastes result from nuclear power plants, fuel reprocessing plants, and hospitals and research laboratories using radio-isotopes. Soft-drink bottling plants produce highly alkaline wastes with high BOD from the washing of bottles, which involves removal of cigarette butts, paper, and other debris left in the bottles by previous users. A complete review of industrial water wastes is found in the book by Nemerow [12].

FISH KILLS

A very interesting source of information about U.S. water pollution—especially as it affects aquatic life—is the annual report on fish kills [15]. Although these reports are not complete, they indicate that tens of millions of fish are killed each year by a wide variety of different pollutants from many different sources, municipal and industrial.

In 1968, for example, over 15 million fish were reported killed in 42 states. Over 6 million were killed by malfunctioning, poorly operated, or inadequate municipal sewage treatment plants; over half of them were killed in a number of separate incidents in Dog River, Ala., from an overloaded treatment plant at Mobile. The largest single kill, over 4 million fish, came on the Allegheny River at Bruin, Pa., from chemicals released into a tributary stream when a petroleum refinery's lagoon overflowed into a pond whose walls broke.

Other causes reported in 1968 included crop poisons from aerial spraying, copper sulfate applied to control algae in a ditch, discharge of zinc cyanide from the plating room of a chain company, discharge of acid water from the waste piles of a coal mine, detergent discharged from a chemical laboratory, chlorine escaping from a private swimming pool, gasoline from a pipeline break, endrin sprayed on a tomato field, alkyd base paint from the storm drain of a chemical company, fluid chicken manure from a poultry farm that reached the water after a waterline break, leak of wash water containing caustic soda and sulfuric acid from a steel company plant, dairy waste, anhydrous ammonia refrigerant from a dismantled refrigeration unit, drainage from a solid waste disposal operation, oil from a tank car overturned in a railroad wreck, a fuel tank leak at an oil bulk station, and a thermal pollution incident.

TREATMENT AND DISPOSAL OF INDUSTRIAL WATER WASTES

Most of the water treatment methods used by industries are the same primary, secondary, and tertiary treatment methods available to municipalities. Many industries have water wastes of fairly uniform or predictable quality and since the exact pollutants are known, it is often possible to choose the best treatment for those particular pollutants. Biological treatment is well suited for wastes from dairy and food industries, for example, while chemical treatments have worked well for wastes from the metal plating industry. As new methods are invented or old methods improved, an industry may find that it can change to a better treatment.

The vast demand for water by many large industrial users has led to an emphasis on water reuse. One steel plant was designed to use only 4.6 m³ of water per metric ton of steel produced compared to an industry-wide average of 27.1 m³, and one oil refinery decreased its daily use 92.3% by reuse [16].

Some methods are used by industry for treating or disposing of waste waters that have no present counterpart in municipal treatment. Some make use of solar evaporation to help reduce pollution problems but this involves loss of water [16]. Many industries recover valuable by-products from their waste water.

Spray irrigation is a technique that has been used by dairy, food processing, tanning, paperboard mill, and other industries [16]. A notable example is the system operated by the Campbell Soup Company's food processing plant at Paris, Tex. [17, 18]. This plant produces 12,000 m³ (3.2 million gal) of organic waste water daily, which is spread by 700 sprinklers over 200 hectares (500 acres) of grassland to fertilize the grasses, which are then harvested as a by-product hay crop and sold to local farmers. The land was cleared and prepared over the period from 1960 to 1964 at a cost of about $2500/hectare ($1000/acre), and the operating costs are about $0.05/1000 gal (3.8 m³). The entire system is automatic and consists of about 10 km of underground force main, 16 km of portable aluminum pipe, and 55 km of control tubing.

Another disposal method that is gaining in popularity among industries producing concentrated, unusable, inorganic wastes is deep well injection [19, 20]. In the period since the early 1930's the petroleum drilling industry has made great use of brine injection wells to dispose of oil field brines, and there may be 40,000 such wells in the U.S. today [20]. During the 1960's a number of other industries became interested in this method, and today there are over 100 such wells in use by chemical and pharmaceutical companies, the depths ranging from a few hundred to 4500 m. A deep well for injection of toxic wastes requires certain geological features that do not exist in some parts of the country. There must be a porous, sedimentary rock stratum, usually sandstone, limestone, or dolomite, that can accept large amounts of the liquid wastes. There must be impermeable strata of rock, such as shale, above and below the injection stratum. Finally the injection stratum must lie below groundwater tables and contain no commercial brine or mineral deposits [20]. The dangers associated with deep well injection of concentrated toxic wastes are likely to keep this method from ever being a completely satisfactory solution to industrial water pollution.

There may be still other problems associated with these wells. The U.S. Army's Rocky Mountain Arsenal northeast of Denver, Colo., put a 3671-m deep well into operation in March 1962 for the disposal of wastes from chemical manufacturing operations, but injection was suspended 4 years later when it was suggested that it might be responsible for an increased frequency of earth tremors in the Denver area that began after the well was put into use [21]. Although the statistical evidence is very strong—the tremors did not appear to occur in the 1950's or earlier, the epicenters of the tremors were located close to the well, and the number of tremors appeared to correlate with the volume of fluid injected—it has not convinced everyone. Healy and his colleagues have concluded [21] that the fluid injection did trigger the earthquakes by reducing the frictional resistance to faulting and urged that steps be taken to minimize the continuing earthquake hazard by removing fluid from the well. Although deep well injection may continue to become a more widespread method of waste disposal, it is clear that each such well must be carefully considered.

Heavy chemical industries have also resorted to the foolish practice of disposing of highly toxic wastes by putting them in large drums and dropping them into the sea [22]. Sooner or later the drums become corroded and the wastes spill into the water. The worst aspect of this practice is that many drums have been dropped into shallow, intensely fished waters like the North Sea, where plans have had to be made to recover the drums and burn their contents.

REFERENCES

1. International Joint Commission, *Report on the Pollution of Boundary Waters.* Washington, D.C.: U.S. Govt. Printing Office, 1951.
2. International Joint Commission, *Final Report.* Washington, D.C.: U.S. Govt. Printing Office, 1918.
3. International Joint Commission, *Pollution of Lake Erie, Lake Ontario and the International Section of the St. Lawrence River.* Washington, D.C.: U.S. Govt. Printing Office, 1970.

4. "Pollution from Vessels Threatens to Negate Clean Water Goals." *Environ. Sci. Tech.* **2:**93–98 (1968).

5. Degler, Stanley E., *Oil Pollution: Problems and Policies*. Washington, D.C.: Bureau of National Affairs, Inc., 1969.

6. Holcomb, Robert W., "Oil in the Ecosystem." *Science* **166:**204–206 (1969).

7. Horn, M. H., et al., "Petroleum Lumps on the Surface of the Sea." *Science* **168:**245–246 (1970).

8. Heyerdahl, Thor, "Atlantic Ocean Pollution and Biota Observed by the 'Ra' Expeditions." *Biol. Conserv.* **3:**164–167 (1971).

9. *Environmental Quality. The First Annual Report of the Council on Environmental Quality*. Washington, D.C.: U.S. Govt. Printing Office, August 1970.

10. Knapp, Carol E., "Membrane Processing Upgrades Food Wastes." *Environ. Sci. Tech.* **5:**396–397 (1971).

11. Mercer, Walter A., "Industrial Solid Wastes—The Problems of the Food Industry." In *Proceedings of the National Conference on Solid Waste Research, December 1963*. Chicago, Ill.: American Public Works Assn., 1964. Special Report No. 29.

12. Nemerow, N. L., *Liquid Waste of Industry*. Reading, Mass.: Addison-Wesley Pub. Co., Inc., 1971.

13. Myers, Arnie, *Pollution! A Survey of Air and Water Pollution Problems in British Columbia*. Vancouver, B.C.: The Vancouver Sun, 1965.

14. *Petrochemical Effluents Treatment Practices*. Ada, Okla.: Robert S. Kerr Water Research Center, February 1970. Water Pollution Control Research Series Program No. 12020.

15. *Pollution-Caused Fish Kills*. Washington, D.C.: U.S. Govt. Printing Office. Published annually by the Environmental Protection Agency (formerly by the Federal Water Pollution Control Administration).

16. Abel Wolman Associates, *Water Resource Activities in the United States: Present and Prospective Means for Improved Reuse of Water*. Washington, D.C.: U.S. Govt. Printing Office, 1960.

17. Law, James P., Jr., et al., *Nutrient Removal from Cannery Wastes by Spray Irrigation of Grassland*. Washington, D.C.: U.S. Dept. of the Interior, 1969.

18. "Campbell's Waste Water Irrigates Texas Prairie." *N.Y. Times,* May 10, 1970, p. 12-F.

19. Warner, Don L., *Deep-Well Injection of Liquid Waste*. Cincinnati, Ohio: U.S. Dept. of Health, Education, and Welfare, 1965. Public Health Service Publication No. 999-WP-21.

20. "Deep Well Injection is Effective for Waste Disposal." *Environ. Sci. Tech.* **2:**406–410 (1968).

21. Healy, J. H., et al., "The Denver Earthquakes." *Science* **161:**1301–1310 (1968).

22. Greve, P. A., "Chemical Wastes in the Sea: New Forms of Marine Pollution." *Science* **173:**1021–1022 (1971).

11

AGRICULTURAL POLLUTION

Augeas, King of Elis, had a herd of three thousand oxen, whose stalls had not been cleansed for thirty years. Hercules brought the rivers Alpheus and Peneus through them, and cleansed them thoroughly in one day. —Bulfinch's Mythology [1]

The old legends of the labors of Hercules say nothing about the reactions of the people living downstream or of the effects on the fish in the rivers, but the magnitude of the task was enormous. The stables presumably contained $3000 \times 30 \times 365 = 33$ million oxen-days' accumulation of wastes, and since a large ox would produce about 4.5 kg of manure daily on a dry basis, Hercules had some 150,000 metric tons of wastes to dispose of!

United States agriculture has always managed to produce an ample supply of food for the U.S. population. Advances in agricultural technology have been so great that today only about 4.5 million persons are employed on farms, compared to 9.9 million in 1950, 12.5 million in 1930, and even more before that; the U.S. population is about twice what it was 50 years ago, while the farm population is only one-third what it was then. In terms of food costs, Americans spend only an average of 18% of their disposable income for food, compared to 40% and more in some other developed countries.

American agriculture has done well for a variety of reasons. Mechanization, improved plant varieties and seeds, artificial fertilizers, chemical pesticides, and improved methods of preserving and transporting food have all played a role. In the meantime prime agricultural land has often been taken out of production to permit the construction of new highways and residential developments, population growth has steadily reduced the per capita acreage (see Figure 2-2), air and water pollution has affected agriculture adversely, and agricultural methods have themselves contributed to environmental pollution.

The effects of pollution on agriculture have been dealt with from time to time in other chapters. In this chapter some of the environmental problems associated with agriculture are discussed in some detail. Pesticides are dealt with at greater length in Chapter 12.

FARM ANIMAL WASTES

Over the centuries, farm animal wastes have been regarded as an important source of soil fertility by all peoples. Fields fertilized with manure generally show higher yields than those without manure, and even when the manure treatment is ended, the favorable influence persists for years. In an experiment performed at the Rothamsted (England) Experiment Station, barley yields on untreated land over a 50-year period were only 20 to 25% of those on land continuously manured; land manured for 20 years showed yields decreasing to 50% in 10 years and 40% after 30 years [2]. In the U.S. today, however, it is estimated that only one-third of the value of farm manure is being realized [2].

The problem with manure as a fertilizer from the U.S. point of view is that its relative concentrations of the plant nutrients nitrogen, phosphorus, and potassium are much lower than those of artificial chemical fertilizers. The analysis of a fertilizer is generally expressed by three percentages referring to nitrogen (N), phosphorus (P), and potassium (K) content. Farm manure rates about 0.5-0.1-0.4 (0.5% N, 0.1% P, 0.4% K) on the average; in order to supply a given quantity of these nutrients to a piece of land, 20 or 30 times as much manure as commercial fertilizer (say 10-10-10) might have to be applied [2]. Although the commercial product has its cost, most farmers have found it easier and simpler to apply and the use of commercial fertilizers in the U.S. has soared since the end of World War II.

Farm animal wastes cannot be avoided, however, and their presence poses serious problems of odor and water pollution in addition sometimes to public health problems arising from their role in transmitting disease. The magnitude of the potential animal waste problem dwarfs that of the human waste problem, as shown below. Domestic animals produce about 1 billion metric tons of fecal wastes each year and about another 400 million metric tons of liquid wastes; the inclusion of used bedding, paunch manure from slaughterhouses, and dead animal carcasses makes the total annual production of solid wastes close to 2 billion metric tons [3]. A dairy cow will produce about 1.5 kg of wet manure for each liter (1.06 qt) of milk, and every 1 kg of livestock weight gain is accompanied by 6 to 25 kg of wet manure [4].

The amount and nature of the wastes produced will depend on the type of animal and the diet (including feed additives) on which it is fed. Most of the wastes consist of undigested foods, largely cellulose and fibers. Data on wastes are difficult to collect because of the variations in their nature, the differing moisture contents (the elapsed time until measurements are made can be important), the amount of bedding present, etc. The usual parameters for the description of municipal sewage—BOD, COD, suspended solids, etc.—were developed to characterize highly liquid wastes consisting mostly of water and are not as applicable to animal wastes, which are quite solid. The total weight values for BOD and suspended solids, however, can be compared with those from the human population to give population equivalent values.

Table 11-1 gives the typical daily materials balance for a 400-kg beef animal [4]. The total feed ration would be 36 kg and the total body weight gain per day would correspond to 1.25 kg. Of the 24-kg excreted wastes, 17 kg would be feces and 7 kg would be urine. Solid excrement would contain about one-half or more of the nitrogen, about one-third of the potassium, and nearly all the phosphorus. Urine contains about half the plant nutrients, then, in a concentrated form; in addition, the plant nutrients in the urine are soluble and directly available to plants or readily become so [2]. Solid and liquid manure are an abundant source of the plant nutrients N, P, and K and also contain sulfur, calcium, magnesium, iron, zinc, boron, copper, and other trace elements [4].

TABLE 11-1. Typical daily materials balance for a 400-kg beef animal [4]. Daily weight gain would be 1.25 kg.

	Feed ration, kg	*Excreted wastes, kg*
Water	**27**	**18.5**
Dry matter	**9.25**	**5.5**
Organic matter, minerals	8.8	5.0
N	0.21	0.17
P_2O_5	0.09	0.07
K_2O	0.15	0.11
Total	**36.25**	**24.0** [a]

[a] 17 kg feces, 7 kg urine.

TABLE 11-2. Data on farm animal wastes [4, 5].

	Hog	*Hen*	*Turkey*	*Cattle*
Typical size (kg)	45	2	2–8	450
Wet manure (kg/day)	2.5	0.11	0.34	29
Total solids (%)	16	29	26	16
Composition (dry basis):				
N (%)	4.5	5.6	5.2	3.7
P_2O_5 (%)	2.7	4.3	2.8	1.1
K_2O (%)	4.3	2.0	1.9	3.0
Volatile solids (%)	85	76	78	80
5-Day BOD (kg/day of O_2)	0.135	0.007	0.02	0.45
Total dry solids (kg/day)	0.40	0.03	0.09	4.6

Representative data on farm animal wastes for hogs, chickens, turkeys, and cattle are given in Table 11-2. As mentioned earlier, actual measurements show wide variations, and Reference 4 may be consulted for more information about these variations.

It is possible to compare the approximate BOD and total solids from farm animals with the corresponding quantities for human beings, who produce about 75 g 5-day BOD and 250 g of total solids per capita per day. Table 11-3 shows the approximate population of various farm animals in the U.S. at any given time, the approximate equivalent number of human beings for each type of animal in terms of BOD and total solids, and the human population equivalent. It will be seen that on a BOD basis the animal population in the U.S. produces over four times as much waste as the human population (now about 210 million), while on a total solids basis farm animals produce about 12 times as much.

TABLE 11-3. Human population waste equivalents of U.S. farm animals in the early 1970's. The number of humans equivalent to each animal has been computed assuming each human produces 75 g BOD and 250 g total solids each day.

Animal	Average no. (millions)	Human equivalents per animal		Human equivalents (millions)	
		BOD	Total solids	BOD	Total solids
Cattle	110	6.0	18.4	660	2020
Swine	60	1.8	4.4	110	260
Hens	400	0.1	0.3	40	120
Turkeys	100	0.3	1	30	100
Sheep	17	0.6	3	10	50
Horses	3	3	13	10	40
Total	690			860	2590

Animal wastes have not been as great a problem in the past as these numbers might suggest because they have traditionally been returned to the soil so that their plant nutrients and organic matter would maintain and improve the quality of the soil. The organic matter in livestock manure improves the tilth of the soil and its capacity to hold water, aids in the aeration of the soil, lessens the amount of erosion produced by wind and water, and has a beneficial effect on soil microorganisms [2, 4]. Returning manure to the land—as the Amish in this country are still careful to do—is a form of recycling.

Animal wastes are a major problem once they are permitted to enter water supplies, when confinement areas are cleaned or when runoff carries the wastes into waterways. Today, farm animals in the U.S. may produce about 8 million metric tons of nitrogen annually but only about 500,000 metric tons are actually getting into waterways.

The problem has been aggravated, however, by the growth of large feedlots during the 1950's and 1960's. Confinement feeding on these feedlots has been increasing as farmers try to cut costs in the face of declining profit margins per animal. Both the number of large commercial livestock feeding operations and the average number of animals per production unit have been rising. In 1966 the number of feeder cattle removed from pastures and grass-lands and confined (with typically 5 to 10 m² per animal) numbered 10 million—up 66% in 8 years and 120% in 15 years. The number of cattle on feedlots at any given time is now well over 10 million. A feedlot of 30,000 head (and many are much larger) may be the BOD equivalent of a city of 200,000 people so the waste problems are severe, whether handled as solid wastes or water wastes.

Large confinement areas also exist for hogs and poultry. A new con-finement area for hogs at Hampton, Iowa, was described in the *Des Moines Register* as "40 acres of wall-to-wall pigs." Poultry farms with hundreds of thousands of birds may also have the wastes of a fair-sized city. Duck farms on Long Island, New York, have been a problem for many years because of their wastes; 33 farms in Suffolk County were estimated to produce wastes equivalent to about 130,000 people [6]. Discharge from these duck farms led to serious bacteria problems that forced the closing of the shellfish industry in Moriches and Bellport Bays in 1967, but the shellfish beds are reopening as the duck farms begin to install their own waste treatment facilities.

With the federal government and the states setting water quality standards, pressures on farmers to dispose of their wastes properly will increase and manure disposal is likely to become an important production expense. There is a great need for research and development, especially since ordinary municipal sewage-treatment methods are not generally suitable for animal wastes. Raw municipal sewage is continuous and might, typically, have 200 mg/l BOD, 400 mg/l COD, and 40 mg/l nitrogen, while feedlot runoff, which contains slugs of waste, might be 1000 mg/l BOD, 8000 mg/l COD, and 700 mg/l nitrogen. Feedlot runoff, when it occurs, can produce heavy pollution in receiving streams and cause large fish kills, such as occurred in Kansas in 1967 when spring rains carried many tons of cattle feedlot wastes into rivers.

Current interest in animal waste control centers around the use of aerobic and anaerobic biologically active systems and the spreading of wastes on the land, although such methods as incineration, dehydration, and com-posting are also being studied [4]. In Europe, biological methods have found some use but the chief emphasis has been on returning the manure to the land, and research and development work has centered around better methods of hauling and handling the wastes. The availability of inexpensive commercial fertilizers, which are easier to handle because the nutrients are more concen-trated, has led U.S. farmers to seek ways of disposing of animal wastes rather than reusing them, and European farmers appear to be moving in the same direction.

Anaerobic digestion systems have proved successful in the laboratory and in the field under some conditions [4]. Anaerobic lagoons are built with a small surface area, as deep as construction and groundwater conditions will permit. Cold weather can interfere with proper operation and solids with a high nitrogen content (such as chicken manure) may lead to odors from ammonia production. Methane gas is produced by anaerobic digestion, although its production becomes minimal when the temperature of the system is below about 13 or 14°C. *The Mother Earth News* (No. 10) carried a story about an English farmer who runs his automobile on methane gas produced from chicken manure in a special digestion tank; since this is completely equivalent to using natural gas, which is relatively nonpolluting, his automobile produces very little air pollution.

The purpose of anaerobic lagoons is to remove and destroy much of the organic matter and they can be combined with aerobic systems to give an effluent that is of adequate quality for discharging into rivers. The anaerobic system accepts slugs from a feedlot operation, for example, and partially degrades and gasifies them, while the aerobic unit degrades the remaining soluble and particulate matter.

Aerobic units are generally shallow oxidation ponds or lagoons that actually have anaerobic conditions near the bottom. Although they work well, they require very large surface areas—perhaps 8 hectares (20 acres) for 1000 head of cattle and 2 hectares (5 acres) for the same number of hogs [4]. Sometimes the amount of land needed can be reduced by the use of mechanical or diffused aeration systems.

Manure disposal on the land is not fundamentally complicated but confinement feeding leads to difficulties in using this method. A poultry unit housing 100,000 hens may produce 11 metric tons of manure daily, and its application to the soil at a rate of 18 metric tons/hectare (8 short tons/acre) annually means that some 220 hectares (540 acres) will have to be found

FIG. 11-1. Aerial view of an Iowa beef farm and its lagoon. (*Courtesy of Iowa State University Extension Service.*)

for disposal. The European philosophy seems to be that the manure will have to be disposed of anyway so one might as well return it to the soil and make use of its nutrients.

Insofar as nutrients are concerned, it is most efficient to haul manure into the fields while fresh and work it into the soil immediately. Nutrients can be lost by volatilization (especially of nitrogen and organic matter); leaching of soluble components by rain; or by loss of the liquid urine, which contains about half the nutrients [2]. In general, it is preferable to apply fresh manure to the fields daily rather than to let it rot first; however, this is often impractical. In winter the soil is frozen and unworkable much of the time in many parts of the country, and if manure is spread over snow-covered ground, it is likely to pollute waterways as soon as a thaw occurs. During the growing season, crops are on the soil and manure cannot be worked into it.

Several processes are being investigated which make use of very fluid (diluted) manure which can be spread thinly on the land, injected into the subsoil, or sprayed on the land. Spray irrigation of grasslands can be used for manure as well as for food processing wastes (discussed in Chapter 10). Many feedlots are building detention ponds (which are sometimes required by law) to catch runoff from rainstorms, and the water can be pumped out and sprayed over fields. One hog-feeding facility in Oregon which has about 15,000 hogs at any given time periodically washes the pens with water which is flushed into lagoons, from which it is pumped out after dilution into a sprinkler system onto 200 hectares of grainfields, hayfields, and pastures [7]. On another small farm in Nebraska, wastes from 200 to 300 head of cattle are drained into a lagoon, from which they may be pumped out (even when up to 25% solids) to irrigate and fertilize cornfields [8].

SOIL EROSION

Sediments—soil and mineral particles washed into streams by storms—were discussed briefly in Chapter 8, where it was pointed out that the amount of soil erosion occurring in the world today may be two or three times what it was before man's intervention in nature. In the U.S. some 3½ billion metric tons of sediments are washed into tributary streams each year, about one-fourth of them eventually being transported into the sea [3]. At least half of this sediment is coming from agricultural lands, with water erosion being a dominant problem on 72 million hectares of cropland and a secondary problem on another 20 million at least. (Total U.S. cropland is about 120 million hectares.) Eroded materials can also be contributed by urban, industrial, and highway construction sites and by road banks, and urban areas can be major sediment sources [9].

Sediments have been both a blessing and a curse to agriculture [3]. They have added to the fertility of the soil where deposited, as in the Nile River Valley in Egypt, and they have settled in and destroyed irrigation

FIG. 11-2. Examples of serious erosion. (Top) This huge gully was washed out in a period of a few years. (Lower) Water erosion of these cornfields has made farming very difficult. (*Courtesy of Iowa State University Extension Service.*)

systems such as those of the ancient Sumerians. Sediment can settle on trout spawning beds and suffocate the eggs. It can clog gills of adult fish. It interferes with domestic and industrial uses of water and produces extra expenses for clarification of the water. It can lead to reduced dissolved oxygen levels and thereby adversely affect aquatic life. It affects the recreational values of lakes, reservoirs, and streams and shortens the life of farm ponds by producing an increased rate of filling through siltation.

The 3½ billion metric tons of sediment delivered to U.S. waterways annually corresponds to the equivalent of 1.6 million hectares of topsoil 18 cm (7 in.) deep [3]. Of the total sediment, some 75% originates from agricultural and forest lands, which in the process lose perhaps 2.6 million metric tons of N, 1.7 million metric tons of P, and 32.5 million metric tons of K [3].

In recent years it has been realized that sediments carry many inorganic and organic materials with them, often adsorbed to the surfaces of silt particles. In particular, phosphorus adsorbed on silt and carried to the bottom of lakes and rivers may play an important role in eutrophication, and many water-insoluble pesticides (such as the chlorinated hydrocarbons) are carried into waterways by sediment. Control of soil erosion consequently looms as one of the most important present-day concerns in the control of environmental pollution.

Control methods exist that can cut erosion rates down to 10 to 50% of current rates without affecting agricultural productivity. These methods act to reduce the kinetic energy of raindrops and flowing water and keep the soil particles in place. The presence of vegetation, either living or dead, reduces erosion. If the soil is covered with grasses or trees, the vegetation will intercept raindrops and prevent them from imparting their energy to soil

FIG. 11-3. Aerial view of terraced farmland just after a very heavy rain. The water is standing on the terraces and has not coursed downhill eroding the land. (*Courtesy of Iowa State University Extension Service.*)

FIG. 11-4. Blowing dust from farm-lands has piled up to a depth of half a meter near a roadside fence. (*Courtesy of Iowa State University Extension Service.*)

particles. Much of the rainfall will be absorbed in the ground instead of flowing over the surface of the land and this, too, leads to reduced erosion. It is well-known that in periods of heavy rainfall, flooding will not occur on forest lands but will occur on nearby land which has been developed commercially.

Control methods are especially desirable on sloping lands since the amount of erosion increases rapidly with increasing slope. In addition to planting suitable vegetation, the land may be shaped to reduce the erosion. Even in ancient times good agricultural practice included the building of terraces that followed the contour of the land and forced downward-flowing water to encounter a series of flat pieces of land. Terraces can take a long time to build and can make it difficult to apply mechanized agricultural techniques. In the U.S. the trend is toward shaping the land to make terraces

FIG. 11-5. Dust problems from wind erosion can be reduced by leaving crop residues on the surface. A strong wind will blow away much of the topsoil on the plowed soybean field on the left, while the field on the right is much better protected. (*Courtesy of Iowa State University.*)

straighter and thus easier to farm. Erosion can also be reduced by growing different crops in alternate strips ("strip cropping"), using broad, gently sloping waterways covered with a thick carpet of grass, and by preventing overgrazing of grassland by livestock, which destroys the vegetative cover.

Wind erosion is a related problem. Strong winds can carry soil particles into the air and thus produce the air pollution problems associated with airborne particulate matter. The great U.S. dust storms of the 1930's were produced in this manner; one storm in May of 1934 may have removed close to 300 million metric tons of soil. Many fields filled up with sand dunes and many homesteads were covered up. These problems are alleviated by a good vegetative cover that can be maintained on the land. Most serious wind erosion is likely to accompany periods of drought where maintenance of vegetative cover is difficult. Residue from the previous crop left on the surface of the soil will reduce soil loss from wind erosion. The growing of hedges and other windbreaks (junipers, ponderosa pines, etc.) has also helped in the Great Plains [3].

PLANT RESIDUES

Plant residues from crops and orchards can constitute environmental pollution when they harbor plant diseases and pests or when they are burned and emit smoke and hydrocarbons, since both situations correspond to an unfavorable alteration of man's environment. Agricultural burning is a fairly important air pollution problem, as the data of Table 3-2 showed. It is estimated to be responsible for:

8.3 percent of carbon monoxide emissions.

8.5 percent of particulate matter emissions.

5.3 percent of hydrocarbon emissions.

1.5 percent of nitrogen oxide emissions.

About 250 million metric tons of agricultural refuse (crop residues, scrub, brush, weeds, grass, and other vegetation) are burned annually in the U.S. [10]. Much of this burning is for sanitation purposes since plant debris can carry diseases or pests to succeeding crops. Burning of grass residue is the only practical method of controlling blind seed disease by the Oregon grass seed industry [3]. Each year about 4% of the potato crop is lost to late potato blight, which is believed to be primarily due to potato residues. Agricultural burning and the resulting air pollution problems are likely to continue until better methods of controlling the plant diseases are available.

Plant debris can be used as a mulch; e.g., it can be spread over the ground under orchard trees. Mulching on a large scale can be effected on croplands by leaving residue from the previous crop on the soil surface.

Mulching is known to reduce wind and water erosion and it sometimes (but not always) leads to improved crop yields. Plant residues are also used to some extent for other purposes, such as bedding for poultry and livestock or in the manufacture of corrugated cartons [3].

Some 23 million metric tons of logging debris are left in forests annually [3]. This debris can act as a reservoir for tree disease and insects, just as elms killed by Dutch elm disease and oaks killed by oak wilt can transmit disease to healthy trees. Controlled burning is the only feasible disposal method at present, costing about $1/ton in comparison with $12/ton for chipping of the debris [3]. It causes air pollution problems, of course, and has interfered with recreational activities and airport operations. Unburned debris from logging operations is a very serious fire hazard, which is known to cause larger-than-average forest fires.

AGRICULTURAL CHEMICALS

In recent years there has been an increased use of agricultural chemicals, notably pesticides and fertilizers. These chemicals have been effective in controlling diseases and pests that affect crops, livestock, and even humans, and they have enabled U.S. farmers to secure high crop yields at relatively low cost. They contribute to air and water pollution, however, and have adverse effects on fish, wildlife, and, upon occasion, human beings. The extent to which pesticides have been studied and debated in the last 2 decades makes them worthy of special attention and they are discussed in Chapter 12.

Agricultural fertilizers sold commercially generally include one or more of the main plant nutrients—nitrogen (N), phosphorus (P), and potassium (K). Although nitrogen may be obtained from certain natural sources such as animal wastes or the great sodium nitrate ($NaNO_3$) deposits in Chile, the majority of the nitrogen in U.S. commercial fertilizers is synthetic ammonia (NH_3) or one of its derivatives, such as ammonium nitrate (NH_4NO_3) or ammonium sulfate [$(NH_4)_2SO_4$]. Nitrogen is sometimes produced in organic form, as in urea [$CO(NH_2)_2$] and calcium cyanamide ($CaCN_2$). Most of the phosphorus in fertilizers comes from rock phosphate, whose principal component is fluorapatite [$(Ca_3(PO_4)_2)_3 \cdot CaF_2$]. Finely ground rock phosphate is generally treated at the manufacturing site with a roughly equal amount of sulfuric acid to produce a water-soluble mixture of monocalcium phosphate [$Ca(H_2PO_4)_2$] and calcium sulfate ($CaSO_4$) known as superphosphate. (Phosphorus concentrations are generally expressed as P or P_2O_5 rather than superphosphate.) Potassium fertilizers originate from the mining of surface and underground deposits of potash-bearing minerals and rocks; commercial fertilizer materials commonly used are potassium chloride (KCl) and potassium sulfate (K_2SO_4) but concentrations are usually expressed as K or K_2O (potash) [2].

The availability and low price of commercial fertilizers has led to

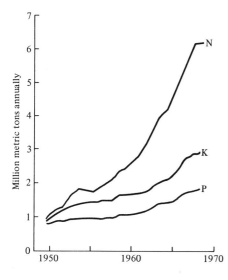

FIG. 11-6. U.S. consumption of nitrogen (N), phosphorus (P), and potassium (K) fertilizers, measured in terms of their content of the elements [11]. Data is for July 1 through June 30 crop years.

rapid increases in their usage since World War II, as shown in Figure 11-6. Problems can arise from excessive application rates since the fertilizers can be transported into groundwater by leaching or into waterways by natural drainage and storm runoff. Nitrates are of special concern but they generally find their way into waterways from feedlots, not from excessive or improperly timed fertilizer use. Nitrates in drinking water can cause methemoglobinemia in babies ("blue babies") because in an infant's stomach the nitrate (NO_3^-) is converted into nitrite (NO_2^-), which acts on the blood hemoglobin to form methemoglobin. The same process can occur in the stomach of ruminants so that livestock can also be affected by nitrate poisoning. The U.S. Public Health Service feels that drinking water supplies used by infants should not contain more than 10 ppm nitrate, and the standard for adults is set at 45 ppm [3]. Groundwater supplies obtained from wells in agricultural areas sometimes exceed one or both of these limits, usually due to feedlot contamination.

Plant nutrients also contribute to eutrophication, as discussed in Chapter 8. Phosphorus is often implicated in algal blooms in U.S. waterways but there is great disagreement about its role. Phosphates have a great affinity for soil and are generally carried into waterways adsorbed to sediment, in a state unavailable for plant growth. Dissolved phosphates are generally much lower in concentration than adsorbed phosphates; the latter can amount to 200 ppm of topsoil and 1000 ppm of the clay fraction of soils [3].

The importance of agriculture in providing plant nutrients to waterways has not been well established; the estimates shown in Table 8-6 indicate that agriculture is probably an important source of these nutrients but not as important as natural sources and municipal sewage. The situation in some areas may not be typical of the U.S. average, however, and as fertilizer usage increases the importance of this type of agricultural pollution is likely to increase. Far more study is needed to determine the extent to which nitrogen and phosphorus in waterways comes from fertilizer and other agricultural sources and the extent to which they occur in forms actually available for plant growth [3].

There are some alternatives to the use of inorganic fertilizers on agricultural crops. One is the return to the methods of years gone by, including use of organic fertilizers (such as animal wastes) and crop rotation (such as planting legume crops occasionally to increase the nitrogen in the soil): A small percentage of American farmers still use such methods, although this "organic farming" can be unfeasible for farmers without livestock. Another possibility is that the government might find it necessary to institute controls on agricultural chemical usage. At present the government has a great deal of control over land usage for agricultural crops and fertilizer control provides another alternative. One mathematical study [12] showed that the effects of fertilizer restrictions on six major crops (corn, wheat, barley, oats, grain sorghum, and cotton) would include higher total food costs, higher total farm income, lower government agricultural payments, and the need for returning diverted acreage to crop production. Complete elimination of all fertilizer use would increase per capita food costs about 25%. The U.S. has enough land to manage without fertilizers if such restrictions ever appear desirable, but there would be a noticeable increase in food costs.

MISCELLANEOUS AGRICULTURAL POLLUTION

There are several other types of pollution that occur in agricultural areas.

ANIMAL DISEASE AGENTS. Infectious agents transmitted by air, water, and soil can affect livestock, poultry, and humans. Some of the diseases produced are listed in Table 8-5. It appears to be possible to distinguish animal contamination of water from human contamination by making counts of both fecal coliform (see Chapter 8) and fecal streptococcal bacteria because the former predominate in human wastes and the latter in animal wastes [3]. The fecal coliform/fecal streptococci ratio is usually greater than 4 for human contamination and less than 0.5 for contamination by cattle, swine, sheep, and poultry; thus a ratio above 2.5 indicates primarily human contamination, one below 1.0 indicates primarily animal contamination, and one in between indicates a relatively even mixture. Great advances in control of veterinary diseases has accompanied the advances in public health of the 20th century and formerly important animal diseases such as bovine tuberculosis and hog cholera are now fairly well controlled.

DEAD ANIMALS. The disposal of dead animals is another problem. The average mortality in hen production is nearly 1% monthly, for example, and 30 million dead birds have to be disposed of annually in the United States [13]. Poultry meat decomposes readily and produces objectionable odors, while the feathers are rather inert and stable. Incineration can be difficult and can produce air pollution. Disposing of carcasses by delivering them to rendering plants is probably the best solution. Sometimes, though, dead animals are thrown in old wells, where they may lead to groundwater pollution.

RURAL DOMESTIC WASTES. The domestic wastes in rural areas must also be considered. Generally the waste waters are disposed of in the ground through the use of pit privies, cesspools, or septic tanks. Sometimes the absorptive capacity of the soil is inadequate for proper disposal. The main concern is to prevent contamination of groundwater supplies or at least to prevent contamination of wells from which drinking water is obtained.

FOOD PROCESSING WASTES. Industry in agricultural areas can also cause serious pollution. Of particular note would be industries closely associated with agriculture, such as the food processing industry, which is discussed to some extent in Chapters 5 and 10. These industries can produce both air pollution (such as by cotton gins and alfalfa mills) and water pollution.

SALINITY OF IRRIGATION WATER. Inorganic salts present in the soils and geologic materials of arid regions can move into waterways and be deposited on farms that use irrigated water, and the salinity of the soil can build up unless effective drainage is provided for. Salinity has already impaired about one-fourth of the 12 million hectares of irrigated land in the western U.S. and become a concern over another one-fourth [3]. The highly productive Imperial Valley of California uses Colorado River water containing about 750 ppm dissolved salts, and a farmer who uses 1.5 m of water during the irrigation season is applying about 11 metric tons of salts per hectare; improperly drained fields produce a crop of "Imperial Valley snow"—a white crust of sodium sulfate.

CONCLUSION

In conclusion, it should be noted that the greatest agricultural pollution problem at present is probably soil erosion by water and wind, especially since many other agricultural pollutants (such as pesticides and phosphates) are transported by sediment.

The animal waste problem is already important in some areas, such as near large feedlots, and could be a problem of overwhelming importance if no animal wastes were returned to the soil.

It should be recognized that our agricultural problems are aggravated by our eating habits. Americans are largely meat eaters, each consuming about 83 kg (183 lb) of meat annually (New Zealanders rank first in the world with annual average per capita consumption of 102 kg of meat). Agricultural problems would be lessened if we were vegetarians since the use of grains and grasses and other plants for animal food, which is then used for human food, is a rather inefficient process. Putting a given acreage into soybeans for human food would yield over 10 times as much protein as putting it into corn to be fed to livestock and eventually converted into meat protein. (Total protein mass is not really a good indication of biological value, of course, since proper nutrition requires that a number of essential amino acids all be present in the diet, and some foods lack one or more of them [14].) Many countries obtain little food from animal sources; meat consumption in China and India is very close to zero.

REFERENCES

1. *Bulfinch's Mythology.* New York: Modern Library. p. 119.
2. Millar, C. E., L. M. Turk, and H. D. Foth, *Fundamentals of Soil Science,* 4th ed. New York: John Wiley & Sons, Inc., 1965.
3. Wadleigh, Cecil H., *Wastes in Relation to Agriculture and Forestry.* Washington, D.C.: U.S. Dept. of Agriculture, March 1968. U.S.D.A. Miscellaneous Publication No. 1065.
4. Loehr, Raymond C., *Pollution Implications of Animal Wastes—A Forward Oriented Review.* Ada, Okla.: Robert S. Kerr Water Research Center, U.S. Dept. of the Interior, 1968.
5. Taiganides, E. Paul, "Agricultural Solid Wastes." In *Proceedings of the National Conference on Solid Waste Research, December 1963.* Chicago, Ill.: American Public Works Assn., 1964.
6. Salpukas, Agis, "L.I. Shellfish Beds Reopen as Duck Farms Curb Waste." *N.Y. Times,* March 8, 1970, p. 76.
7. Underwood, Clarence, "Irrigating with Animal Waste." *Soil Conservation* **34:** 81–82 (1968).
8. "Cattle Feeders Avoid Pollution by Using Wastes in Irrigation." *Soil Conservation* **34:**84–86 (1968).
9. Gross, M. Grant, "New York Metropolitan Region—A Major Sediment Source." *Water Resources Research* **6:**927–931 (1970).
10. *Nationwide Inventory of Air Pollutant Emissions 1968.* Washington, D.C.: U.S. Dept. of Health, Education, and Welfare, August 1970. N.A.P.C.A. Publication No. AP-73.
11. *Statistical Yearbook.* Published annually by the United Nations Statistical Office.
12. Mayer, L. V., and S. H. Hargrove, *Food Costs, Farm Incomes, and Crop Yields With Restrictions on Fertilizer Use.* Ames, Iowa: Iowa State University, 1971. Center for Agricultural and Economic Development Report No. 38.
13. Hart, S. A., and W. C. Fairband, "Disposal of Perished Poultry." In *Poultry Industry Waste Management—Second National Symposium.* Lincoln, Neb.: University of Nebraska, 1964.
14. Lappé, Frances Moore, *Diet for a Small Planet.* New York: Ballantine Books, Inc., 1971.

12
PESTICIDES

The Canker (worm) *is a shrewd disease when it happeneth to a tree; for it will eate the barke round, and so kill the very heart in a little space. It must be looked unto in time before it hath runne too farre; most men doe wholly cut away as much as is fretted with the Canker, and then dresse it, or wet it with vinegar or Cowes pisse, or Cowes dung and urine, &c. untill it be destroyed, and after healed againe with your salve before appointed.* —John Parkinson, *Paradisi in Sole Paradisus Terrestris* ("An Earthly Garden of Paradise in the Sun") (1629 A.D.).

Farmers and gardeners are engaged in a never-ending struggle to avoid the ravages of insects and other pests that damage vegetation. Public health officials are constantly trying to control numerous pests that are vectors of human disease. The number and quantities of chemical compounds used in these campaigns have increased sharply in the past 2 decades. "It appeared for a time that DDT would be a panacea for most of our ills caused by insects, but flaws have been found in the utilization of this compound," a scientist wrote a few years after the introduction of DDT [1]. Further flaws have been discovered since and there is great controversy about whether or not the benefits produced by various uses of pesticides outweigh the disadvantages and possible long-term risks. In this chapter the history of pesticide usage will be discussed, then the types and effects of the important pesticides, and finally some possible alternatives to chemical pesticides.

HISTORY OF PESTICIDE USAGE

Prior to the 19th century, use had been made of lye, lime, soap, turpentine, tobacco (for its nicotine), pyrethrum powder [made from pyrethrum (chrysanthemum) flowers], mineral oil, and arsenicals as pesticides, but the use was never great.

The first real use of a chemical compound as a commercial pesticide was that of Paris green (also known as copper acetoarsenite or copper acetate metaarsenate), whose approximate composition is $Cu(CH_3COO)_2 \cdot 3Cu(AsO_2)_2$, for control of the Colorado potato beetle in the western United States about 1868. Within a few years it was also being used against

189

cankerworms and codling moths on apple trees. Several other arsenic compounds were soon in use; in the early 20th century, lead and calcium arsenates [$Pb_3(AsO_4)_2$ and $Ca_3(AsO_4)_2$] became the preferred choices. In the late 1930's close to 50,000 metric tons of arsenates were being used annually in the U.S., but this dropped to 40,000 by 1950, 8000 by 1955, and 4000 by 1967.

During the late 19th and early 20th centuries several other insecticides and fungicides won approval. Examples included bordeaux mixture (a mixture of copper sulfate and lime), sulfur, oil, nicotine, and rotenone (from derris root).

A major advance was made in 1939 with the discovery of the insecticidal properties of DDT (for *d*ichloro-*d*iphenyl-*t*richloroethane) by Paul Müller, a research scientist with the Swiss chemical firm of J. R. Geigy. DDT itself had been known since 1874. After the demonstration during World War II of the value of DDT in the control of typhus (a rickettsial infection transmitted by lice and fleas) and malaria (a protozoan infection of the mosquito that can be transmitted to humans), Müller was awarded the Nobel Prize for Physiology and Medicine in 1948.

DDT was the first of a number of chlorinated hydrocarbons to be developed as pesticides. Another important group of pesticides, the organic phosphates, were developed as by-products of German research on nerve gases about the time of World War II. The herbicides 2,4-D and 2,4,5-T were also discovered (in the U.S.) during World War II [2].

The past 30 years have seen the development of a wide variety of chemical pesticides discussed in the following sections. There has also been progress in the area of nonchemical methods of pest control, some of them quite old compared to the chemical pesticides; these are discussed in a separate section.

IMPORTANT TYPES OF PESTICIDES

Pesticides can be classified in a number of different ways: (1) by their *target* [insecticides, herbicides, fungicides, rodenticides, algaecides, molluscicides, acaricides (miticides), nematocides, etc.]; (2) by their *chemical nature* (natural organic compounds, inorganic compounds, chlorinated hydrocarbons, organophosphates, carbamates, etc.); (3) by their *physical state* (dusts, dissolved solutions, suspended solutions, volatile solids, etc.); or (4) by their *mode of action* (stomach poisons, contact poisons, fumigants, etc.). In this

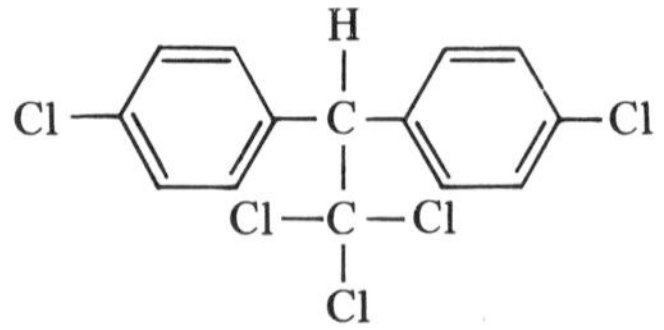

FIG. 12-1. "DDT" or 2,2-bis(*p*-chlorophenyl)-1,1,1-trichloroethane.

section a classification based on the target is adopted, with several subclassifications by other means.

A. INSECTICIDES. Insecticides are usually classified as stomach poisons (including systemic poisons), contact poisons, and fumigants [3]. Stomach poisons are those that are taken internally through the mouth of the insect (and are thus effective against those with biting mouth parts, such as caterpillars) and then absorbed into the body through the digestive tract. Systemic poisons are those that are absorbed into the plant upon application and then ingested by the insect when it eats the plant or its sap. Most of the newer synthetic insecticides are contact poisons, which affect the insect upon contact, entering through the body wall or the respiratory systems and affecting the nerve or respiratory system or the blood stream. Fumigants are gaseous chemicals that kill insects and other pests by entering the respiratory system; these are usually employed in closed spaces.

1. *Stomach poisons.* The most important stomach poison insecticides are the arsenic and fluorine compounds, although several other types are in use. *Arsenicals* are generally arsenites (AsO_2^-) or arsenates (AsO_4^{3-}). Arsenites are easily soluble in water and are very toxic to animals and plants; arsenates are less phytotoxic (toxic to plants) and their poisonous effects generally follow reduction to arsenite. A wide variety of arsenicals are in use today: acid lead arsenate ($PbHAsO_4$), basic lead arsenate [$Pb_4(Pb)H(AsO_4)_3 \cdot H_2O$], calcium arsenate [$Ca_3(AsO_4)_2$], Paris green (discussed earlier), arsenious oxide (As_2O_3), and various other compounds. Arsenic is in the same group of the periodic table as phosphorus and can substitute for phosphorus in certain biochemical reactions with devastating effects, especially since the most common energy source in intermediary metabolic reactions is the P-O-P chemical bond of ATP (adenosine triphosphate) [4]. Arsenic can also combine with —SH groups in enzymes and can cause gross coagulation of proteins [4]. In insects the primary effect of arsenicals is probably the inhibition of respiratory SH enzymes.

Inorganic fluorine compounds have also been used as stomach poison insecticides in certain situations but are not very important today. Major members of this group are cryolite (sodium fluoaluminate, Na_3AlF_6), sodium fluoride (NaF), and sodium and barium fluosilicates (Na_2SiF_6 and $BaSiF_6$). Sodium fluoride is not so suitable as the other compounds because it is toxic to plants; it is also very soluble in water and therefore does not release fluoride as slowly as the other compounds, which are effective over longer periods of time. The fluorides appear to exert their toxic action on pests by complexing with and inhibiting many enzymes that contain metals such as iron, calcium, and magnesium [4].

Other stomach poisons used as insecticides include mercury compounds (such as $HgCl_2$ and $HgCl$), boron compounds (such as borax, $Na_2B_4O_7$, and boric acid, H_3BO_3), antimony compounds, thallium compounds (such as Tl_2SO_4 and CH_3COOTl), yellow phosphorus, and formaldehyde (HCHO).

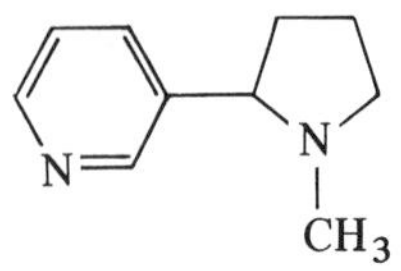

FIG. 12-2.   Nicotine.

2. Contact poisons. Most of the synthetic organic insecticides developed in recent years fall into the category of contact poisons which includes the chlorinated hydrocarbons, the organophosphates, the carbamates, and several other important groups, including the natural organic compounds obtained from plants.

Certain botanical compounds have insecticidal properties and are extracted from plants for use as contact poisons by man. Pyrethrum insecticides, which have been known for several centuries, are extracted from the dried powdered flowers of certain chrysanthemums; they are safe to plants and only slightly toxic to animals but skin contact or inhalation may lead to allergic reactions in certain sensitive persons; they cause rapid paralysis in insects, apparently affecting nerves or muscle, but their mode of action and selectivity to insects are a mystery [4]. Nicotine (see Figure 12-2) is a very toxic heterocyclic amine obtained from the tobacco plant and its relatives, generally marketed as nicotine sulfate. It produces tremors, convulsions, and eventually paralysis in insects; the mean oral lethal dose in adult humans is about 60 mg (1 mg/kg), but a few milligrams may produce severe illness and occasionally death (a typical cigarette might have 15 to 30 mg in its tobacco but only a few milligrams are absorbed from the smoke) [5]. Rotenone (see Figure 12-3), which can also act as a stomach poison, is extracted from the roots of derris and other leguminous plants; it interferes with the biochemistry of oxygen utilization and is very toxic to insects, fish, and pigs but is of low toxicity to most mammals [4]. A number of other organic insecticides are discussed in Reference 3.

Another family of contact poison insecticides, the *chlorinated hydrocarbons* includes DDT (Figure 12-1), its metabolites DDD (also known as TDE) and DDE; BHC (for "benzene hexachloride" although it is really hexachlorocyclohexane); heptachlor; dieldrin; endrin; aldrin; methoxychlor; toxaphene (manufactured by chlorination of camphene compounds); mirex; chlordane; tetradifon; and many others (see Figure 12-4). DDT is character-

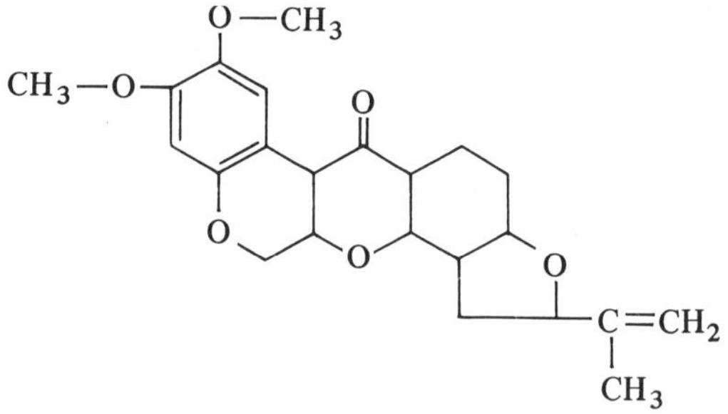

FIG. 12-3.   Rotenone.

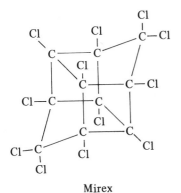

DDD (TDE)

DDE

Methoxychlor

BHC

Mirex

Dieldrin and endrin

FIG. 12-4. Representative chlorinated hydrocarbons. Mirex is a nonplanar molecule resembling a distorted box. Dieldrin and endrin are stereoisomers; i.e., their three-dimensional configurations are different.

ized by a very low vapor pressure (not true of all the chlorinated hydrocarbons) and is very insoluble in water, with a limit of solubility of about 1.2 ppb [4]. DDT is quite soluble in oils, from which it can be absorbed readily through the skin to accumulate in the fatty tissues of the body; it is released slowly when the stored fat is called upon as a source of energy. DDT and other chlorinated hydrocarbons all appear to have their primary effects upon the central nervous system in vertebrates and invertebrates but there is no good understanding of the mechanism of action [4]. In the environment, DDT is degraded or metabolized into DDD, DDE, DDA [(ClC$_6$H$_4$)$_2$CHCOOH], Kelthane [ClC$_6$H$_4$)$_2$C(OH)CCl$_3$], and similar products. Aldrin is known to be converted into dieldrin, an unusually persistent insecticide. The chlorinated hydrocarbons are for the most part persistent broad-spectrum insecticides; i.e., their residues persist in the environment for long periods of time (a few months or years) and they affect a large number of different species. These are the qualities that made them so desirable, but that are also responsible for their undesirable side effects, discussed later in this chapter.

Organophosphorus compounds are the most toxic insecticides, dangerous to insects and mammals alike. Many of them (parathion, thimet, TEPP, tetram, phosdrin, paraoxon, pyrazoxon, HETP, di-syston, and others) are in the "super toxic" category of human poisons for which human fatal doses are less than 5 mg/kg, along with arsenic, cyanides, heroin, strychnine, digitoxin, LSD, and others [5]. As small a total dose as 2 mg of parathion has been known to kill children. Several of these compounds are shown in Figure 12-5. It is widely accepted that the mode of action of organophosphates in vertebrates and invertebrates is the inhibition of the enzyme cholinesterase with consequent disruption of nervous activity caused by accumulation of acetylcholine at the nerve endings [4]. Symptoms include nausea, vomiting, diarrhea, cramps, sweating, salivation, blurred vision, and muscular tremors. In general, organophosphates are more easily biodegraded than the chlorinated hydrocarbons and are thus not as persistent in the environment; however, some of them (such as paraoxon) are quite stable and can persist for years.

A fourth group of contact poisons, the *carbamates*, are derivatives of carbamic acid (NH$_2$COOH), such as carbaryl ("Sevin") shown in Figure 12-6. They are broad-spectrum insecticides with relatively low toxicity to

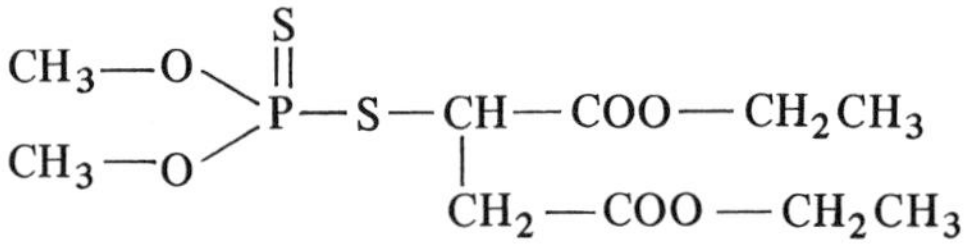

Malathion

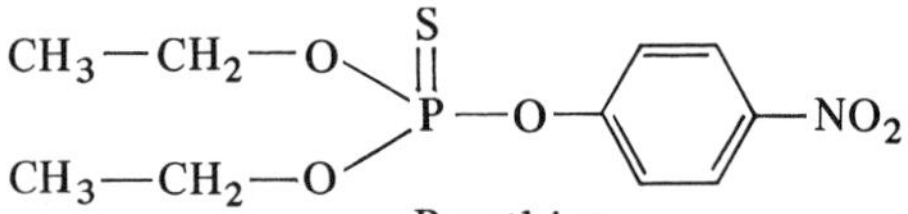

Parathion

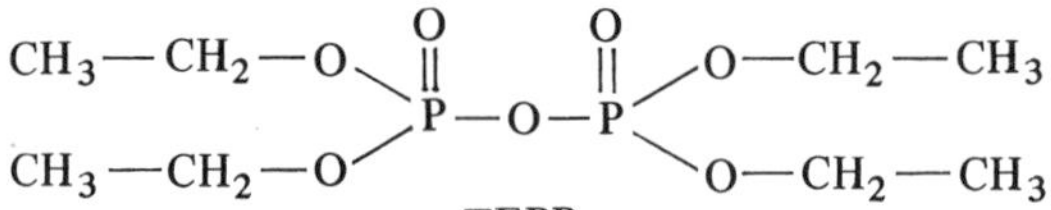

TEPP

FIG. 12-5. Some common organophosphate insecticides.

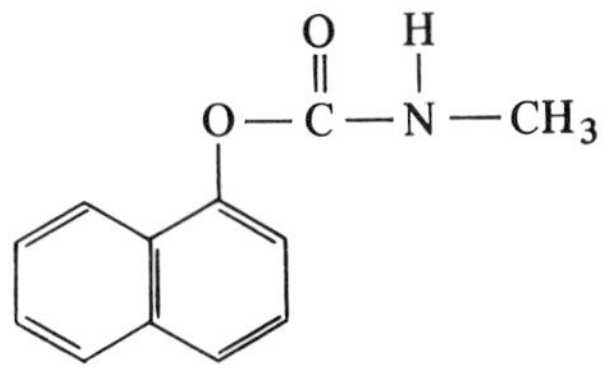

FIG. 12-6. Carbaryl ("Sevin").

mammals and even to some insects. Like the organophosphates, they are inhibitors of cholinesterase and produce similar effects on the nervous system.

3. *Fumigants.* Fumigants are gases (or the vapors from heated liquids or solids) that are used in sealed spaces to kill pests of stored products or nursery stock; sometimes they are used in the open under tents. Popular fumigants include hydrogen cyanide (HCN) released from calcium cyanide (Cyanogas) by atmospheric moisture, methyl bromide (CH_3Br), carbon disulfide (CS_2), carbon tetrachloride (CCl_4), nicotine, naphthalene, and a number of other compounds [3].

B. FUNGICIDES. Fungicides are used to prevent or eradicate fungal diseases of plants. Protective fungicides used to prevent fungi include sulfur, organic mercury compounds (such as methyl mercury used on seeds), and copper compounds. Eradicant fungicides include lime sulfur, organic mercury, formaldehyde, and several dinitro compounds.

C. HERBICIDES. Herbicides are chemicals that kill plants. Nonselective herbicides such as sodium arsenite, sodium chlorate, sulfuric acid, and some oils kill all plants, while selective herbicides kill only certain types of plants, such as ferrous sulfate for dandelions. The most commonly used herbicides have been 2,4-D (for 2,4-dichlorophenoxyacetic acid) and 2,4,5-T (for 2,4,5-trichlorophenoxyacetic acid) and their esters, growth-regulating substances that kill many broad-leaved plants while not injuring many narrow-leaved plants; thus they are valuable weed killers. The human median lethal dose has been estimated at 300 to 700 mg/kg of 2,4-D, and several deaths have resulted from use of the herbicide. Figure 12-7 shows the chemical formulas for these herbicides.

D. RODENTICIDES. Rodenticides are used against rodents such as rats and mice. Some have a direct, immediate action, such as strychnine, sodium fluoroacetate ($CH_2FCOONa$, "Compound 1080"), phosphorus, thallium, and ANTU (alpha-naphthylthiourea). Others are anticoagulants, such as the commonly used warfarin. Of special interest is norbormide, a very selective toxicant for rats that is much less toxic to other animals. The LD_{50} (median lethal dose) of norbormide for rats is 5 to 15 mg/kg taken orally, while for mice it is 100 to 1000 mg/kg and for most lab animals it is well above 1000 mg/kg [5]. This poison seems to constrict the small peripheral blood vessels in rats to an extent that does not occur in other species.

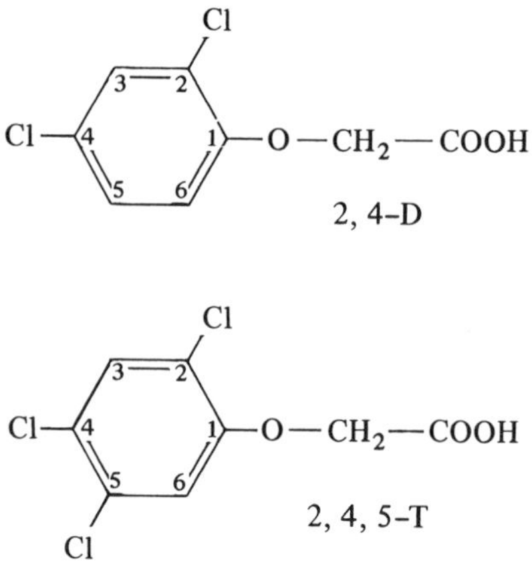

Cl—[benzene ring with positions 3, 2, 4, 5, 6; Cl at position 4, Cl at top]—O—CH₂—COOH

$$Cl \text{—} \underset{5\quad6}{\overset{3\quad2}{\bigcirc}}^{Cl} \text{—} O\text{—}CH_2\text{—}COOH$$

2, 4-D

$$Cl \text{—} \underset{5\quad6}{\overset{3\quad2}{\bigcirc}}^{Cl} \text{—} O\text{—}CH_2\text{—}COOH$$

2, 4, 5-T

FIG. 12-7.   "2,4-D" and "2,4,5-T."   Cl

EFFECTS OF PESTICIDES

The effects of pesticides on pests and other wildlife and on man have been studied extensively during the past 2 or 3 decades [4, 6, 7, 8]. Some are quite desirable but others have proved to be very undesirable and even alarming. In this section we shall review some of the most important questions that have been raised about pesticides.

EFFECTS ON PESTS. Perhaps 1% of the insect species in the U.S. can be considered harmful to man because of their adverse effects on crops, livestock, and poultry. The number of economically important insect pests might be 10,000 and the number of plant species affected 1000, but the 200 major pests cause almost all the damage. About 80% of the insecticides used annually in the U.S. are used against the 100 major insect pests [8]. In parts of the world in which pesticides are not common, agricultural production can be reduced by large amounts; for example, it has been estimated [8] that the fraction of total production lost to insects in Asia is 41% for rice, 18% for cotton, 14% for maize, 13% for vegetables, and 6.5% for wheat, oats, barley, and rye. In the U.S., somewhere between 5 and 15% of total agricultural production is believed to be lost, with the damage estimated at about $5 billion annually [3]. The use of pesticides has reduced this damage by more than the cost of the pesticides, although it is not always certain that the use is limited to quantities that are economically justified.

A number of pests are vectors of human diseases (see Table 8-5), and pesticides have been of great use in controlling some of these diseases. In recent years the majority of the DDT manufactured in the U.S. has gone into public health programs around the world, especially into the campaign against malaria. The World Health Organization estimates that the anti-malaria campaign has saved over 5 million lives and prevented over 100 million illnesses in less than a decade; the annual malaria death rate in India was reduced from 750,000 to 1500. In Ceylon the incidence of malaria in the early 1950's was over 2 million cases a year, but DDT spraying reduced it to almost zero before the use of DDT was stopped in 1964; by 1968 it had risen back to over 1 million and DDT spraying was reinstated despite the realization that wildlife would be adversely affected. In the U.S., deaths from malaria in 1940, 1945, 1950, 1955 and 1960 were 1442, 443, 76, 18, and 7, respectively, while those for tularemia (a bacterial infection of rodents and humans, transmitted by insects) were 189, 122, 15, 9, and 4, respectively [9]. These and similar downtrends were accelerated by the use of DDT and other pesticides, although they had begun earlier with other improvements in public health programs.

One problem that has been encountered with pesticides is the development of resistant strains of pests. An insecticide that kills a large fraction of the population of some insect will after several generations be less effective due to Darwinian selection of the resistant individuals in the population. This phenomenon was noted even with the older pesticides (arsenicals, cyanide, etc.) but has become extremely important with the advent of the synthetic organic pesticides. As early as 1947 it was realized that DDT was losing its effectiveness in controlling houseflies [4]. The DDT resistance of houseflies has been widely studied and is apparently due to the presence in resistant strains of higher concentrations of an enzyme (called DDT-dehydrochlorinase) that dehydrochlorinates the DDT [4]. Exposure to one pesticide sometimes leads to cross resistance—resistance not only to the pesticide used but also to other pesticides, including even chemically dissimilar compounds. Resistance appears to occur for every toxicant and in all parts of the world and is a very well-established phenomenon (see Chapter 15 of Reference 4 for an extensive discussion).

Resistance to rodenticides can also occur, even though the lifetimes of rodents are much longer than those of insects and the selection process is therefore much slower. Great Britain has recently had to cope with the spread of warfarin resistance among rats, and attempts to contain resistant rats within fixed areas have not proved very successful [10]. A related problem, touched on in Chapter 17, is the resistance of bacteria to antibiotics.

Another problem with pesticides has been the harm sometimes caused to beneficial species, such as bees. In addition, pesticides can cause great changes in an ecosystem, sometimes actually increasing the amount of harm caused by a pest. The mathematics of a two-species system, one species being a predator (or parasite) of the other, was worked out by the noted mathematician Volterra, who showed that in the absence of disturbances the numbers of the two groups would fluctuate periodically [11]. Volterra also showed that if for some reason the same proportion of the two species were suddenly destroyed, different fluctuations would result with the average (over time) number of the predators being decreased and the average number of the prey being increased. This has been observed in practice, a pesticide sometimes increasing the average numbers of a pest because some natural enemy has also been affected. Likewise, the reduction of a natural enemy has sometimes increased the numbers of an insect of no economic importance to the point at which it is of economic importance.

EFFECTS ON HUMANS. Every year some persons die because of pesticides, although the number is much less than from other accidental causes or even from other types of poisoning, such as carbon monoxide poisoning (see Chapter 3). In 1956 there were 152 deaths caused by pesticides (94 among children under 9 years of age) in the U.S., of which 104 were due to compounds older than DDT and 35 from newer synthetic pesticides, the rest being unidentified [12]. Almost all the pesticides in use can be fatal in sufficiently large amounts but the organophosphates are the most toxic [5]. While the organophosphate parathion can be lethal at an oral dose of 10 to 20 mg and the mean lethal adult dose is probably 300 mg (roughly 4 mg/kg of body weight), it takes 10 mg/kg of DDT or dieldrin to produce untoward symptoms and the mean lethal oral doses are perhaps 20 to 70 mg/kg for dieldrin and 250 mg/kg for DDT [5].

No exact lethal doses can be fixed, of course, since there are many variables that can affect human response to pesticides—age, sex, diet, state of health, and many others.

Upon occasion, large numbers of persons have died from eating foods contaminated accidentally with pesticides, although no such cases have been reported in the U.S. These cases have included 330 deaths in Turkey from eating seed grain treated with hexachlorobenzene, 80 deaths in Colombia from eating flour contaminated with parathion, and 17 deaths in Mexico from eating sugar containing parathion. In 1969 in the U.S. there was a widely reported incident in which a New Mexico family was poisoned by organic mercury present (as a fungicide) in seed grain fed to hogs that the family had slaughtered for food [13].

Chlorinated hydrocarbons are known to be present in the body fat of human beings, the main source being small amounts present in food supplies. Studies have shown that the daily content of DDT and DDE in meals in the U.S. might be about 0.04 to 0.5 mg, while concentrations of DDT and derived materials in body fat might be 5 to 12 ppm [14]. There have been no detectable effects from these concentrations, and no correlations noted with any disease rates or other adverse health effects. No detectable health effects were noted in volunteers who took 35 mg of DDT daily for 21 months, even though their body fat concentrations increased greatly [14].

The absence of short-term effects in human beings receiving greater-than-average exposure to these compounds is no guarantee of their long-term safety, of course. Persons exposed in their occupation or in other ways to radiation and asbestos, to mention two prominent examples, often showed no detectable effects until after 20 or 30 years had elapsed, sometimes even when exposures did not continue during the interval. Since chlorinated hydrocarbons were first used commercially only a quarter of a century ago, there could very well be important effects that have not yet showed up, although there is no basis at present for expecting such effects. Some disturbing effects on wildlife have been noted, as discussed below.

Long-term effects that might occur could be related to carcinogenic, mutagenic, and teratogenic properties—the abilities to produce cancers, mutations, and congenital *in utero* malformations. Sufficiently large quantities of DDT, dieldrin, aldrin, heptachlor, and mirex have all been found capable of inducing tumors in mice, including malignant tumors [8]. Rotenone and carbaryl, on the other hand, have not been able to induce tumors in mice or rats. Several pesticides have been discovered to be capable of producing mutations or chromosomal aberrations in plants but much more study is needed along these lines [8]. Teratogenicity is also an open question but the tragic experience with thalidomide in the early 1960's suggests the need for more care and testing for this property. It has already been established that 2,4,5-T is teratogenic and fetocidal (capable of killing the fetus) in some mice when administered subcutaneously or orally [15].

Pesticides may be capable of causing other effects. The question of small effects on reproduction has been raised. One interesting medical report [16] dealt with impotence (for up to a year) that occurred in four of five agricultural workers (the fifth was single and was never questioned) who had been exposed to a variety of toxic chemicals, including malathion, dieldrin, and 2,4-D in April and August 1967 when their troubles began; each worker was unaware of his colleagues' impotence, and it was accidental that one doctor became aware of the cases.

The Commission on Pesticides and Their Relationship to Environmental Health summed up their report by stating:

> The field of pesticide toxicology exemplifies the absurdity of a situation in which 200 million Americans are undergoing life-long exposure, yet our knowledge of what is happening to them is at best fragmentary and for the most part indirect and inferential. While there is little ground for forebodings of disaster, there is even less for complacency [8].

EFFECTS ON WILDLIFE. The effects of pesticides on wildlife are better known and often alarming. Pesticides are often concentrated in food chains (see the examples in Chapter 1), leading to levels high enough in birds and fish to damage their health or reproductive abilities. One example has been the killing of robins by DDT used against Dutch elm disease, the pesticide being concentrated in the chain from leaf to litter to earthworms (to which apparently no harm is done) to robins. The dangers to birds and mammals of acute pesticide poisoning (as opposed to the more subtle effects discussed below) are not known but there is evidence that it could be important [17]. In an experiment with Japanese quail, it was found that when diets of from 2 to 250 ppm dieldrin were fed, deaths occurred over a period of time but tended to occur earlier with the higher-dosage diets; brain residues of dieldrin correlated well with death, typically being 4 or 5 ppm and higher [17].

Chlorinated hydrocarbons have been implicated in the decline of several bird populations through interference with their reproduction [18, 19]. The famous peregrine falcon, which survived the warfare of farmers, hunters, and egg collectors over the centuries, began a sudden, drastic decline in population in both Europe and North America in the early 1950's, and similar trends occurred with grebes, woodcock, bald eagles, ospreys (fish hawks), kestrels (sparrow hawks), and other species at about the same time. The cause was related to a drop in the reproduction of the birds, rather than to their killing. Breeding was delayed or no eggs were even laid; the eggs that were laid were thin and easily broken; and fledgling mortality was high [19]. Reproductive failure was greatest in areas in which persistent pesticides, especially the chlorinated hydrocarbons, were used most widely.

During the 1960's it was demonstrated that of the many possible factors (diseases, parasites, predators, killing by humans, or environmental pollutants) that might be involved only the concentration of chlorinated hydrocarbons by these raptors (carnivorous birds, standing at the top of long food chains) could explain their simultaneous decline in numerous parts of the world. Sublethal concentrations of chlorinated hydrocarbon residues are now known to be capable of interfering with the calcium metabolism of the birds, perhaps through inhibition of the enzyme carbonic anhydrase. As a result, insufficient calcium is made available during the process of formation of the eggshell (which typically occurs in the last 20 h before laying), and the eggshells are very thin. (The calcium in the eggshells is present as $CaCO_3$.) This occurs even when the pesticide is injected into the birds very shortly before egg laying and is therefore not due to an insufficient supply of stored calcium in the body [19, 20]. The administration of pesticide early in the breeding cycle has another effect—delay of egg laying; this is believed to result from inhibition of liver enzymes by the pesticides with accompanying depression of estrogen levels in the blood. It is important to realize that these are not mere hypotheses but have actually been observed in experimental laboratories [20, 21].

The decrease in the thickness and weight of the eggshells of raptors occurred immediately after the introduction of chlorinated hydrocarbons into general usage. Several studies have been made of museum collections of eggs obtained over a period of many decades, and they all show the same trend (see Figure 12-8): a rather stable mean thickness and weight until the late 1940's and early 1950's, when a sharp drop occurred, well correlated with the introduction of the chlorinated hydrocarbons [22, 23].

The chlorinated hydrocarbon pesticides are not the only chemicals that might be having adverse effects on wildlife. The polychlorinated biphenyls ("PCB") used as plasticizers and in the manufacture of paints, resins, electrical insulators, and other products are now being widely distributed over the world and are being concentrated in the food chains of marine ecosystems [24]. In one experiment it was found that mallard ducklings fed PCB in the diet for 10 days at levels from 25 to 100 ppm suffered no apparent clinical

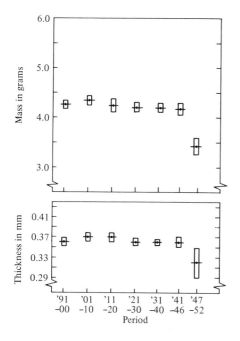

FIG. 12-8. Measurements of the mass and thickness of 614 California peregrine falcon eggshells collected since 1891. The solid horizontal bars mark the mean values and the rectangles mark the 95% confidence limits for the mean. From Hickey, J. J., and D. W. Anderson, "Chlorinated Hydrocarbons and Eggshell Changes in Raptorial and Fish-Eating Birds." *Science* **162**:271–273 (11 October 1968). (*Copyright 1968 by the American Association for the Advancement of Science.*)

intoxication but when challenged with duck hepatitis virus they suffered significantly higher mortality than unexposed birds [25]. The importance of these compounds in natural ecosystems remains to be established.

Fish also exhibit acute responses to pesticides. One notable example, still a matter of some controversy, was the series of major fish kills that began in the late 1950's along the Mississippi and Atchafalaya Rivers in Louisiana [26]. Catfish and other fish exhibited hemorrhaging and death in great numbers despite the fact that the water did not give unusual measurements of parameters such as pH, dissolved oxygen, or temperature, and there was no evidence of pathogens. Investigations that attempted to reproduce the symptoms with viruses, toxic metals, etc., were fruitless until it was discovered that chloroform extracts of bottom sediments would reproduce them exactly. Eventually traces of endrin were identified as the source of the intoxication. It was known that catfish in the vicinity exhibited acute effects when endrin was sprayed over the Louisiana cane sugar fields; although this spraying seemed to be a logical source for the endrin, levees were keeping the cane field water out of the river systems in question and the fish kills in the rivers were not occurring at or near spraying time. Federal officials identified the source as the endrin plant of a chemical company at Memphis, Tenn., several hundred kilometers upstream, although the company denied being responsible; when the company stopped the discharges, however, the fish kills ended [26].

The question of sublethal effects in fish from pesticide contamination has not been investigated very much but behavioral changes are believed to occur. One experiment on brook trout indicated that DDT affected the central nervous system and interfered with responses to temperature [27].

The possibility that chlorinated hydrocarbon pesticides may affect photosynthesis and growth in marine phytoplankton has also been raised by experiments [28, 29]. It has been found that DDT residues present in marine phytoplankton collected in Monterey Bay, Calif., increased by three times in the period from 1955 to 1969 [30].

DEFOLIATION. Herbicides used to control plants can have both desirable and undesirable effects. The most common herbicides, 2,4-D and 2,4,5-T and related compounds, are selective against woody vegetation and are extensively used against unwanted trees, shrubs, and woody vines. Use of excessively volatile herbicides has often killed or injured vegetation in neighboring areas. The use of 2,4-D by corn growers and the highway department in western Iowa caused grape production in vineyards in the area to drop 96% during the 1960's, for example. Many homeowners have had valued plants harmed after their next-door neighbors used herbicides. On the other hand, use of herbicides to control woody plants can be ecologically beneficial in increasing the productivity of rangeland and pastureland [31].

The most widely debated example of defoliation in recent years— because of its moral and political implications—has been the U.S. Army's herbicide program in South Vietnam [32, 33, 34]. The program was experimental in 1961 and operational in 1962, after which it grew rapidly to a peak in 1967, declined over the next few years, and was terminated in 1971.

The chief herbicides used against forests were "Agent Orange," a mixture of 2,4-D and 2,4,5-T, and "Agent White," a mixture of 2,4-D and picloram; the herbicide most commonly used against crops was "Agent Blue," a solution of cacodylic acid, an arsenic compound [34]. Herbicides used against croplands constituted about 10% of the total. Some 2000 km² of land have been sprayed to destroy crops, causing the loss of enough food to feed 600,000 persons for a year [34]. In addition, significant quantities of herbicides seem to be transported by the winds over broad areas of cropland, producing adverse psychological impacts on the population in addition to defoliation [33].

The Herbicide Assessment Commission of the American Association for the Advancement of Science, which visited Vietnam in 1970, discovered that one-fifth to one-half of South Vietnam's mangrove forests—some 1400 km²—had been destroyed, the entire plant community generally being killed outright by a single spraying [34]. For unknown reasons, no new plant life is present several years after spraying, except for worthless shrubs and undesirable ferns. Restoration of the forests will apparently take at least several decades.

About 35% of South Vietnam's 57,000 km² of tropical hardwood forests have been sprayed one or more times, killing merchantable timber equal to the country's current demand for 31 years. Damaged forest lands are now being invaded by tenacious bamboo and grasses that may take over the area for many decades [34]. (Another result of the war is that shrapnel in the timber has become a serious problem for the Vietnamese lumber industry, causing sawmills to lose 1 to 3 h daily due to damage of saw blades [33].)

The ecological effects of the military defoliation program have been extensive and generally harmful, and many Vietnamese—especially the Montagnard tribesmen of the Central Highlands—have been deprived of food. Many ecologists and other scientists have argued that this program is much more inhumane than it would appear to be on the surface since the civilian population will be adversely affected for decades to come.

ALTERNATIVE METHODS OF PEST CONTROL

A number of alternative pest control methods have been developed or are being developed that may involve less environmental pollution than the chemical pesticides presently in use. The Agricultural Research Service of the U.S. Department of Agriculture has been heavily involved in the investigation of alternatives to chemical pesticides since the early 1950's, largely through the efforts of E. F. Knipling, director of the A.R.S. Entomology Research Division [35]. This is in contrast to industrial research groups, which have emphasized the development of new but less hazardous chemical pesticides for the past quarter century.

The alternative methods, some of them rather old, generally involve biological controls of various types or selective chemicals [36–39].

AUTOCIDE. Insect sterility was developed through the efforts of E. F. Knipling, who in the 1930's had theorized that the screwworm fly so troublesome to livestock in the southern United States might be eliminated by raising large numbers of flies in the laboratory, sterilizing them, and then releasing them (during the season in which the screwworm fly population was at its lowest) to mate with the natural population—without offspring [40]. The noted geneticist H. J. Muller suggested the use of X rays to sterilize the laboratory flies. The success of the method was proved in 1954, when it was used to eliminate the screwworm fly from the island of Curaçao, chosen because it is 80 km from other land and therefore insulated from invasion by migrating flies. In 1958, an 18-month campaign was begun to eradicate the fly from the southeastern part of the U.S.; it proved successful after the release by airplane of a total of 2.75 billion sterilized flies. Since then the screwworm fly has been suppressed in the Southwest, and sterile flies are released from time to time at a buffer zone along the Mexican border and wherever the flies reappear [38].

The same technique is being used along the California-Mexico border to control the Mexican fruit fly and is being tested in California and Arizona against the pink bollworm [38]. It has also been used against other fruit flies and may be developed for use against the cabbage looper, the boll weevil, the corn earworm, and several other insect pests [39].

While ordinary chemical pesticides are most useful when the pest population is high and have their effectiveness reduced when the pest population has been decreased (especially if some resistance is occurring), autocide is most useful when the pest population is very low and is therefore often used after chemical pesticides or natural causes have reduced the population. The method also requires careful mass rearing and sterilization in the laboratory.

CHEMOSTERILIZATION. A similar method involves the sterilization of natural pest populations by application of a chemosterilant, leading to a reduced reproduction rate. It has sometimes been difficult to get the pests to take the chemosterilant or the effect on reproduction might be only temporary; these constitute disadvantages of this method. One study [41] has suggested that the feral pigeon common in urban areas could be controlled by the use of synthetic grit containing the synthetic steroid mestranol, which is capable of inhibiting reproduction in birds.

SEX ATTRACTANTS. Chemicals known as pheromones are secreted into the environment by animals and elicit specific reactions in receiving individuals of the same species [42]. Sex attractants offer some promise as pest control techniques since they may be used with traps baited with insecticides or may be used to permeate the atmosphere to prevent males from locating females. The first sex attractant was isolated from the female gypsy moth in 1960 and several hundred have since been identified, some very complex molecules (*cis*-7,8-epoxy-2-methyloctadecane for the gypsy moth [43]) and some rather simple (phenol for the grass grub beetle [44]). A synthetic lure, methyl eugenol, was used to eradicate the oriental fruit fly on the island of Rota in the Marianas; the attractant was combined with an insecticide on pieces of fiberboard distributed by airplane [39]. Sex attractants are being tested for a number of insects.

JUVENILE HORMONES. Juvenile hormones and ecdysones are internal secretions of insects that regulate their growth and metamorphosis from larva to pupa to adult. These and related compounds offer another promising approach to pest control [45]. Although they are vitally necessary for proper development at certain stages in the insect's life cycle, they must be absent at other periods or else abnormal development will occur—eggs fail to hatch or immature insects die without reproducing. Exposure to these hormones (or to synthetic compounds of similar effect) at the appropriate time can thus aid in controlling the insect. These compounds appear to have no effect on other types of life and therefore may not have dangerous effects on humans, fish, and wildlife, although this has not been proved. They have the important advantage that insects cannot develop resistance to them since they are essential to the insect's life cycle [45].

It is believed that the juvenile hormone of each insect is a specific chemical compound, and isolation and identification of these compounds is being accomplished. Figure 12-9(a) shows the chemical structure of the juvenile hormone of the Cecropia moth (the adult stage of the silkworm). Synthetic compounds with identical activity, and therefore just as useful as the true juvenile hormone, have also been discovered. Although these affect only insects, as far as is known, they generally affect beneficial or harmless insects as well and hence have the disadvantages inherent in any broad-spectrum pesticide [45].

The possibility remains that chemical compounds may be found that affect only particular insect species. One such compound was discovered at Harvard University by Karel Slama and Carroll M. Williams. Slama had brought some European linden bugs (*Pyrrhocoris apterus*) from Czecho-slovakia but it was impossible to raise them to the normal, sexually mature adult stage; the larvae continued to grow as larvae or molted into adult-like forms with larval characteristics. These difficulties had not arisen in Europe and the source of the juvenile hormone interfering with proper development was traced to paper toweling in the rearing jars, occurring only when North American paper was used. The hormone originated in the balsam fir, the principal source of paper pulp in North America, and the active compound responsible is shown in Figure 12-9(b). This "paper factor" interferes with the development of only the one family of insects, Pyrrhocoridae, which contains some important cotton pests, and may thus be a useful pesticide [45].

Evergreen trees like the balsam fir synthesize and emit many different terpenoid materials (see Chapter 3), for no apparent reason, but the "paper factor" and other compounds may have been evolved as a defense against natural insect enemies of the trees and may include useful pesticides [45].

NATURAL PREDATORS AND PARASITES. As early as 1888 the ladybug beetle *Rodolia cardinalis* was imported and used successfully to control the cottony-cushion scale of citrus plants. In Mexico the citrus black fly is being

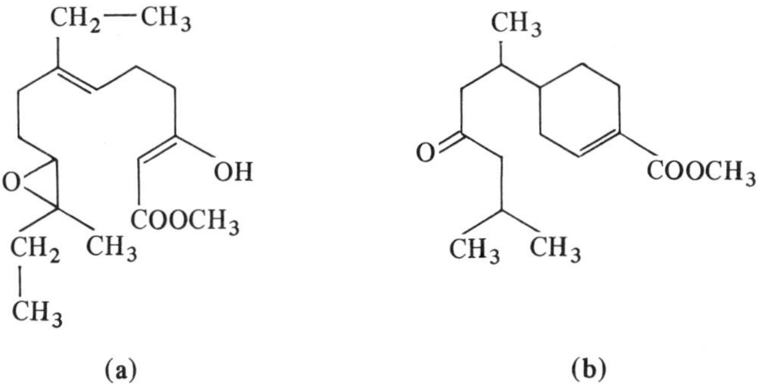

(a)   (b)

FIG. 12-9. (a) The juvenile hormone of the Cecropia moth [45]. (b) The "paper factor" found in balsam fir, a potent juvenile hormone for the insect family Pyrrhocoridae [45].

controlled by the release and establishment of hymenopterous parasites. The establishment of several parasites of the spotted alfalfa aphid, together with the development of resistant species of alfalfa, may constitute a permanent solution to the alfalfa aphid problem [38].

About 520 different species of predators and parasites have been deliberately imported into the U.S.; 115 have become established; and 20 of these provide significant control of destructive pests [39].

Bacteria and viruses that affect insects offer another possibility for control. The insect pathogen *Bacillus thuringiensis* was identified in 1927 and has been used against a variety of insect pests, including the cabbage looper, the diamondback moth, the alfalfa caterpillar, the corn borer, and the tobacco hornworm. Its toxin has been isolated and bacterial toxins may be even more useful as insecticides than the pathogens, at least if commercially economical synthesis becomes possible. Several viruses have also been developed for use against such pests as the corn earworm (also called the cotton bollworm) but since there has been no precedent for registration of such a pesticide, the U.S.D.A. and the F.D.A. are proceeding cautiously before giving approval, even though no effects seem to occur in animals or humans when they are exposed to these viruses [38].

RESISTANT CROP VARIETIES. These are yet another alternative to chemical pesticides. The outstanding example is wheat resistant to the Hessian fly. The first such variety was introduced in 1942, and some 22 now exist. Although occasionally one variety will lose its effectiveness against the Hessian fly in some area, the problem is not so serious as the development of insect resistance to chemical pesticides, and another resistant variety can be substituted [38]. A variety of alfalfa that resists the spotted alfalfa aphid has also proved successful, as mentioned above in the discussion of parasites. A number of other agricultural crops and trees with insect resistance have been developed and released. The juvenile hormone produced by the balsam fir mentioned above affords protection against an insect that is now extinct or has learned to avoid the balsam fir entirely [45].

The development of a resistant variety of crop may take 10 or 20 years but the benefits can be enormous since the use of pesticides is then avoided year after year for long periods of time. Even partial resistance, reducing (but not eliminating) the need for pesticides, can be valuable.

CULTURAL PRACTICES. Used by farmers before more sophisticated pesticide methods, cultural practices can be of great help in reducing the damage caused by pests. Included are proper sanitation, destruction of crop residues, deep plowing, rotation of crops and animals on the land, planting of crops in strips, and proper timing of planting to escape major pest damage. For example, the corn rootworm is easily controlled by crop rotation—avoiding planting corn on the same acreage 2 years in a row. These are difficult to use as effectively today as in the past since U.S. agriculture is based on the monoculture system, with each crop and livestock produced in the region of the country best suited as far as soil and climate are concerned [39].

QUARANTINE. Another practice that has been helpful in preventing invasions of exotic pests and diseases is quarantine. U.S.D.A. inspectors at ports of entry have intercepted Mediterranean fruit flies as many as 148 times in 1 year but there have been four invasions by the "Medfly" in 14 years, one of them requiring $10 million to eradicate [39].

INTEGRATED CONTROL PROGRAMS. Programs for important pests are being developed by entomologists. They involve a variety of pest control methods in proper sequence at the proper times to make the best use of natural controls and create the least hazard to humans, wildlife, and non-target species [39]. The cost of developing and using such a program is generally regarded as rather high but perhaps only because the benefits to the environment are being neglected.

CONCLUSION

The major pesticides in use today are the synthetic chemical pesticides developed in the past quarter century. Although they have been of great help in reducing the damage caused by pests that affect agricultural crops and livestock and are vectors of human disease, they have often had undesirable effects on beneficial species and they have often lost their effectiveness as pest resistance has evolved. (Projections of the use of specific insecticides generally involve information about the development of insect resistance year by year [8].)

The question of the value of different levels of pesticide use is an important one but one which is difficult to resolve at present [9]. Studies have shown that the use of pesticides is economically justified in many situations but in some cases farmers may be overusing pesticides; i.e., their total benefits may be greater than their total costs but their marginal benefits may be less than their marginal costs. Farmers generally obtain their information about pesticides from the chemical dealer (rather than from agricultural extension agents) and are sometimes clearly oversold on the use of particular chemicals [46, 47].

From the point of view of society as a whole, it is also necessary to ask which uses are most important and which are frivolous. Disease control would certainly be a high priority use of pesticides; perhaps we should continue DDT usage in malaria control programs, although we should actively search out safer substitutes. Agricultural uses probably rank somewhat lower in importance—at least in the developed countries with perennial agricultural surpluses. The importance of keeping roadsides free of weeds is much harder to appreciate, however, especially for someone who prefers natural roadsides to unnecessarily formal and well-mowed highway rights-of-way.

A number of alternative pest control methods are available but most have not been afforded the amount of research and development that chemical pesticides have enjoyed from chemical firms. Some of these methods have (or could have) some of the same disadvantages as chemical pesticides but others might not. These alternative methods should be given high priority, with public funding if necessary. These methods are already being investigated by companies such as the Zoecon Corporation, which is developing the economic potential of hormones and other selective chemicals.

In the meantime, if chemical pesticides are used, it is desirable to replace broad-spectrum, persistent pesticides by those that are more selective and less persistent in the environment whenever possible. Some of the latter pesticides (such as the organophosphates) are more toxic and it is clearly necessary to exercise great care in their application and storage.

It would also be wise to ask whether or not we are being a little too fastidious in some of our eating habits and thereby making use of pesticides to an excessive extent. No one would be pleased to learn about rodent hairs or other remains in his or her jam because of the eating habits of rodents and their ability to transmit diseases to human beings. But the presence of bits of larvae that fed on the fruit cannot reasonably be objected to except out of fastidiousness. Insects that feed on fruits and other crops relished by man can also serve as food. Moses, in Chapter 11 of Leviticus, permitted some insects to be eaten by the Israelites, although others were forbidden. John the Baptist lived upon locusts and wild honey in the desert. The ancient Greeks and modern American Indians relished cicadae. Roman epicures ate beetle larvae. Today, some people eat snails, grasshoppers, ants, wasp grubs, and such creatures that other people regard as disgusting.

Why not eat insects [48]? Table 12-1 gives a grasshopper recipe to start the reader on a new culinary adventure.

TABLE 12-1. Recipe for common large grasshoppers [48].

---

1. Pluck off heads, legs, and wings.
2. Sprinkle with pepper, salt, and chopped parsley.
3. Fry in butter with a little vinegar.

REFERENCES

1. Friend, Roger B., "Insect Control by Chemicals." *American Scientist* **40**:136 (1952).
2. Hamner, Charles L., and H. B. Tukey, "The Herbicidal Action of 2,4 Dichlorophenoxyacetic and 2,4,5 Trichlorophenoxyacetic Acid on Bindweed." *Science* **100**:154–155 (1944).

3. Davidson, R. H., and L. M. Peairs, *Insect Pests of Farm, Garden, and Orchard,* 6th ed. New York: John Wiley & Sons, Inc., 1966.

4. O'Brien, R. D., *Insecticides. Action and Metabolism.* New York: Academic Press, 1967.

5. Gleason, M. N., et al., *Clinical Toxicology of Commercial Products—Acute Poisoning,* 3rd ed. Baltimore: Williams & Wilkins Co., 1969.

6. Kraybill, H. F. (ed.), "Biological Effects of Pesticides in Mammalian Systems. A conference, New York, May 1967." *Annals of the New York Academy of Sciences* **160**:1–422 (1969).

7. Miller, M. W., and G. C. Berg (eds.), *Chemical Fallout.* Springfield, Ill.: C. C. Thomas, Pub., 1969.

8. *Report of the Secretary's Commission on Pesticides and Their Relationship to Environmental Health.* Washington, D.C.: U.S. Dept. of Health, Education, and Welfare, December 1969.

9. Headley, J. C., and J. N. Lewis, *The Pesticide Problem: An Economic Approach to Public Policy.* Washington, D.C.: Resources for the Future, 1967.

10. Greaves, J. H., "Warfarin-Resistant Rats in Britain." *Agricultural Science Review* **8**(1):35–38 (1970).

11. Volterra, Vito, *Leçons sur la Théorie Mathématique de la Lutte pour la Vie* [*Lessons on the Mathematical Theory of the Struggle for Survival*]. Paris: Gauthier-Villars, 1931.

12. *Cleaning Our Environment. The Chemical Basis for Action.* Washington, D.C.: American Chemical Society, 1969. p. 230.

13. Curley, A., et al., "Organic Mercury Identified as the Cause of Poisoning in Humans and Hogs." *Science* **172**:65–67 (1971).

14. Hayes, W. J., Jr., "Monitoring Food and People for Pesticide Content." In *Scientific Aspects of Pest Control.* Washington, D.C.: National Academy of Sciences—National Research Council, 1966. Publication 1402. pp. 314–342.

15. Courtney, K. D., et al., "Teratogenic Evaluation of 2,4,5-T." *Science* **168**: 864–866 (1970).

16. Espir, M. L. E., et al., "Impotence in Farm Workers Using Toxic Chemicals." *British Medical Journal* **1**:423–425 (1970).

17. Stickel, W. H., et al., "Tissue Residues of Dieldrin in Relation to Mortality in Birds and Mammals." In Reference 7, pp. 174–204.

18. Wurster, C. F., Jr., "Chlorinated Hydrocarbon Insecticides and Avian Reproduction: How Are They Related?" In Reference 7, pp. 368–389.

19. Peakall, David B., "Pesticides and the Reproduction of Birds." *Scientific American* **222**(4):72–78 (April 1970).

20. Peakall, David B., "*p,p'*-DDT: Effect on Calcium Metabolism and Concentration of Estradiol in the Blood." *Science* **168**:592–594 (1970).

21. Porter, R. D., and S. N. Wiemeyer, "Dieldrin and DDT: Effects on Sparrow Hawk Eggshells and Reproduction." *Science* **165**:199–200 (1969).

22. Hickey, J. J., and D. W. Anderson, "Chlorinated Hydrocarbons and Eggshell Changes in Raptorial and Fish-Eating Birds." *Science* **162**:271–273 (1968).

23. Ratcliffe, D. A., "Decrease in Eggshell Weight in Certain Birds of Prey." *Nature* **215**:208–210 (1967).

24. Risebrough, R. W., "Chlorinated Hydrocarbons in Marine Ecosystems." In Reference 7, pp. 5–23.

25. Friend, M., and D. O. Trainer, "Polychlorinated Biphenyl: Interaction with Duck Hepatitis Virus." *Science* **170**:1314–1316 (1970).
26. Graham, Frank, Jr., *Since Silent Spring*. Boston: Houghton Mifflin Co., 1970.
27. Anderson, J. M., and M. R. Peterson, "DDT: Sublethal Effects on Brook Trout Nervous System." *Science* **164**:440–441 (1969).
28. Wurster, C. F., Jr., "DDT Reduces Photosynthesis by Marine Phytoplankton." *Science* **159**:1474–1475 (1968).
29. Menzel, D. W., et al., "Marine Phytoplankton Vary in Their Response to Chlorinated Hydrocarbons." *Science* **167**:1724–1726 (1970).
30. Cox, J. L., "DDT Residues in Marine Phytoplankton: Increase from 1955 to 1969." *Science* **170**:71–73 (1970).
31. Barrons, Keith C., "Some Ecological Benefits of Woody Plant Control with Herbicides." *Science* **165**:465–468 (1969).
32. Tschirley, Fred H., "Defoliation in Vietnam." *Science* **163**:779–786 (1969).
33. Orians, G. H., and E. W. Pfeiffer, "Ecological Effects of the War in Vietnam." *Science* **168**:544–554 (1970).
34. Boffey, Philip M., "Herbicides in Vietnam: AAAS Study Finds Widespread Devastation." *Science* **171**:43–47 (1971).
35. Kramer, Joel R., "Pesticide Research: Industry, USDA Pursue Different Paths." *Science* **166**:1383–1386 (1969).
36. Ordish, George, *Biological Methods in Crop Pest Control*. London: Constable & Co., Ltd., 1967.
37. Kilgore, Wendell W., and Richard L. Doutt (eds.), *Pest Control. Biological, Physical, and Selected Chemical Methods*. New York: Academic Press, 1967.
38. Holcomb, Robert W., "Insect Control: Alternatives to the Use of Conventional Pesticides." *Science* **168**:456–458 (1970).
39. Irving, George W., Jr., "Agricultural Pest Control and the Environment." *Science* **168**:1419–1424 (1970).
40. Knipling, E. F., "The Eradication of the Screw-Worm Fly." *Scientific American* **203**(4):54–61 (October 1960).
41. Sturtevant, Joan, "Pigeon Control by Chemosterilization: Population Model from Laboratory Results." *Science* **170**:322–324 (1970).
42. Shorey, H. H., and L. K. Gaston, "Pheromones." In Reference 37, pp. 241–265.
43. Bierl, B. A., et al., "Potent Sex Attractant of the Gypsy Moth: Its Isolation, Identification, and Synthesis." *Science* **170**:87–89 (1970).
44. Henzell, R. F., and M. D. Lowe, "Sex Attractant of the Grass Grub Beetle." *Science* **168**:1005–1006 (1970).
45. Williams, Carroll M., "Third-Generation Pesticides." *Scientific American* **217**(1):13–17 (July 1967).
46. Van den Bosch, R., "The Toxicity Problem—Comments by an Applied Insect Ecologist." In Reference 7, pp. 97–112.
47. Shea, Kevin P., "Cottons and Chemicals." *Scientist and Citizen* **10**(9):209–219 (November 1968).
48. Holt, Vincent M., *Why Not Eat Insects?* Published in 1885, reprinted in 1967 by E. W. Classey Ltd., Hampton, England.

13

SOLID WASTES

Keep an old colander over a pan or pail and throw the refuse as it accumulates in this, turning it into a separate pail as the colander fills and after it has stood long enough to drain thoroughly. At night it may be placed on old tins, kept for the purpose, and dried in the oven. Every morning, after breakfast is over and before the fire is to be prepared for the day's work, the dried accumulations of the previous day may be safely burned, all drafts being opened. This, if rightly done, is an excellent sanitary measure. In a well-ordered kitchen, it should be as great a disgrace to have sour and vile-smelling refuse about as to send rancid butter or moldy bread into the dining room. —*New York Times*, February 25, 1894, p. 18.

Today the disposal of garbage and other solid wastes seems much simpler: urban dwellers merely place them in bags or other containers, put these in garbage cans, and have them picked up by a public or private collection agency. Only when one passes an open dump or sees a landfill might the real magnitude of the U.S. solid waste problem become apparent. This chapter discusses some of the characteristics of solid wastes, the quantities produced, and present and proposed disposal methods, with special consideration of avoidance of solid wastes.

TYPES, CHARACTERISTICS, AND QUANTITIES OF SOLID WASTES

Information about solid wastes has been comparatively scanty for a number of reasons. The need for this information was not recognized a number of years ago and many communities made no weight or other measurements. Measurements are difficult to make because of the complexity of the material, and techniques for sampling and making laboratory measurements have not been standardized. The quantities and character of solid wastes also vary greatly with geographical location, food patterns, etc. Data collection has begun to improve, however, with the holding of national conferences [1, 2] and with the growth of the solid wastes program of the Public Health Service and now of the Environmental Protection Agency.

A typical classification of solid wastes is the following [3, 4]:

1. Garbage—putrescible (decomposable) wastes from food, slaughter-houses, canning and freezing industries, etc.

2. Rubbish—nonputrescible wastes, either combustible or noncombustible. Combustible wastes would include paper, wood, cloth, rubber, leather, and garden wastes. Noncombustibles would include metals, glass, ceramics, stones, dirt, masonry, and some chemicals.

3. Ashes—residues (such as cinders and fly ash) of the combustion of solid fuels for heating and cooking or the incineration of solid wastes by municipal, industrial, and apartment house incinerators.

4. Large wastes—demolition and construction rubble (pipes, lumber, masonry, brick, plastic, roofing and insulating materials), automobiles, furniture, refrigerators and other home appliances, trees, tires, etc.

5. Dead animals—household pets, birds, rodents, zoo animals, etc. There are also anatomical and pathological wastes from hospitals.

6. Sewage treatment process solids—screenings, settled solids, sludge.

7. Industrial solid wastes—chemicals, paints, sand, explosives, etc.

8. Mining wastes—"tailings," slag heaps, culm piles at coal mines, etc.

9. Agricultural wastes—farm animal manure, crop residues, etc.

Some of these wastes are the result of control techniques for air and water pollution, such as fly ash recovered from flue gases and sludge from sewage treatment plants. Others are capable of producing air or water pollution quite readily, such as organic matter that can decay to produce odors or farm animal wastes that can enter water supplies. Much of the solid waste produced in the U.S. is collected or otherwise disposed of but much is permitted to litter the landscape—highway litter, dog wastes in cities, discarded oil drums on the Alaskan tundra [5], etc.

The sources of solid wastes are (1) municipal—street sweepings, sewage treatment plant wastes, wastes from schools and other institutions; (2) domestic—garbage, rubbish, and occasional large wastes from homes (and, in years past, large amounts of ashes from the use of coal); (3) commercial—from stores and offices; (4) industrial—from manufacturing plants; (5) mining—from coal mining, strip mining, etc.; and (6) agricultural.

Solid wastes could be measured in terms of their mass or their volume. Their densities generally vary over a wide range—typically 0.05 to 1.0 g/cm^3—and mass (or weight) measurements are felt to be more useful. Solid wastes are rather low in density because of the large amounts of air present in them and it is usually not difficult to compact them to several times their original density, as is done in sanitary landfill operations or by home compactors.

Several estimates have been made in recent years of the total amount of solid wastes generated in the United States but the most accurate are probably those of the National Solid Waste Survey for 1968 being compiled by the federal government. An interim report [6] gave the total amounts and per capita daily amounts shown in Table 13-1. More recent information [7] suggests that mineral wastes may total over 1.5 billion metric tons and agricultural wastes over 2 billion metric tons but these are quantities that are rather difficult to define. Household, commercial, municipal, and industrial solid wastes total 326 million metric tons annually, which amounts to 4.5 kg (10 lb) per capita per day. The breakdown of the average solid waste actually collected is shown in Table 13-2, which does not include solid wastes disposed of by the household or industry that generates the waste. Several decades ago the rule of thumb figure for refuse collected in urban areas was about 2 lb (almost 1 kg) per person per day but the present figure is already almost three times as much and is expected to be four times as much by 1980 [7].

TABLE 13-1. Total solid wastes produced in the U.S. annually and daily per capita production [6]. Uncollected wastes are those disposed of by the producer rather than collected by a public or private agency.

Type of waste	Total mass (10^6 metric tons/year)	Per capita (kg/day)
Household, commercial, municipal	**226**	**3.1**
Collected	172	2.4
Uncollected	54	0.7
Industrial	**100**	**1.4**
Mineral	**1000**	**14**
Agricultural	**1860**	**25**
Farm animal wastes	1360	19
Crop residues	500	7
Total (rounded)	**3200**	**44**

TABLE 13-2. Average solid waste *collected* in the U.S. in kilograms per day per person according to the 1968 National Solid Wastes Survey [6]. The "household and commercial" category is estimated to be about three-fourths household and one-fourth commercial.

	Urban	Rural	National
Household and commercial	1.97	1.56	1.88
Industrial	0.29	0.17	0.27
Demolition and construction	0.10	0.01	0.08
Street and alley	0.05	0.01	0.04
Miscellaneous	0.17	0.04	0.14
Total	**2.59**	**1.78**	**2.41**

The composition of collected refuse has changed over the years, with the quantity of ashes decreasing as coal heating of households has been replaced by gas, oil, and electric heating and with the amount of paper increasing dramatically along with the growth in consumer packaging (discussed in more detail later in this chapter). The annual average municipal refuse composition in New York in 1939 included 43% ashes, 22% paper, 17% garbage, 6.8% metal, and 5.5% glass [3], while a study [8] of a number of midwestern communities in the early 1960's showed 42% paper, 12% garbage, 10% ashes, 8% metal, and 6% glass; the remaining percentages in each case were largely wood, grass, leaves, and dirt. The 1960's undoubtedly produced further changes from the growth of packaging, especially non-returnable beverage containers of metal and glass.

CURRENT DISPOSAL METHODS

A number of different collection and disposal methods are in use in the United States and will be discussed in this section. Fairly recent data are available from the National Solid Wastes Survey [6]. At present, the approximate fractions of household, commercial, and municipal wastes actually collected that are disposed of by various methods are:

77% in open dumps (including landfills covered less frequently than once a day).

13% in sanitary landfills.

8% in municipal incinerators.

2% by miscellaneous methods (composting, hog feeding, ocean dumping, salvage operations).

HOG FEEDING. At one time, hogs were extensively used for garbage disposal. In colonial times hogs wandering in the streets served as scavengers of the wastes in the gutters. As urban areas developed, hog farms nearby found it desirable to collect urban garbage for feeding to hogs. A 1941 survey [9] of the 412 U.S. cities with population of 25,000 or more in the 1940 census found that 2 million metric tons of the 7 million metric tons of garbage produced and collected in those cities was used for feeding hogs. Experiments showed that a gain of 100-kg live weight could result from the feeding of 6 metric tons of garbage, and that the resulting pork would amount to 55 kg. Complete use of all U.S. urban garbage for feeding hogs would have provided pork equivalent to only about 2% of 1940 U.S. production, however, so that this use was not a major threat to other sources of hog feed.

It was known that the feeding of raw garbage to hogs led to high incidences of trichinosis in humans, but the practice continued until the mid-1950's. During the years 1953 to 1955 the rapid spread of a virus disease of hogs, vesicular exanthema, largely through raw garbage, led the U.S.

Public Health Service and state health departments to institute regulations forbidding the feeding of raw garbage to hogs. Although heat treatment (such as exposing the raw garbage to live steam for a half hour or more) cured the problem, the feeding of garbage to hogs decreased in the late 1950's and 1960's and is presently insignificant.

OPEN DUMPS. Disposing of solid wastes in open dumps is the most common method used in the U.S. Much of the uncollected refuse is disposed of privately in a similar manner. Open dumps produce health and air pollution problems and are not an acceptable method of disposal. They can cause public health problems by encouraging the growth of populations of flies (which can transmit typhoid fever, cholera, dysentery, tuberculosis, anthrax, and other diseases), rats (which can transmit plague, murine typhus fever, leptospirosis, rabies, rickettsialpox, and other diseases), cockroaches, mosquitoes (which can transmit malaria, yellow fever, dengue, mosquito-borne encephalitis, and filariasis), and other pests [10, 11]. Air pollution problems arise when the dumped wastes are burned in order to reduce their volume and conserve space or when spontaneous combustion or arson leads to fires.

SANITARY LANDFILLS. The sanitary landfill became common after World War II, although its origins date back to pre-World War I days, at least for the disposal of garbage. In sanitary landfill operations, refuse is spread in thin layers that are compacted by heavy bulldozers before another layer is spread. After the refuse is perhaps 3 m deep, it is covered by a thin layer of clean earth, which is again compacted. At the end of the day, the fill is topped with up to another meter of compacted earth. (In some landfills, which are not properly called sanitary landfills, covering is less frequent than once a day.) Figure 13-1 shows the cross-sectional areas of two types of sanitary landfills.

The advantages of a sanitary landfill, as opposed to an open, uncovered dump, are (1) the public health problems are minimized because flies, rats, and other pests are unable to breed in the covered refuse; (2) there is no air pollution from burning and none from dust or odors; and (3) fire hazards are very small. There is danger of groundwater or surface water pollution, however, if the landfill site is improperly chosen or if it is dug out too deep.

Inside the filled region, there can be aerobic reactions to the extent that oxygen is present and anaerobic reactions (with production of methane, H_2S, CO_2, and water), both involving bacterial action. The inside temperature may rise to 70°C in a few days from the heat produced. The amounts of air and water present influence the speed of decomposition, with wet landfills sometimes showing rather complete decomposition of organic matter within a few years, while dry landfills show little decomposition after several decades. Gas production can sometimes be dangerous. In the early 1960's, in Chicago, methane gas produced in a 20-year-old landfill, which was normally vented

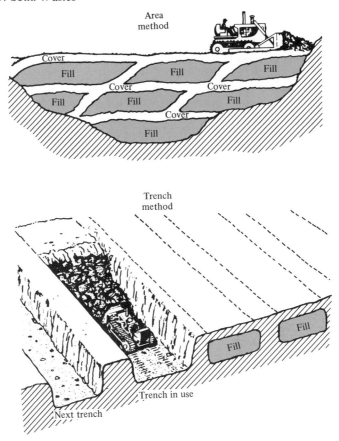

FIG. 13-1. Cross-sectional views of area and trench-type sanitary landfills. (*Courtesy of Caterpillar Tractor Co.*)

harmlessly into the atmosphere, was trapped underground by several feet of snow, forcing it to enter a sewer system where it caused an explosion. Methane has also been trapped by buildings built on landfills.

Landfill sites are sometimes regarded as valuable (but not by persons who preferred the site in its natural state). San Diego County, Calif., has converted swamps, marshes, used-out gravel pits, and canyons into parks, playgrounds, parking areas, airports, and other facilities [12]. Difficulties may arise from methane production or from settling of the land. A swampy pit in the East Bronx of New York was used in 1959 as the site of a number of three-story brick row houses that began to settle and show cracks within 6 months, apparently because the piles had not been driven deep enough (leading to settling) and the foundations had not been properly placed on the piles (leading to cracking) [13]. The residents had to leave, losing their investments, and the Building Commissioner obtained a court order to have the buildings demolished as unsafe.

The rule of thumb for the amount of land required in a sanitary landfill to dispose of urban wastes is 1 acre (0.4 hectare) per year for each 10,000 persons, assuming the refuse is filled in to a depth of about 3 m. This amount of land may be difficult to find. In the mid-1960's the nine-county San Francisco Bay Area in Calif. was disposing of 2.6 million metric tons of refuse annually in 77 disposal sites of from less than 1 hectare to about 200 hectares; 38 sites, with about 20% of the disposal volume, were publicly owned and the others were privately owned [14]. The available volume was estimated to suffice until about 1977, and another 28,000 hectare-m would be needed by the year 2000. Some 64% of the land being used for refuse disposal was adjacent to San Francisco Bay—36% in the form of tidelands (surrounded by impervious dikes) and 28% classified as marshlands. Other large cities are also short of disposal space; New York's landfill sites will suffice until about 1975. Choices of new landfill sites are likely to encounter opposition from landowners and conservationists, to judge from recent experience.

INCINERATION. There are several hundred municipal incinerators in use in the U.S., accounting for about half the tonnage burned, and there are thousands of small, privately owned trash burners [7]. True incineration involves the burning of solid wastes at high temperatures, to leave ashes, glass, metal, and unburned combustibles amounting to perhaps one-fourth the original weight, which must then be disposed of in a landfill or other dump. Air pollution is often a problem, and New York has had to legislate upgrading of incinerators in apartment buildings because of their emissions into the atmosphere. Incinerator technology has been developing in recent years, with air pollution control a particular concern.

OCEAN DUMPING. Some coastal cities dump solid wastes into the ocean. New York formerly dumped wastes off the New Jersey shore but the State of New Jersey in 1933 won a Supreme Court decision forbidding this practice. New York still dumps wastes over an area several square kilometers in size 20 km out into the Atlantic Ocean; this area has been described by critics as a "dead sea" of muck and black goo, largely because of the sewage sludge disposed there.

COMPOSTING. Another interesting idea for the utilization of municipal refuse is composting, which is practiced on a large scale in some European countries [15, 16] but has not been tried extensively in the U.S. Composting involves fermentation of refuse into a product, compost, which supplies valuable humus for the soil. The composting is generally accomplished by heaping the refuse and moistening it, then letting it ferment for about 6 months (the decomposition is faster if the refuse is first ground up into smaller particles) [17]. The fermentation takes place at 50 to 80°C, which is apparently too high for pathogens to survive so that disease is not of concern. Sewage sludge can also be added to the compostable material.

Compost is valuable as a soil conditioner, as no artificial product is capable of adding humus to the soil. Its fertilizer value is very low—typically it contains 0.5% nitrogen, 0.4% phosphorus, and 0.2% potassium [16]. A small market for compost has been developed in Europe but it is in intensive, luxury-type agriculture (bulb and flower growing, hillside vineyards, etc.) and not in basic agriculture. Composting is most common in The Netherlands, where it is used to dispose of about one-sixth of all municipal refuse.

Occasionally American business firms have tried composting municipal refuse to produce a marketable soil conditioner but the lack of a ready market and the competition of agricultural chemicals have prevented their success. American solid wastes, with less organic matter and more glass, pottery, plastics, metals, rubber, leather, etc., than European solid wastes, would seem less profitable to compost than the European solid wastes. Nevertheless, a number of composting plants have recently been built in the United States. Even though composting may not prove profitable for private enterprise, municipalities may find it to be economically competitive with other disposal methods, in addition to being more satisfactory to conservationists and environmentalists.

SALVAGE. A little salvaging of solid wastes still takes place in this country but it accounts for only a small percentage of the total weight of all solid wastes produced. Some "recycling," or salvaging of materials before they become solid wastes, also takes place, as discussed in more detail later in this chapter.

THE COSTS OF SOLID WASTES

The social costs of solid wastes are not easy to determine. Open dumps may be responsible for a small amount of disease and incinerators add to the air pollution costs of a community. Most disposal methods involve some aesthetic displeasure. In rare circumstances the costs can be much higher. Some 145 persons, mostly children, were killed at Aberfan, Wales, on October 21, 1966, when an avalanche of 2 million tons of rain-soaked coal wastes slipped 150 m down a mountainside and buried a school and over a dozen cottages. The slag heap was 120 m high and had been begun in 1870; the villagers had made complaints in 1964 to the effect that the heap was dangerous and should be investigated but the National Coal Board had taken no action.

Litter along American roadsides constitutes a solid waste problem with considerable psychological impact, judging from the interest in antilitter and highway beautification programs. Occasionally there are massive efforts to pick up litter along highways, generally resulting in the recovery of several hundred kilograms per kilometer. Since there are approximately 5 million km of roads in this country, highway litter might amount to roughly one million metric tons at present—a small but important part of the solid waste problem, and one which is generally believed to be growing.

The major cost of solid waste disposal in the U.S. at present comes from the collection, transportation, and transfer of the wastes, not from the final stages of disposal. The National Solid Wastes Survey [6] found that the average per capita annual expense for all communities was $6.81, of which $5.39 was for collection ($0.53 capital costs and $4.86 operating costs) and $1.42 for disposal ($0.25 capital costs and $1.17 operating costs). This survey showed that 80% of total expense is for collection and only 20% for final disposal, which demonstrates our emphasis on collection services and our neglect of proper disposal practices.

The disposal costs of various methods, per metric ton of refuse disposed of, have been estimated as [4, 6, 18]:

11¢ to 28¢ for open dumps.

28¢ to 55¢ for controlled burning dumps.

40¢ to 83¢ for ordinary refuse filling.

77¢ to $1.65 for sanitary landfills, and

$3.30 to $6.60 for incineration.

These figures are subject to change as technology changes and as the cost of land for sanitary landfills increases.

The 1970 cost of solid waste management programs in the U.S. was estimated [19] by the Council on Environmental Quality as $5.7 billion, or about $27 per person annually. The Council estimated that in order to satisfy federal environmental standards the 1975 annualized costs (costs for amortization of capital investment, interest, operation, and maintenance) would have to be $7.8 billion, or 37% greater. Such costs would thus amount to much less than 1% of gross national product.

PACKAGING

One solid waste that is worth looking at in closer detail is packaging. The Midwest Research Institute recently made a careful study [20] of the packaging industry over the period from 1958 to 1966 and made projections of its growth through 1976. Per capita annual consumption of packaging materials rose from 183 kg (404 lb) in 1958 to 238 kg (525 lb) in 1966 and is projected to rise to 300 kg (661 lb) in 1976. In 1966 packaging cost the American public $25 billion (3.4% of the gross national product)—$16 billion for the packaging materials themselves and $9 billion for value added by the packaging manufacturer, plus about $200 million for machinery to use in the manufacturing. Of the 1966 total of 23.5 million metric tons of packaging materials, about 90% were discarded instead of being reused, accounting for over 13% of the residential, commercial, and industrial wastes produced in the U.S. [20].

There are several reasons for the phenomenal growth of the packaging industry. One is the continuing rise in self-service merchandising, which requires packages that will help to sell the product by themselves. Another is the use of paper and plastic packages for small items (batteries, razor blades, picture-hanging hooks, and countless others—look over any grocery or department store) that might otherwise be easily shoplifted. Another is the trend toward the use of nonreturnable bottles and cans for beverages.

Glass containers produced for U.S. consumption totaled 20.2 billion in 1958 and 29.4 billion in 1966, with a breakdown in 1966 as follows [20]:

Returnable beverage containers	2.7 billion
Nonreturnable beverage containers	9.3 billion
Food products	10.8 billion
Drugs and cosmetics	5.8 billion
Industrial and household chemicals	0.8 billion
Total	29.4 billion

The actual number of items sold in glass containers totaled an astounding 71.8 billion—about 1/day/person. With 26.7 billion (29.4 billion less 2.7 billion) being in nonreturnable containers, it will be noted that returnable containers were sold a total of 45.1 billion times—they accounted for 63% of the fillings despite the fact that only 9% of the glass containers manufactured were returnable. This is due to the fact that the average returnable glass container was used 19 times over an average lifetime of almost exactly 1 year. This explains why the glass industry would like to drive out the use of the returnable bottle—about 2½ times as many bottles will have to be produced even if the number of fillings does not increase.

The number of nonreturnable bottles produced has been rising sharply in recent years, bringing higher profits to leading glass manufacturers such as Owens-Illinois and Anchor-Hocking. Figure 13-2 shows the growth in the number of glass bottles produced in the U.S. in the last 2 decades, divided into returnable beverage (beer and soft drink) bottles, nonreturnable beverage bottles (including wine and liquor bottles), and all other bottles (which are almost all nonreturnable, of course). The major growth has evidently been in nonreturnable beverage bottles.

The same is true of other nonreturnable beverage containers. Table 13-3 shows the number of beer and soft-drink containers made and used in 1958 and 1966 and the projections for 1976 [20]. By 1969 the total number of such containers had already reached 43.8 billion [7]. Milk containers show a similar trend, with returnable glass containers being insignificant in numbers today, having been displaced by nonreturnable paperboard and plastic containers.

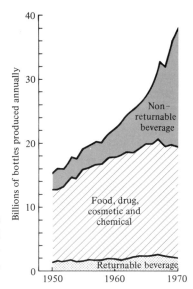

FIG. 13-2. Glass bottle production in the U.S. during the 1950's and 1960's, classified according to returnable beverage bottles, non-returnable beverage bottles (soft drink, beer, wine, and liquor), and all other types [21].

TABLE 13-3. Production and use of beer and soft-drink containers in 1958 and 1966 and projections for 1976 [20]. Entries (except those in the last row) are in billions.

Type	1958	1966	1976
Nonreturnable bottles			
soft drink	0.192	1.980	13.5
beer	1.239	5.031	8.6
Nonreturnable cans			
soft drink	0.409	5.612	17
beer	8.337	12.947	19
Nonreturnable total	**10.177**	**25.570**	**58.1**
Returnable bottles			
soft drink	1.240	1.922	1.20
beer	0.388	0.577	0.46
Returnable total	**1.628**	**2.499**	**1.66**
Total containers	**11.805**	**28.069**	**59.76**
Total fillings	52.921	65.213	79.5
Fillings per container	4.48	2.32	1.33

The types and amounts of packaging consumed in 1958 and 1966 and the projections for 1976 are shown in Table 13-4. These materials accounted for 13% of municipal refuse in 1966 and are projected to account for 21% by 1976. Their composition and nature will greatly affect the collection and disposal of municipal solid wastes. Most packaging materials are not very degradable. Paper, the most degradable of the major categories, has been known to persist unchanged for over 60 years in landfills. The tin in "tin cans" is a thin coating that protects the steel from corrosion and provides a surface that is attractive and easily soldered and decorated; the tin accounts for only 0.5% of the weight of the can but its use makes the steel unsalvageable. Some wastepaper is presently salvaged for reuse in paper and paperboard production; although the amount reused rose from 6.6 million metric tons in 1946 to 8.0 million in 1956 and 9.25 million in 1966, the fiber content of paper products derived from wastepaper decreased from 35% in 1946 to 26% in 1956 to 21% in 1966.

TABLE 13-4. Packaging material consumption (in millions of metric tons) in the U.S. in 1958 and 1966 with projections to 1976 [20]. Per capita figures in kilograms are given in parentheses.

	1958		1966		1976	
Paper and paperboard	30.0	(85.9)	45.5	(115.9)	67.0	(150.6)
Metal	11.4	(32.7)	13.0	(32.9)	15.3	(34.3)
Glass	10.8	(30.7)	14.9	(37.9)	21.6	(48.5)
Plastics	0.6	(1.9)	2.0	(5.1)	5.7	(12.8)
Wood	6.5	(18.7)	7.3	(18.7)	8.0	(18.0)
Textile bags	0.5	(1.5)	0.5	(1.2)	0.3	(0.6)
Miscellaneous [a]	4.3	(12.1)	10.4	(26.4)	15.5	(34.9)
Total	64.1	(183.4)	93.6	(238.1)	133.4	(299.8)

[a] Cushioning materials, coatings and applied materials, pallets and skids, etc.

JUNKED AUTOMOBILES

Another highly visible solid waste problem that has attracted a great deal of interest in recent years is the problem of junked automobiles. Figure 13-3 shows the annual number of motor vehicles scrapped, which has risen from a level of 2 to 3 million annually in the 1930's and late 1940's to over 7 million in 1965 [22]. Most of these vehicles are reprocessed, although it has been estimated that the number of junked automobiles still present in the environment may be 15 or 20 million [23].

Auto wreckers generally take scrapped vehicles, remove the parts that can be used or rebuilt for reuse, burn the hulk to get rid of upholstery and other burnable material (creating air pollution problems), and convert the remaining metal to scrap by baling, shearing, or shredding. The scrap metal is usually compressed into "No. 2 bundles," which may have typically 0.5% copper, an undesirable impurity as far as the steel industry (which can tolerate only about 0.1% copper) is concerned. Large shredders costing several hundred thousand dollars exist that can be used to shred the hulks. Magnetic separators can then be used to remove the ferrous metals, producing a higher-grade scrap. This is an expensive process and the costs of transporting an old automobile hulk to the shredder can be prohibitive. The economics of getting rid of old automobiles is such that the junk car dealer usually

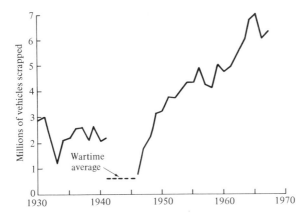

FIG. 13-3. Motor vehicles scrapped annually in the U.S. [22].

cannot afford to pay anything for the car or may not even want it at all. As a consequence, large cities have had to cope with increasing numbers of vehicles that have simply been abandoned by their owners. In New York the number of abandoned cars that had to be towed away rose from 2500 in 1960 to 21,943 in 1965 to 72,961 in 1970.

Various suggestions have been made to alleviate the problem. Cars could be designed with less copper content (by using aluminum wiring) or with the copper easier to remove. Steel mills could be designed or redesigned to accept automobile scrap. New uses could be developed for automobile scrap. All new car owners could be charged a disposal fee (of say $25) that could be redeemed when the car was properly disposed of or simply used to pay general disposal costs. (This could also be applied to other products, and the City Club of New York has urged a fee of $0.02/lb on all nonfood items to pay for the eventual disposal costs.)

Automobiles are not the only vehicles that are junked. The number of junked airplanes in the nation's airports has led government officials to request cleaning up operations. On the west side of Staten Island in New York, there is a 200-acre tract owned by the Witte Marine Equipment Company for the disposal of ship scrap. The company acquires surplus tonnage, dismantles it, and sells most of it as scrap; until the depressed scrap market of the 1960's hit the company, it produced up to 50,000 tons of scrap annually. The site has been described as a "vast panorama of rotting hulks, splintering barges, piles of rusting anchor flukes and stacks of twisted steel" [24].

BETTER SOLID WASTE MANAGEMENT

In recent years new and better solid waste management methods have been suggested and/or developed [4]. During the 1960's the federal government began to fund solid waste research and development through the Bureau of Mines in the Department of the Interior and the Bureau of Solid Waste Management in the Department of Health, Education, and Welfare, and this work is now being supervised by the Environmental Protection Agency. A number of ideas are discussed in this section.

UTILIZATION. In some cases solid wastes can be put to direct use [25]. Fly and bottom ash from industrial installations now total about 30 million metric tons annually, of which only some 5 to 10% is used commercially, largely as a cement substitute in concrete for dams, highways, and other major construction. This is much less than in some foreign countries; for example, some 50% of the fly ash in France and 40% in Great Britain finds commercial application and does not need to be treated as a solid waste. New uses are being developed for fly ash in the U.S. and other countries: to make brick, to dewater industrial waste water sludge, as a land cover, etc. [26–28].

Another use for ordinary municipal solid wastes would be the production of heat or electricity during incineration. A U.S. company is developing a fluidized-bed incinerator which burns solid wastes at high pressure producing hot gases which power a turbine and generate electricity [7]. Municipal solid wastes have a heating value (heat per unit mass) about one-third that of coal, and a typical American community might be able to obtain 10% of its electric power requirements by burning its own refuse. Several European cities have such facilities, and Montreal has recently put into operation an incinerator (see Figure 13-4) that burns about 1100 metric tons of refuse daily and uses the heat to produce steam [29]. St. Louis, Mo., is shredding part of its municipal refuse and adding it to pulverized coal fueling the Meramec plant of the Union Electric Co.

RECOVERY. In some circumstances, valuable materials can be recovered from solid wastes, lessening disposal problems and financial costs [25]. Over the years, salvaging of materials from solid wastes has been practiced at times, abandoned when it became uneconomical, and sometimes revived. Items that have had resale value at some time include rags, newspapers, cardboard, bottles, rubber, tin cans, ferrous metals, nonferrous metals, and glass [30]. At one time some cities (such as Washington, D.C., in the 1920's) passed their refuse onto a continuously moving belt from which salvageable materials could be picked off. Increasing labor costs, depressed scrap markets, and a general trend toward lack of concern about the possible waste of natural resources has made recovery rare. In recent years there have been suggestions for at least partial recovery using magnetic separation, screening, gravity separation, and other methods in addition to hand sorting [4, 25].

RECYCLING. During World War II many cities collected flattened tin cans from residents and shipped them to detinneries for recovery of tin and

FIG. 13-4. An aerial view of the Montreal des Carriers plant at which municipal refuse is incinerated to produce steam. (*Courtesy of Research-Cottrell, Inc.*)

ferrous metal [30]. Waste paper drives by Boy Scouts, the Salvation Army, and other groups—including environmental action groups in recent years—have kept paper out of municipal solid wastes, permitting it to be reused. These are examples of recycling—the salvage of valuable products that are never permitted to become a solid waste. Aluminum, bottling, and can companies or groups have set up "recycling centers" in some large communities, paying typically $0.10/lb ($220/metric ton) for aluminum (all-aluminum cans, frozen food trays, etc.), $0.015/lb ($33/metric ton) for metal cans, $0.01/lb ($22/metric ton) or more for used bottles, etc. Newspapers still bring a few dollars a ton (the price varies with time and with geographical location), and each ton corresponds to the wood from about 17 trees. The U.S. presently recycles about one-fifth of its paper, while Japan recycles one-half. In the U.S. in the mid-1960's the fractions of annual consumption that were derived from secondary (recycled) metal were about 20% for aluminum, 42% for copper, 45% for iron and steel, and 25% for zinc [31]. About 12% of the total consumption of new rubber was reclaimed rubber, which amounted to 265,000 metric tons in 1966. About 30% of the 100 million motor vehicle tires discarded annually in the U.S. are reclaimed, and tire companies are developing plants to recover carbon black, gas, oils, and other products by destructive distillation of old tires.

Recycling conserves natural resources but in too many cases it is not

FIG. 13-5. Recycling of paper into paperboard. **(a)** Newspapers collected in Boy Scout paper drives are carried by a forklift truck to a hydro-pulper to begin the cleaning and reconditioning process. **(b)** In the hydro-pulper the paper stock is diluted with water. **(c)** The paper fibers are reconditioned by screening, steaming, and heat procedures. The paper stock is thickened when it forms on the cylinders and the water drains through the screen mesh. **(d)** The wastepaper is eventually made into new paperboard ready to be cut and folded into cartons. (*Courtesy of the Container Corporation of America.*)

economical (according to our customary standards of what is economical) to recycle materials compared to the exploitation of virgin resources. There is no reason why government policies that have encouraged the exploitation of natural resources might not be changed to encourage recycling instead [32].

AVOIDANCE OF SOLID WASTES. The real reason for the phenomenal growth of solid waste tonnages in the United States has been the trend to the "throw-away" society: throw-away bottles, throw-away cans, disposable diapers, disposable medical supplies, increased packaging, planned obsolescence, etc. Until people realize that apparent conveniences involve less apparent inconveniences, present and future, and espouse a program of environmental stewardship, the solid waste problem will continue to increase. The U.S. badly needs efforts and talents directed at reversing the trend toward needless consumption.

REFERENCES

1. *Proceedings of the National Conference on Solid Waste Research, December 1963*. Chicago, Ill.: American Public Works Assn., 1964.
2. *Proceedings of the National Conference on Solid Wastes Management, April 1966*. Davis, Calif.: University of California, 1966.
3. Rogus, C. A., "Refuse Quantities and Characteristics" in Reference 1.
4. Engdahl, R. B., et al., *Solid Waste Processing. A State-of-the-Art Report on Unit Operations and Processes*. Washington, D.C.: U.S. Dept. of Health, Education, and Welfare, 1969. Public Health Service Publication No. 1856.
5. "Oil Drums Litter Coast in Alaska." *N.Y. Times,* January 3, 1971, p. 39.
6. Black, Ralph J., et al., *The National Solid Wastes Survey: An Interim Report*. Washington, D.C.: U.S. Dept. of Health, Education, and Welfare, 1968.
7. *Environmental Quality. The First Annual Report of the Council on Environmental Quality*. Washington, D.C.: U.S. Govt. Printing Office, August 1970.
8. Bell, John M., "Characteristics of Municipal Refuse" in Reference 1.
9. Stolting, W. H., *Food Waste Materials. A Survey of Urban Garbage Production, Collection, and Utilization*. Washington, D.C.: U.S. Dept. of Agriculture, 1941.
10. Anderson, R. J., "Public Health Aspects of the Solid Waste Problem" in Reference 1.
11. Hanks, T. G., *Solid Waste/Disease Relationships. A Literature Survey*. Cincinnati, Ohio: U.S. Dept. of Health, Education, and Welfare, 1967. Public Health Service Publication No. 999-UIH-6.
12. Goode, C. S., "Utilization of Sanitary Landfill Sites" in Reference 1.
13. Roberts, Steven V., "Residents Leave 10 Sinking Houses." *N.Y. Times,* December 1, 1966, p. 49; Anonymous, "Errors in Sinking Bronx Homes Laid to 2 by City Inquiry Board." *N.Y. Times,* April 9, 1967, p. 68.
14. Hickey, J. H., "The Problem in Detail" in Reference 2.
15. Jensen, Michael E., *Observations of Continental European Solid Waste Management Practices*. Washington, D.C.: U.S. Dept. of Health, Education, and Welfare, 1969. Public Health Service Publication No. 1880.

16. Hart, Samuel A., *Solid Waste Management/Composting. European Activity and American Potential.* Cincinnati, Ohio: U.S. Dept. of Health, Education, and Welfare, 1968. Public Health Service Publication No. 1826.
17. Weststrate, I., "Composting of City Refuse" in Reference 1.
18. Black, Ralph J., "Sanitary Landfills" in Reference 1.
19. *Environmental Quality. The Second Annual Report of the Council on Environmental Quality.* Washington, D.C.: U.S. Govt. Printing Office, August 1971.
20. Darnay, Arsen, and William E. Franklin, *The Role of Packaging in Solid Waste Management 1966 to 1976.* Washington, D.C.: U.S. Dept. of Health, Education, and Welfare, 1969. Public Health Service Publication No. 1855.
21. *Glass Containers.* Published annually by the Glass Container Manufacturers Institute, Inc., 330 Madison Ave., New York, N.Y. 10017.
22. *Automobile Facts and Figures.* Published annually by the Automobile Manufacturers Assn., Inc., 320 New Center Building, Detroit, Mich. 48202.
23. U.S. Bureau of Mines, *Automobile Disposal. A National Problem.* Washington, D.C.: U.S. Dept. of the Interior, 1967.
24. Bamberger, Werner, "Ships Become Scrap on S.I.'s 'Dark Side'." *N.Y. Times,* April 12, 1970, p. 90.
25. Drobny, N. L., H. E. Hull, and R. F. Testin, *Recovery and Utilization of Municipal Solid Waste.* Washington, D.C.: U.S. Environmental Protection Agency, 1971. Public Health Service Publication No. 1908.
26. "Fly Ash Utilization Climbing Steadily." *Environ. Sci. Tech.* **4:**187–189 (1970).
27. Rippon, J., and E. G. Barber, "PFA—a 20th Century By-Product." *Science Journal* **6**(8):74–79 (1970).
28. *Fly Ash Utilization. A Summary of Applications and Technology.* Washington, D.C.: U.S. Govt. Printing Office, 1970.
29. Malin, H. Martin, Jr., "Plants Burn Garbage, Produce Steam." *Environ. Sci. Tech.* **5:**207–209 (1971).
30. Institute for Solid Wastes of American Public Works Assn., *Municipal Refuse Disposal.* Chicago, Ill.: Public Administration Service, 1970.
31. *Cleaning Our Environment. The Chemical Basis for Action.* Washington, D.C.: American Chemical Society, 1969.
32. Grinstead, Robert R., "The New Resource." *Environment* **12**(10):2–17 (December 1970).

14

THERMAL POLLUTION

When the temperature of the water is slowly raised to 45°C, the circulation stops forever. The force of the vital action is conquered. The plant dies. The most favorable temperature for the life and circulation of Chara *(the freshwater algae* Nitella flexilis*) appears to be between 12 and 25°C; below and above these two limits the life and circulation of* Chara *exist only through the power of the vital action, which always finally ends by being conquered.* —Translated from MM. Dutrochet and Becquerel, "Observation sur le *Chara flexilis." Comptes rendus* **5:**775–784 (1837).

It has long been known that plants and animals thrive best in certain temperature ranges and that changes in the temperature of a body of water will affect the types and numbers of organisms in the aquatic ecosystem. The use of river and lake water for industrial cooling purposes in the U.S. can raise the temperature of the water enough to produce major changes in the ecosystems. In some cases the water has become so hot that fish are unable to live in it.

In this chapter we shall discuss the limitations imposed on the efficiencies of steam-electric generating plants by the second law of thermodynamics and point up the magnitude of the thermal pollution problem in the U.S. today and in the following few decades. Some effects of thermal pollution will be considered and methods of using waste heat or dispersing it will be discussed.

THE SECOND LAW OF THERMODYNAMICS

According to the second law of thermodynamics (thermodynamics is the branch of physics dealing with the properties of heat), no process that converts heat into mechanical work can be completely efficient and any such process must waste a certain fraction of the heat. A device that converts heat into mechanical work is known as a "heat engine," and its efficiency is defined as the ratio of the work obtained to the heat extracted for conversion into work. It is a consequence of the laws of thermodynamics that the efficiency of a heat engine cannot be greater than (and is actually always less than) that of a particular idealized heat engine known as the Carnot engine, which takes the working substance of the engine through a particular combination of reversible isothermal and adiabatic processes. The efficiency of a Carnot engine operating between a lower temperature T_L and an upper temperature T_H is $e = 1 - (T_L/T_H)$ where both temperatures are measured with respect to absolute zero [1].

As an example of a heat engine it is instructive to consider the actual operation of a steam-electric generating plant, the type that produces over 80% of the electricity generated in the U.S. Figure 14-1 shows (schematically!) the various components of a fossil-fuel electric plant of the type that burns coal, fuel oil, or natural gas. The purpose of the plant is to convert the heat produced by the combustion of the fossil fuel into mechanical work, which is then converted into electrical energy. (No economically competitive way to convert heat directly into electricity exists at present.) The steps involved are the following:

1. Heat is produced by combustion of the fuel and used to convert the water in the boiler into steam; some of the heat unavoidably escapes so this step is not 100% efficient by any means. The steam is produced in a boiler at very high pressures (typically 1.2×10^7 to 2.4×10^7 N/m², or 120 to 240 atm) and heated to very high temperatures (typically 540 to 590°C).

2. The steam produced is passed through the blades of a turbine, where it is permitted to expand adiabatically. An expanding gas does work on its environment, and the steam causes the turbine blades to rotate and transmit power to the turbine shaft. At the same time the steam cools.

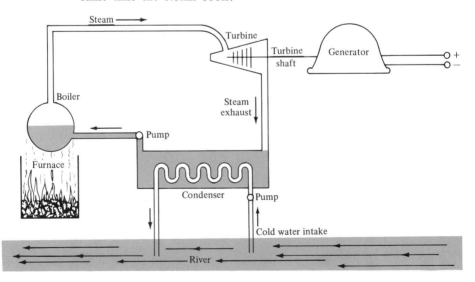

FIG. 14-1. Schematic plan of a steam-electric plant.

3. The power transmitted to the turbine shaft is used to drive an electric generator and thereby produce electrical energy, usually with an efficiency above 95% (for this step alone). Actually a voltage is developed across the terminals, and the power can be transmitted by high-voltage overhead or underground lines to homes, commercial establishments, and factories.

4. The "spent" steam leaving the turbine travels to the condenser, where it is cooled and condensed. The purpose of the condensation is to reduce the volume of the working substance (the water) so it can be reheated under pressure in the boiler to permit the whole cycle to be repeated. The simplest method of cooling the steam is to pass cold water from a river, lake, or ocean through pipes in the condenser so that heat can be exchanged between the working substance and the cooling water, which is then exhausted back into the water source. The elevated temperature of the returned cooling water constitutes "thermal pollution," at least to the extent to which it produces undesirable effects.

Although some electricity has been produced, part of the heat eventually was lost to the cooling water. The total efficiency of the plant is the amount of electrical energy produced divided by the amount of heat produced by combustion. If the temperature of the steam leaving the boiler is 580°C (853 K) and the temperature of the condensed water is 30°C (303 K), the efficiency would necessarily be less than $1 - (303/853) = 0.645$ (or 64.5%), the efficiency of a Carnot engine operating between 303 K and 853 K. Actually, the steam cycle just described is not a Carnot cycle but a less efficient cycle known as the Rankine cycle, and some heat is lost around the boiler as well. The best fossil-fuel generating plants now in use have an efficiency of about 40% and the upper limit attainable has been estimated at 45 to 46% [2]. In order to increase the efficiency it is necessary to increase the steam temperature and pressure; during the 20th century, higher and higher steam temperatures and pressures have come into use as the design of generating plants has improved. Table 14-1 shows the average efficiencies

TABLE 14-1. Average efficiencies of all U.S. steam-electric generating plants in use and corresponding ratios of waste heat to electrical energy [3].

Year	Average efficiency	Waste heat/electrical energy
1930	17.24%	4.8
1940	20.81%	3.8
1950	24.33%	3.15
1960	31.72%	2.15
1970	ca 34 %	ca 1.9

of steam-electric generating plants in actual use and the amount of waste heat energy rejected or lost per unit of electrical energy. Around the turn of the century, the average efficiency was probably only about 8% so it is clear that progress in raising efficiency has been quite marked.

Nuclear power plants differ from fossil-fuel plants in several ways. They do not have furnaces but heat energy is produced in the reactor core through controlled nuclear fission (see Chapter 15). Most of the nuclear power plants in use today are of one of two designs:

1. The boiling water reactor (produced in the U.S. by General Electric). In this type of reactor, the reactor core in which the fission takes place is surrounded by water, which comes in as liquid water and leaves as steam. The pressure of the steam is typically 68 atm (6.8×10^6 N/m²) and its temperature 285°C. A boiling water reactor rated at 800 MW(e) (which means it produces 800 megawatts electrical power) might have a reactor vessel 20 m high and 6 m in diameter.

2. The pressurized-water reactor (produced by Westinghouse Electric in the U.S.). In this type of reactor, the water passing through the reactor does not boil and is not the working substance. This water is under 150 atm pressure and is heated to 315°C; it is passed through a steam generator to produce the steam that is the working substance.

Nuclear power plants differ from fossil-fuel power plants largely in having a reactor in place of a furnace since they too use steam as the working substance. Since they are operated with lower steam temperatures and pressures for safety purposes, they have lower efficiencies, typically about 33% implying a waste heat/electrical energy ratio of 2.0 instead of the 1.5 characteristic of the best new fossil-fuel plants with 40% efficiency. This situation is not likely to remain the case for many more years. In fact the gas-cooled reactors (which are similar to pressurized-water reactors but which use a gas such as helium or carbon dioxide in place of the pressurized water), such as the one at Peach Bottom, Pa., have 40% efficiencies right now.

Table 14-2 shows explicitly the eventual fate of the heat energy produced in a steam-electric generating plant assuming 33% overall efficiency, including 97% generator efficiency. Heat losses have been taken as 15% in a fossil-fuel plant and 5% in a nuclear plant, which are representative values [3]. The fossil-fuel plant will eventually have to remove 51% of the original heat energy in the cooling water, while the nuclear plant removes 61%. With the same in-plant heat losses given above, the ratio of waste heat in cooling water/electrical energy, for a plant with efficiency e, is $0.85/e - 1$ for fossil-fuel plants, and $0.95/e - 1$ for nuclear power plants [3]. A 33% efficient nuclear power plant will then reject $(0.95/0.33) - 1$

TABLE 14-2. Energy balance in a steam-electric generating plant.

	Fossil-fuel plant	Nuclear plant
A. Assumed overall efficiency	33%	33%
B. Assumed generator efficiency	97%	97%
C. Assumed heat losses in plant	15%	5%
D. Heat energy produced	1.00[a]	1.00[a]
E. Heat losses in plant (C × D)	0.15	0.05
F. Heat delivered to steam (D − E)	0.85	0.95
G. Energy delivered to turbine (A × D/B)	0.34	0.34
H. Electrical energy produced (A × D)	0.33	0.33
I. Heat removed at condenser (F − G)	0.51	0.61

[a] Figures from here to bottom of column represent arbitrary units of energy.

= 1.9 times as much waste heat to the cooling water as electrical energy produced, while a 40% efficient fossil-fuel plant rejects (0.85/0.40) − 1 = 1.1 times as much—42% less than the nuclear plant. The situation as of the early 1970's is then that new nuclear power plants produce considerably more thermal pollution per kilowatt-hour than new fossil-fuel plants, although this may not remain the case for very many more years.

In order to determine the extent to which large electric power plants are capable of raising the temperature of water bodies, suppose a generating facility involving three 1000-MW(e) power plants were to be planned near the mouth of the Missouri River (to choose a large river for our example). [Plants of 1000 MW(e) capacity are now being built, and facilities involving several of them are being planned.] Heat would have to be rejected to the cooling water from the river at a rate of about 3300 MW for a 40% efficient fossil-fuel installation and about 5700 MW for a 33% efficient nuclear installation when the facility is operating at full capacity. The average flow of the Missouri River near its mouth is 1800 m^3/s but the average yearly minimum flow is only 120 m^3/s. Even if the electric power facility could use *all* the river's flow through its condensers, the temperature rise (calculated using the specific heat of water, 1 cal/cm^3-°C = 4.184 MJ/m^3-°C) would be as shown in Table 14-3. This example supposes a very large installation on a very large river but the same magnitude of temperature rise would occur for moderate-sized installations on a smaller river.

TABLE 14-3. Temperature rise in the Missouri River if the total flow were used to cool steam at a 3000-MW(e) power installation operated at capacity. (See text for discussion.)

	Fossil-fuel plant	Nuclear plant
At average flow	0.44°C (0.8°F)	0.76°C (1.4°F)
At minimum flow	6.6°C (12°F)	11°C (20°F)

About 80% of the cooling water required by industry (see Table 8-4) is used by electric generating plants, whose future requirements will be enormous if electric power production continues to double every decade (see Figure 2-4). At present, about 20% of U.S. electrical energy is produced by hydroelectric plants rather than steam-electric plants, and some electric power plants use lakes or ocean water for cooling or make use of cooling towers (discussed later in this chapter).

EFFECTS OF THERMAL POLLUTION

Numerous effects of temperature on living organisms have been recorded in the biological literature [4]. The temperatures of water bodies generally vary with the seasons but there are sometimes natural fluctuations that occur over and above the seasonal variations, and these are usually reflected in the growth and numbers of aquatic organisms. Fluctuations of 2°C in the annual means of north Pacific waters have been associated with different geographical ranges of fish, for example.

Fish and other aquatic life can live only within certain temperature ranges, and the range in which well-being exists is narrower than the range in which survival is possible. These ranges vary from one individual to another; many tests have been carried out to define the LD_{50} tolerance limits, defined as those within which 50% of a given species will survive for some specified period of time. For example, these LD_{50} tolerance limits for large-mouth bass acclimated to 20°C (68°F) are 5°C (41°F) for 24-h exposure and 32°C (89.6°F) for 72-h exposure; for the same fish acclimated to 30°C the corresponding limits are 11°C and 34°C [3].

Favored temperature ranges exist for other species as well. In an unpolluted stream, diatoms grow best at 18 to 20°C; green algae, at 30 to 35°C; and blue-green algae, at 35 to 40°C [3]. Thermal discharges to a waterway may thus favor the growth of blue-green algae over green algae, with resulting damage to the ecosystem. Blue-green algae are a poorer food source and are believed in some cases to be toxic to fish. Thermal discharges are usually favorable to bacteria and pathogens as well. If the food supply is adequate, raising the temperature of natural waters, even in summer, will increase the bacterial multiplication rate, although if the food supply is limited, a more rapid die-off may be produced [3].

Some acclimation to elevated temperatures is usually possible in the case of fish, at least if the temperature changes occur sufficiently slowly. Rapid temperature changes produce "thermal shock" and sometimes almost immediate death; a sudden temperature rise of 16.7°C (30°F) has been found [5] to kill stickleback in 35 s and chum salmon in 10 s.

Temperature is, of course, of vital importance to physiology, controlling reproductive cycles, digestion rates, respiration rates, and the many chemical activities taking place in the body. Higher temperatures generally correspond

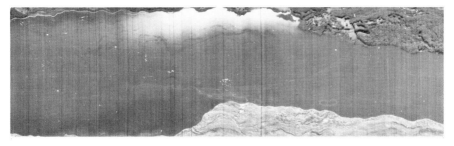

(a)

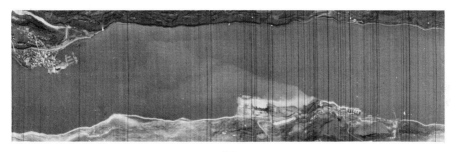

(b)

FIG. 14-2.   Images recorded by infrared scanning system of waterways exhibiting natural and man-made temperature fluctuations. The warmer areas show up as areas of light tone, approaching white; the cooler areas are indicated by the darker tones. Because water is a strong absorber of infrared radiation, the temperature measurements are representative of only the near-surface layer (0.1 to 0.2 mm) and correspond to the maximum prevailing water temperatures. **(a)** This image recorded on August 21, 1968 at 9:40 A.M. shows the naturally occurring temperature gradients in a shallow cove on the east bank of the Hudson River about 8 km above Hyde Park, N.Y. At this point the river widens and a series of shallows and small coves are formed on the eastern shoreline. Water flow in these areas is relatively slow so that the sun tends to warm the water to higher temperatures than are found in the main river channel. The water depth of the cove shown in this image is approximately 1 m as compared to 15 m, the depth of the main river channel. **(b)** This image, recorded on the same day at 9:27 A.M., shows the heated discharge of the Danskammer Point Power Station (Central Hudson Gas and Electric Corporation) during flood tide conditions. This power station, which consists of four fossil-fuel units totaling 515 MW(e), is located at Danskammer Point on the west bank of the Hudson River about 7 km above Newburgh, N.Y. A significant tidal effect is experienced in this portion of the Hudson River. The heated discharge is initially directed downstream by the outfall structure. It collects in a shallow cove to the south of the plant in which the influence of the discharge flow predominates and the tidal influence is small. As the discharge water moves out into the river, the tidal influence begins to take effect and directs the discharge upstream as it disperses. This image was recorded 1 h before maximum flood tide at a tidal velocity of about 1.1 km/h. (*Courtesy of the New York State Atomic and Space Development Authority.*)

to increased chemical reaction rates, and the behavior of physiological processes in speeding up by a factor of 2 or 3 times for every 10°C (18°F) rise in temperature is sometimes referred to as the Van't Hoff-Arrhenius law. The fact that temperature rises produce rises in metabolic rates means that an animal will require more food just to maintain body weight; the converse holds for lowered temperatures, and a fish that thrives best at, say, 15°C when the food supplies are ample will thrive best at perhaps 5°C when the food supply is restricted. Higher temperatures also lead to faster growth rates and shorter life-spans.

TABLE 14-4. Some properties of water as a function of temperature [3]. The last column lists the concentration of dissolved oxygen when the water is saturated with oxygen.

| Temperature | | Density | Viscosity | Vapor pressure | Dissolved O_2 |
°C	°F	(g/cm³)	(centipoise)	(mm of Hg)	(mg/l)
0	32	0.99987	1.7921	4.58	14.6
5	41	0.99999	1.5188	6.54	12.8
10	50	0.99973	1.3077	9.21	11.3
15	59	0.99913	1.1404	12.8	10.2
20	68	0.99823	1.0050	17.5	9.2
25	77	0.99707	0.8937	23.8	8.4
30	86	0.99567	0.8007	31.8	7.6
35	95	0.99406	0.7225	42.2	7.1
40	104	0.99224	0.6560	55.3	6.6

Higher temperatures also affect the physical and chemical properties of water. The density of water is a maximum of 1.00000 g/cm³ at 4°C and then decreases slightly at higher temperatures. As the temperature increases, the viscosity of water decreases sharply and the vapor pressure (and thus the evaporation rate) increases sharply. The solubility in the water of gases will also decrease with increasing temperature. Table 14-4 lists some of these properties of water, including the amount of dissolved oxygen present when the water is saturated with oxygen. These properties have important effects on aquatic life. The decreases in viscosity and density cause the settling speed of suspended particles to increase, in accordance with Stokes' law given in Chapter 3; the sediment load of a stream will then tend to settle more rapidly, possibly affecting aquatic food supplies. The lowered solubility of oxygen in water is clearly of great importance since fish require at least 5 mg/l (7 mg/l in spawning areas) for well-being. From Table 14-4 it is clear that above 35°C the dissolved oxygen content may be dangerously low even if the water is saturated with oxygen, which is not generally the case. Since higher temperatures favor bacterial growth and increase the rates of physiological processes, the decomposition of organic and other oxygen-demanding wastes will be speeded up, increasing the rate of oxygen depletion and further aggravating the dissolved oxygen problem.

Despite all the possible problems of thermal pollution, the fish kills that actually occur in this country (see Chapter 10) are usually due to causes other than thermal pollution. In order to protect aquatic life, the National Technical Advisory Committee to the Secretary of the Interior has recommended [6] restrictions on temperature extremes and temperature increases for various types of water in order to prevent fish kills and other damage to aquatic ecosystems; these are summarized in Table 14-5. During the summer months, a lake becomes stratified into an epilimnion or warm upper layer, a thermocline in which there is a rapid transition to lower temperatures, and a hypolimnion or cooler lower layer (typically at about 5°C) (see Figure 14-3). Steam-electric plants on lakes have made use of the colder water of the hypolimnion during the time the lake is stratified, returning it to the epilimnion, but ecologists have warned that this could have serious effects on the ecosystem, such as by transfer of nutrients to the epilimnion and consequent increased growth in aquatic life [7].

TABLE 14-5. Recommendations of the National Technical Advisory Committee on Water Quality Criteria relating to temperature [6].

Freshwater Organisms

Warm Water Biota
1. In no month should the heat added to a stream be in excess of the amount that will raise the water temperature (at the expected minimum daily flow for that month) more than 5°F (2.8°C) or the heat added to a lake increase the epilimnion temperature more than 3°F (1.7°C). Heated effluents should not be discharged into the hypolimnion.
2. Natural daily and seasonal temperature variations should be maintained.
3. Heat of artificial origin should not be permitted to increase the water temperature beyond certain recommended maxima [such as 90°F (32.2°C) for growth of largemouth bass and 84°F (28.9°C) for growth of pike, perch, walleye, and smallmouth bass].

Cold Water Biota
1. Inland trout streams, headwaters of salmon streams, trout and salmon lakes, hypolimnions containing salmonoids and other cold water forms, and the vicinity of spawning areas should not be heated or used for cooling water.
2. Elsewhere, the restrictions for warm water biota above apply.

Marine and Estuarine Organisms

1. Monthly means of maximum daily temperatures in coastal and estuarine waters should not be raised by heat of artificial origin by more than
   (a) 1.5°F (0.83°C) during the summer months June through August (July through September north of Long Island and in the waters of the Pacific Northwest north of California).
   (b) 4°F (2.2°C) during the rest of the year.
2. The rate of temperature change should not exceed 1°F (0.56°C) per hour except when due to natural phenomena.

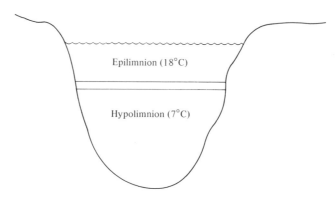

FIG. 14-3. Typical summer stratification of a lake into a warm epilimnion and a cooler, denser hypolimnion.

ALTERNATIVES TO ONCE-THROUGH COOLING

The use of water from a stream, lake, or ocean for cooling purposes, with return to the waterway after passage through the condenser, is referred to as once-through cooling. Several alternative methods of cooling are possible. In this section, artificial cooling lakes and ponds and cooling towers are considered. Of course, the problem of thermal pollution can also be alleviated by improvements in efficiencies of electric generating plants, and these are considered in Chapter 16.

ARTIFICIAL LAKES OR COOLING PONDS. Man-made bodies of water offer one possible alternative. Heated effluent is discharged into the lake's shallow end 1 or 2 meters deep, and the water for cooling purposes is drawn from beneath the surface at the other end of the lake, which may be 15 m deep [8]. A 1000 MW(e) electric power plant will require a surface area of perhaps 400 to 800 hectares (1000 to 2000 acres), although the required area may be greatly reduced (perhaps to 40 hectares) if the heated effluent is sprayed onto the surface of the lake from a height of a couple of meters [3]. Since the heat would eventually be dissipated through evaporation, the cooling pond would have to be replenished continuously, such as by a small stream flowing into the lake.

COOLING TOWERS. Cooling towers transfer heat from cooling water to the atmosphere, generally through evaporation of water [9]. *Evaporative cooling towers* are of two types:

1. Natural draft towers, hyperbolic in shape, in which the hot water is sprayed down through a rising current of air [see Figure 14-4(a)]. The air gains water vapor that has removed heat by evaporation, and the cooled water is collected at the bottom and returned to the

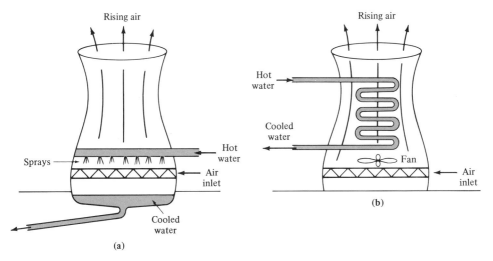

FIG. 14-4. **(a)** Schematic drawing of a natural-draft evaporative cooling tower. **(b)** Schematic drawing of a mechanical-draft "dry" cooling tower.

condenser or to a waterway. The natural draft towers are designed so that air is continuously being drawn in through air inlets at the bottom and exhausted at the top of the tower. Initial costs for such a system might be on the order of $10 million for a 1000 MW(e) plant, and operating costs would be fairly small. The towers are often quite large (the TVA has some 130 m high and 100 m in diameter at the base) and may be aesthetically offensive to some persons. Towers of this type are shown in Figure 14-5.

2. Mechanical draft cooling towers, in which the air flow is forced or induced by fans. Initial costs are somewhat less than for natural draft towers, but operating costs are greater and there can be noise created by the fans. These towers are smaller than the hyperbolic tower and are not as intrusive on the landscape (see Figure 14-6).

Evaporative cooling towers typically cool the water by 10°C or more, but in the process they evaporate 2% or more of the water. More precisely, the fraction of water evaporated is $(\Delta T/580)$ for a cooling of ΔT (in °C) since the specific heat of water is approximately 1 cal/g-°C and its latent heat of vaporization at ordinary temperatures is about 580 cal/g. A 1000 MW(e) plant might have to evaporate 1 m³/s, the exact amount depending on the plant efficiency, the weather conditions, and other relevant parameters. Under some atmospheric and temperature conditions, especially in colder climates, problems of reduced visibility and formation of fog, ice (on roads and lines), and even snow may arise.

There are also *nonevaporative cooling towers*, in which heat is transferred directly to the air without evaporation by the use of heat exchangers. Figure 14-4(b) shows a mechanical cooling tower of this "dry" type.

FIG. 14-5. Hyperbolic natural-draft cooling towers with a total capacity of 250,000 gal/min (16 m³/s) at the Fort Martin Power Station, Fairmont, W. Va. (*Courtesy of The Marley Company, Mission, Kan.*)

At present the "dry" towers are more expensive, approximately three times as expensive in the case of mechanical draft towers but, as mentioned above, the "wet" types may not be appropriate for some climates.

The total costs of cooling towers, including initial and operating costs, are believed to increase consumer bills by 1 or 2% compared to once-through cooling.

USES OF WASTE HEAT

A number of suggestions have been made for the use of the waste heat from steam-electric power generation since such use would alleviate the thermal pollution problem and simultaneously help conserve our fuel resources.

HEATING OF BUILDINGS. Exhausted steam is already used to heat buildings in some places such as institutions having their own power plants. This solution would naturally work best in winter; it would be of no use in summer when the thermal pollution problem is likely to be worse because of the heavy production of electricity for air conditioning. In the past 2

FIG. 14-6. An 11-cell mechanical-draft cooling tower with a total capacity of 197,000 gal/min (12 m³/s) at the Victoria (Tex.) Power Station of Central Power and Light Company. (*Courtesy of The Marley Company, Mission, Kan.*)

decades there has emerged a new concept—"total energy"—of generation of power where it is needed (institution, home, office building), avoiding transmission losses and permitting use of waste heat [10]. Another interesting possibility, as yet undeveloped and perhaps suitable only in areas with high population densities, is to use the waste heat in the heated cooling water to warm homes by means of a heat pump, which is a refrigerator unit used for heating rather than cooling. Water that is not hot enough to use in radiators can still serve as a source of heat for a heat pump.

HEATED SWIMMING POOLS. Heated pools may be installed in some areas; the city of Cleveland, Ohio, did this on an enclosed portion of Lake Erie.

DEICING OF WATERWAYS. Heated effluent in waterways could extend the navigation season in northern climates or permit waterfowl to spend the winter further north than usual. Ecologists warn, however, that this practice could easily damage natural ecosystems.

DESALINATION. Freshening of seawater or other brackish water is being accomplished on a pilot plant scale in some areas. A University of California at Riverside study [11] has suggested that the natural steam and hot water wells in California's Imperial Valley could be developed to yield 20,000 MW(e) of electric power and 6 to 8 billion m³ of desalted water annually, the latter at a cost of $0.10/1000 gal (3.8 m³).

AQUACULTURE. The production of fish or shellfish that thrive in warm water could also benefit from heated waste water, and Long Island oyster farmers have found that their harvesting season is extended by the warming of the ocean by large power plants in the area. Heated water discharged at lower depths in a lake or the ocean will rise, carrying plant and animal nutrients to the surface and improving the fishing, at least over the short term; any such major disturbance of an aquatic ecosystem needs to be carefully studied in advance by ecologists.

WARM-WATER IRRIGATION. A promising idea that is being investigated in several places is warm-water irrigation. The Eugene (Oregon) Water and Electric Board has cooperated with local farmers in such a project in the hopes of stimulating and enhancing plant growth and protecting fruit trees from killing frosts [12].

CONCLUSION

Thermal pollution problems will continue to grow at a sharp rate if electric power production continues to increase as dramatically as it has in the first 7 decades of the 20th century. More efficient methods of generating power and using it are needed, and more emphasis is required for avoiding unnecessary energy consumption.

REFERENCES

1. Halliday, D., and R. Resnick, *Fundamentals of Physics*. New York: John Wiley & Sons, Inc., 1970. Chapter 21, pp. 401–420.
2. Cootner, P. H., and G. O. G. Löf, *Water Demand for Steam Electric Generation: An Economic Projection Model*. Washington, D.C.: Resources for the Future, Inc., 1965.
3. Pacific Northwest Water Laboratory, *Industrial Waste Guide on Thermal Pollution*. Corvallis, Ore.: Federal Water Pollution Control Administration, 1968.
4. Krenkel, P. A., and F. L. Parker (eds.), *Biological Aspects of Thermal Pollution*. Nashville, Tenn.: Vanderbilt University Press, 1969.
5. Snyder, George R., "Discussion." In Reference 4, pp. 318–337.
6. *Water Quality Criteria. Report of the National Technical Advisory Committee to the Secretary of the Interior*. Washington, D.C.: U.S. Dept. of the Interior, Federal Water Pollution Control Administration, 1968.

7. Eipper, Alfred W., "Nuclear Power on Cayuga Lake." In Harte, John, and Robert H. Socolow (eds.), *Patient Earth*. New York: Holt, Rinehart and Winston, Inc., 1971. pp. 112–131.

8. Clark, John R., "Thermal Pollution and Aquatic Life." *Scientific American* **220**(3):18–27 (March 1969).

9. Woodson, Riley D., "Cooling Towers." *Scientific American* **224**(5):70–78 (May 1971).

10. Diamant, R. M. E., *Total Energy*. Oxford: Pergamon Press, 1970.

11. Rex, Robert W., *Investigation of Geothermal Resources in the Imperial Valley and Their Potential Value for Desalination of Water and Electricity Production*. Riverside, Calif.: The Institute of Geophysics and Planetary Physics, University of California, 1970.

12. Carter, Luther J., "Warm-Water Irrigation: An Answer to Thermal Pollution?" *Science* **165**:478–480 (1969).

15

RADIATION

Numerous writers, both in Europe and America, have published accounts of the injurious effects produced on the skin by a too prolonged exposure to the Roentgen rays, the symptoms varying in nature and intensity from "sunburn" to dermatitis, vesication, and ulceration. Loosening of the hair, sometimes carried so far as to result in baldness, was frequently observed.   —Lancet **1:**752 (1897).

Within a few months of Roentgen's discovery of X rays in 1895, they were already being used as a diagnostic tool in medicine to make X-ray photographs ("the new photography") of the human body. In 1896 an article appeared describing their use in pregnancy to determine the position of the fetus [1]. Overexposures to this unknown type of radiation led to skin problems and to falling out of the hair, as the quotation above indicates. Nevertheless, excessive exposures continued for several decades in scientific and medical work, often causing the loss of limbs or of life. The use of X rays as a depilatory proved especially difficult to control. Even after the dangers had been well established, a United States physician Dr. Albert C. Geyser marketed his "Tricho System" for doing away with unwanted hair in beauty parlors throughout the country, attracting thousands of women into a 12-week period of radiation treatment that produced ulcerations, malignant growths, and even death in unknown numbers of victims [2]. Not many years ago, shoe stores used X-ray fluoroscopes to check the fit of shoes on children, delivering large doses of radiation.

Today, with the existence of nuclear power plants, the presence of radioactive fallout in the atmosphere, and the use of radiation and nuclear devices for "constructive" purposes, the sources of radiation in the environment have multiplied, although medical X-ray usage is still the major source of artificial radiation. The following sections will discuss radiation and its effects on life, the natural and artificial sources of radiation in the environment, the problem of setting radiation standards, and the disposal of radioactive wastes.

RADIATION AND ITS EFFECTS ON LIFE

In its broadest sense, radiation is energy being propagated from one place to another. Sound (discussed in Chapter 7) is a wave motion of the air propagating small amounts of energy capable of being heard and also capable of being absorbed by living tissue to produce small amounts of heat. Light is a form of electromagnetic radiation, exhibiting wave properties in some circumstances and particle properties in other. Photons are the particles associated with electromagnetic radiation of all types, being more energetic for visible light than for radio waves but being even more energetic in the form of X rays and gamma rays. Light energy can be absorbed by living things to produce heat and molecular vibrations and is essential for many important biochemical processes, such as photosynthesis.

The radiation that is of concern as pollution is *ionizing radiation*, radiation of sufficiently great energy to ionize atoms and molecules. An atom is ionized when it gains sufficient energy for one or more of its electrons to be separated from the atom; ionization of a molecule might split it into two charged fragments, such as $H_2O \rightarrow H^+ + OH^-$. Sometimes the fragments are uncharged and then they are referred to as "free radicals," as in the example $H_2O \rightarrow H + OH$. Free radicals are generally very reactive chemically.

The most important types of ionizing radiation from the standpoint of pollution (as opposed to the standpoint of the scientific researcher interested in new knowledge of the fundamental properties of matter) are alpha, beta, and gamma radiation:

1. Alpha radiation is due to energetic alpha particles. An alpha particle is a He^4 nucleus, which consists of two protons and two neutrons bound together into a very stable particle by what the physicist refers to as a "strong nuclear interaction." Alpha particles have a positive electric charge whose absolute magnitude is twice that of the electron's charge and are capable of interacting strongly with ordinary matter (including living tissues) by electromagnetic interactions.

2. Beta radiation consists of energetic electrons (or their antiparticles, the positrons). These particles occur in the beta decay of nuclei, which also produces antineutrinos (or neutrinos) which interact only very weakly with matter and are of no importance as radiation. Beta particles, being electrically charged, can interact strongly with matter by the electromagnetic interaction.

3. Gamma radiation consists of very energetic photons, i.e., very short wavelength electromagnetic radiation. Despite being uncharged, photons are capable of very strong electromagnetic interaction with matter.

There are other types of ionizing radiation that are encountered less frequently but can also be dangerous to living systems—protons, neutrons, deuterons, and other nuclei of sufficiently great energy, for example. The particular importance of alpha, beta, and gamma radiation arises from the fact that they are produced in the radioactive decay of certain atomic nuclei, i.e., in the spontaneous or induced disintegration of these nuclei.

When ionizing radiation passes through body tissue, it splits the molecules of the tissue into ions or free radicals, as in the examples of water molecules mentioned above. These fragments may later combine to form new chemical compounds [such as H_2 from $H + H$ or H_2O_2 (hydrogen peroxide) from $OH + OH$] or free radicals (such as HO_2 from $H + O_2$). Ionizing radiation can thus split molecules into useless or reactive fragments and permit the formation of other reactive compounds. These effects can be classified into two groups:

1. *Direct effects*—fragmentation of biologically important molecules, such as DNA molecules in the cell nucleus.

2. *Indirect effects*—fragmentation of biologically less vital molecules (such as water) with the formation of reactive ions or free radicals that can later affect more important molecules and impair their usefulness.

The damage done by radiation depends on the amount of energy that is lost by the radiation and made available for the ionization of matter. The amount of energy lost by the radiation per unit distance traveled (which is not necessarily in a straight line—beta rays, for example, are easily deflected by matter) is referred to as the "linear energy transfer" (LET). The LET of a given type of radiation increases as the energy of the radiation decreases, in part because its speed will be less and the traversal of a unit distance will require a longer period of time.

Table 15-1 shows some average LET values and total tissue penetration distances for several types of ionizing radiation. Photons and neutrons have average LET values approximately equal to those of electrons and protons, respectively, of half as much energy; e.g., a 200-keV neutron has an average LET close to that of a 100-keV proton.

Low-LET radiation does most of its tissue damage through indirect action, while high-LET radiation does most of its damage through direct action. A particle entering tissue with high energy will have low LET initially but as its energy decreases, its LET increases. Such a particle might be useful in treating a deep-seated tumor if the particle's range is just right for it to come to a stop near the tumor [3].

From a biological point of view, radiation effects can also be classified as somatic or genetic. *Somatic effects* are the effects on the body itself and are of direct concern to the person exposed to the radiation. *Genetic effects* are those involving mutations of the chromosomes or genes in sex cells; they pose a potential hazard to the descendants of the exposed person and are of concern to the whole society. Genetic effects are of concern only in persons who are not yet past their reproductive years.

TABLE 15-1. Average LET values and tissue penetration distances for electrons, protons, and alpha particles [3]. 1 keV = 1000 eV (electron volts) = 1.6×10^{-16} joules. 1 μm = 10^{-6} m.

Particle	Energy (keV)	LET (keV/μm)	Tissue penetration distance (μm)
Electron	1	12.3	0.01
	10	2.3	1
	100	0.42	180
	1000	0.25	5000
Proton	100	90	3
	2000	16	80
	5000	8	350
	10,000	4	1400
	200,000	0.7	300,000
Alpha	100	260	1
	5000	95	35
	200,000	5	20,000

In order to discuss the effects of radiation, it is necessary to define the commonly used radiation units. These are listed in Table 15-2. Of particular importance in discussing radiation effects on life are the units of rads and rems, which are often used interchangeably even though their definitions are

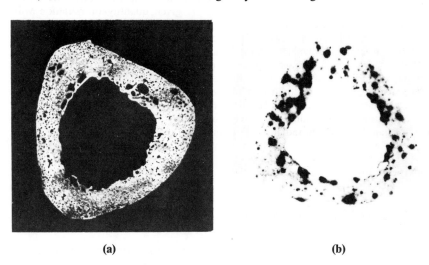

(a)   (b)

FIG. 15-1 A section of bone from the body of a former radium watch-dial painter, who, in order to maintain a fine brush tip, was in the habit of touching the tip with his tongue. **(a)** A photograph showing dark areas of damaged bone. **(b)** An autoradiograph in which the bone "took its own picture" by being held against the film, showing areas exposed by the radium alpha particles. The areas of high alpha activity correspond to the areas of maximum damage in photograph **(a)**. (*Courtesy of the U.S. Atomic Energy Commission.*)

TABLE 15-2. Radiation units.

---

1. *Disintegrations per second.* The number of radioactive disintegrations occurring each second for a given amount of radioactive material. This equals $N\lambda$, where N is the number of atoms present and λ is the radioactive decay constant giving the probability of decay of the atom in question. For Ra^{226}, the naturally occurring isotope of radium, $\lambda = 1.36 \times 10^{-11}$/s, so 1 g of Ra^{226} (which has $6.025 \times 10^{23}/226 = 2.7 \times 10^{21}$ atoms) will produce $2.7 \times 10^{21} \times 1.36 \times 10^{-11} = 3.6 \times 10^{10}$ disintegrations per second.
2. *Curie.* The quantity of any radioactive material that gives 3.70×10^{10} disintegrations per second. This unit was defined so that 1 g of natural radium together with its decay products amounted to 1 curie. Millicurie and microcurie units are also used. The infrequently encountered *rutherford* is defined to be exactly 1 million disintegrations per second.
3. *Roentgen.* Originally defined as the amount of gamma (or X) radiation that produced by ionization 1 e.s.u. of electricity of each sign in 1 cm^3 of dry air at 0°C and atmospheric pressure; equivalently it produced 2.08×10^9 singly charged ions of each sign. At one time the roentgen was believed to correspond to the absorption of 83 ergs/g of air since it was thought that 32.5 eV were required, on the average, to produce one ion-pair; later, 86 ergs/g of air was determined to be a more accurate value. Experiments showed that 1 roentgen of gamma radiation led to the absorption of about 97 ergs/g in soft tissue, and a new unit was defined: The *rep* (for "roentgen equivalent physical") denoted a dose of 97 ergs/g of body tissue.
4. *Rad* (for "radiation absorbed dose"). That quantity of radiation that leads to the absorption of 100 ergs/g of the absorbing material. For body tissue, this closely matches the definition of the rep. A given quantity of radiation will correspond to different numbers of rads in different materials but the number of roentgens will be the same in each case.
5. *Rem* ("roentgen equivalent man"). That quantity of radiation that produces the same biological damage in man as one rep of gamma radiation. Different types of radiation seem to have different biological effects, and they can be characterized by their "relative biological effectiveness" (RBE), which gives the number of rems per rad. The RBE really depends on the effect considered, the location of the radiation source, and other things, but typical values are 1 for gamma and beta radiation of all energies above 50 keV, 2 to 5 for lower energy gamma and beta radiation, 10 for fast neutrons and protons (0.5 to 10 MeV) and natural alpha particles, and 20 for heavy nuclei and fission particles [3].

---

slightly different. Rads are easier to measure but rems are somewhat more meaningful.

A basic understanding of the biological action of ionizing radiation does not yet exist despite a great deal of progress in research [4]. Most experiments have been carried out using laboratory animals, and comparison with humans is difficult to make. Some data have been accumulated over the years from human exposures to large radiation doses obtained accidentally or from nuclear explosions. Table 15-3 shows the approximate short-term effects that might be experienced for whole-body radiation exposures over a short period of time. A whole-body exposure of 1 rad would mean an average

TABLE 15-3. Estimated short-term effects of single-dose, whole-body radiation exposures in man [5].

Less than 25 rads	No observable effect
About 25 rads	Threshold level for detectable effect
About 50 rads	Slight temporary blood changes
About 100 rads	Nausea, fatigue, vomiting
200 to 250 rads	Fatality possible, though recovery is more likely
About 500 rads	Perhaps half the victims would die
About 1000 rads	All the victims would die

absorption of 100 ergs/g over the whole body. Table 15-3 refers to immediate effects, and there is no guarantee that recovery might not be followed many years later by other effects—such as greater incidence of cancer than that occurring in unexposed persons [6]. Since different persons differ in their sensitivity to radiation, as in their sensitivity to infections and numerous other things, the doses corresponding to different effects will vary widely from person to person.

Exposure to a few hundred rads leads to acute radiation illness: nausea, fatigue, and vomiting within a few hours and for a day or two, a decrease in red and white blood cells and blood platelets for a few weeks, then anemia, susceptibility to bacterial infection, and hemorrhaging for some period of time, followed often by death. If the victim survives, he or she will still have a greater-than-average chance of developing leukemia (especially in the first few years after exposure), other forms of cancer, cardiovascular disorders, and eye cataracts [3].

Of particular interest in studies of radiation effects are the continuing research findings of the Atomic Bomb Casualty Commission. The Commission was begun in 1948 as a cooperative venture between the National Research Council of the United States and the National Institute of Japan, and it has been following the medical histories of thousands of survivors (and their children) of the Hiroshima and Nagasaki atom bombs of 1945 [7, 8].

Atom bomb survivors exhibited higher-than-average leukemia rates, which peaked in 1951—6 years after exposure—but which were still higher than usual in 1966 [7]. Mortality rates for persons within 1200 m of the hypocenter in the two towns, after excluding the effects of leukemia, were 15% higher than in unexposed Japanese during the decade 1950 to 1960—a statistically significant increase. No effect was found on mortality of children conceived to survivors after the exposure, however. Very high rates of chromosomal abnormalities have been found among survivors and their children who were *in utero* at the time of the bombings but not in children conceived afterward [7].

Despite frequent criticism of the ABCC studies in Japan, these studies are being continued and are playing an important role in defining the hazards of radiation. They are especially important now when increased cancer rates or other effects might be appearing after a long period of latency.

RADIATION SOURCES IN THE ENVIRONMENT

Radiation sources in the environment are partly natural and partly man-made. Some natural radiation is produced by naturally occurring radioactive elements, such as the ones listed in Table 15-4. The table lists the radioactive isotopes, their approximate abundances in the lithosphere (the earth's crust), their half-lives, and the nature of their radiation [9]. These are an important source of radiation exposure to humans. The amount of whole-body radiation produced annually varies from place to place, depending on such variables as the nature of the soil and the distance to mineral deposits [11]. There are places in Brazil where the soils are so rich in thorium and uranium that whole-body exposures can amount to 2 rem annually, perhaps 40 times the average U.S. radiation exposure from earth. Some natural radioactive isotopes (K^{40}, C^{14}, Th^{232}, U^{238}, etc.) are present in human body tissues themselves, the amounts varying somewhat with geographical location, type of water supply, and other variables.

TABLE 15-4. Some naturally occurring radioactive isotopes, their abundances in the lithosphere, their half-lives, and the nature of their radiation [9]. Ra^{226}, U^{238}, and Th^{232} are all members of natural radioactive series [10] that produce all three types of radiation—alpha, beta, and gamma.

Isotope	Abundance	Half-life (years)	Radiation
Ra^{226}	2×10^{-12} (in soil)	1622	Alpha, gamma
U^{238}	4×10^{-6} (in soil)	4.5×10^9	Alpha
Th^{232}	12×10^{-6} (in soil)	1.4×10^{10}	Alpha, gamma
K^{40}	3 ppm	1.3×10^9	Beta, gamma
V^{50}	0.2 ppm	5×10^{14}	Gamma
Rb^{87}	75 ppm	4.7×10^{10}	Beta
In^{115}	0.1 ppm	6×10^{14}	Beta
La^{138}	0.01 ppm	1.1×10^{11}	Beta, gamma
Sm^{147}	1 ppm	1.2×10^{11}	Alpha
Lu^{176}	0.01 ppm	2.1×10^{10}	Beta, gamma

Another important source of natural radiation is cosmic radiation. Cosmic rays are high-energy charged particles (mostly protons) of extra-terrestrial origin. They are capable of producing other energetic radiation by collisions with oxygen and nitrogen nuclei in the atmosphere. Table 15-5 lists some products of cosmic rays. The radiation dose obtained from cosmic rays is greater at greater heights in the atmosphere and may amount to about 10 rem/year above the atmosphere. Natural radiation from cosmic rays will be higher at higher elevations (e.g., greater in Denver than in coastal cities) and higher in a jet aircraft at 10 km than on the ground.

Table 15-6 lists typical radiation exposures in the United States; these numbers are based on the 1963 report of the Federal Radiation Council. Natural sources contribute perhaps 100 to 600 mrem annually, with the aver-

TABLE 15-5. Products of cosmic radiation, their half-lives, and their concentration in disintegrations per minute per cubic meter of air in the lower troposphere [9]. The concentrations refer to those resulting from cosmic radiation and do not include the effects of nuclear weapons tests.

Isotope	Half-life	Concentration
H^3	12.3 years	10
C^{14}	5760 years	4
Be^7	53 days	1
S^{35}	87 days	0.015
P^{33}	25 days	0.015
P^{32}	14.3 days	0.02

TABLE 15-6. Typical whole-body radiation exposures in the United States [12].

Natural sources	*Millirems/year*
A. External to the body	
1. From cosmic radiation	50.0
2. From the earth	47.0
3. From building materials	3.0
B. Inside the body	
1. Inhalation of air	5.0
2. In human tissues (mostly K^{40})	21.0
Total from natural sources	**126.0**
Man-made sources	
A. Medical procedures	
1. Diagnostic X rays	50.0
2. Radiotherapy X rays, radioisotopes	10.0
3. Internal diagnosis, therapy	1.0
B. Nuclear energy industry, laboratories	0.2
C. Luminous watch dials, television tubes, radioactive industrial wastes, etc.	2.0
D. Radioactive fallout	4.0
Total from man-made sources	**67.2**
Total	**193.2**

age near the lower end of the range. Man-made sources contribute only a few millirems annually apart from medical procedures.

Medical procedures are the major source of man-made radiation exposures. Not too many years ago, medical and dental X-ray examinations would give perhaps 300 rems exposure over part of the body (not the whole body) but modern techniques involve much lower exposures. Still, the average chest X ray gives 200-mrem exposure to part of the body, and the average gastrointestinal tract examination gives 22 rem [13, p. 190]. The average exposures listed in Table 15-6 correspond to having only occasional X rays.

The contribution of nuclear power plants to man-made radiation is quite small. Radiation standards set by the Atomic Energy Commission (AEC) formerly permitted a maximum dose of 500 mrem/year at any point on or beyond the boundary of a nuclear power plant, which would be several times the usual natural background; in 1971 this limit was reduced to 5 mrem/year for light-water-cooled nuclear power reactors [14]. Operating nuclear power plants (except for less than half a dozen older plants) in practice emitted no more than a few percent of the old 500 mrem/year limit, and in the great majority of cases less than 1% of it [13, 15]. Thus the actual exposure of a person remaining around the clock at the least favorable location (in the direction of the prevailing winds at the plant boundary) would be no more than 5 mrem/year, in accordance with the new limits [16]. It is thus believed unlikely that anyone in the U.S. is receiving even 1 mrem/year radiation exposure from nuclear power plants—an amount approximately equal to that received from increased cosmic radiation exposure during a single transcontinental jet flight.

There are a number of other everyday sources of man-made radiation. Older-type luminous watch faces using radium could deliver local doses of up to 2 mrem/h, but tritium (H^3) is used today and the doses are only a negligible portion of the man-made radiation. X rays from black-and-white television sets might contribute another millirem per year and large doses are occasionally reported from other sources—10 mrem/h from houses built with radioactive stone, 100 mrem/h from bathtubs glazed with uranium pigments, etc.

Radioactive fallout from nuclear weapons tests has been of concern to scientists since the mid-1950's. When a nuclear weapon is tested in the atmosphere, there is *local fallout* of radioactive fission products over the immediate area for about a day, then worldwide *tropospheric fallout* for about a month from fission products released into the troposphere, and *stratospheric fallout* worldwide for many years thereafter [17]. This fallout is easily detected at scientific laboratories around the world and its study has greatly increased our knowledge of worldwide atmospheric transport processes. The U.S. and U.S.S.R. have not tested nuclear weapons in the atmosphere since the signing of the Limited Test Ban Treaty in 1963, but France and the People's Republic of China—who have not signed the treaty—are currently testing nuclear weapons in the atmosphere from time to time. The rapid transport of fallout from one hemisphere to the other is indicated by the detection within 22 days at 34° north latitude of radioactive I^{131} and Ba^{140} from a French nuclear test at 21° south latitude [18].

In order to be of concern to man, a radioactive fission product must be produced in sufficient quantities, have a sufficiently long half-life (which is as short as 8 days in the case of I^{131}), be transferable to man, and remain in the body long enough to do damage. The most hazardous radionuclides turn out to be those listed in Table 15-7. C^{14} is produced naturally in the atmosphere by cosmic radiation and is a constituent of all living tissues; weapons tests through 1965 increased the amount in the atmosphere by 70 to

TABLE 15-7. Radionuclides important in fallout.

Element	Isotope	Half-Life
Carbon	C^{14}	5760 years
Strontium	Sr^{89}	51 days
Strontium	Sr^{90}	28.9 years
Iodine	I^{131}	8.1 days
Cesium	Cs^{137}	30.2 years

100% but circulation in the biosphere will reduce this to about 3% by 2040 A.D. unless atmospheric testing continues [17]. Sr^{89} is similar to (but much shorter-lived than) Sr^{90}, which was discussed in Chapter 1 as a pollutant that is discriminated against biologically but is still of great importance. Sr^{90} and I^{131} both reach man through cow's milk, the Sr^{90} going into bones and the I^{131} into the thyroid gland. Cs^{137} reaches human tissues through ingestion of milk and meat but has a biological half-life (i.e., half-life inside the body) of only 70 to 140 days because of its fairly rapid removal from the body through metabolic action.

Figure 15-2 shows the Sr^{90} and Cs^{137} content of milk in New York over the years. The two large peaks in the data occurred just after the ends of the two periods of heavy U.S. and U.S.S.R. atmospheric testing in 1957 to 1958 and 1961 to 1962.

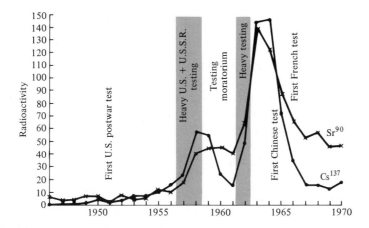

FIG. 15-2 The Sr^{90} and Cs^{137} content of New York milk, 1946 to 1970. The 1946 to 1957 data are from Reference 19, converted to volume units using 1 liter milk = 78 g dry milk solids. The 1958 to 1970 data are from *Radiological Health Data and Reports* [20]. Periods of atmospheric nuclear tests are indicated. The last U.S. atmospheric test was in November 1962 and the last U.S.S.R. atmospheric test was in December 1962. The People's Republic of China carried out 11 tests between October 1964 and the end of 1970. Units are picocuries/4 liters for Sr^{90} and picocuries/liter for Cs^{137}.

Another possible man-made radiation source in the future would be the so-called "peaceful" or "constructive" uses of nuclear explosives for releasing natural gas from underground regions, for building harbors, or even for constructing a new sea level canal between the Atlantic and Pacific Oceans to replace the Panama Canal. The U.S. Atomic Energy Commission has been promoting several such uses as part of the Plowshare Program [21]. Although much of the discussion has centered about the economic advantages of the use of nuclear explosives for certain operations, there has been a growing realization in the United States that the possible radiation dangers must also be seriously considered. Natural gas released with a nuclear explosion is likely to be contaminated by radioactive gases such as Kr^{85}, for example, and large-scale construction processes may lead to radioactive contamination of air and water. Even the use of radioisotopes for medical and scientific purposes involves some risk to the general population; the burnup of an orbiting SNAP (Systems for Nuclear Auxiliary Power) generator in the atmosphere in April 1964 led to detectable amounts of Pu^{238} in rainwater [22]. Although society may very well decide that the risks involved in the peaceful uses of nuclear energy are outweighed by the benefits, it is essential that the public and its elected representatives in federal, state, and local governments understand the risks and act on the basis of this understanding, rather than permit the decisions to be made by any small group of experts whose values may not coincide with those of society.

RADIATION STANDARDS

The important types of radiation were discovered in the late 19th century and quickly put to medical use, as mentioned at the beginning of this chapter. It was many years before groups were set up to consider the problem of radiation standards, however, and many tragic incidents occurred from medical and commercial use of various radiation devices, not to mention the sale of radioactive waters (see Chapter 17).

The International Commission on Radiological Protection (ICRP) was established in 1928 as a body of scientific experts from many different countries. Today it consists of a commission of 13 members from the U.S., the U.S.S.R., Great Britain, France, and other countries, plus four committees dealing with radiation effects, internal exposure, external exposure, and application of the commission's recommendations. In 1929 the National Council on Radiation Protection and Measurements (NCRP) was set up in the U.S. under the sponsorship of the National Bureau of Standards to work with the ICRP. Today it consists of several dozen members from government, industry, universities, and private medical practice. In 1959, Congress set up the Federal Radiation Council (FRC) to recommend radiation protection guidelines based on information from the ICRP, the NCRP, the AEC, and other sources. When its functions were transferred in 1970 to the new Environmental Protection Agency, it consisted of the secretaries of the Departments of Agriculture; Commerce; Defense; Health, Education, and Welfare; Labor; and the Interior and the Chairman of the Atomic Energy Commission.

The radiation standards that have been set over the years have generally been lowered as new information has been obtained about the effects on life. In 1946, as the atomic energy industry was just beginning, the NCRP was using a radiation protection standard of 0.1 rem/day, which would amount to 36.5 rem/year exposure, for adults. This would correspond to perhaps 200 or 300 times the natural radiation background exposure. Because the technology was available, this standard was lowered to 0.3 rem/week (15 rem/year).

The recommended guidelines of the FRC were published in 1960 and 1961, and they can be summarized in somewhat simplified fashion as follows [15]:

1. A dose not exceeding 5 rem/year whole-body exposure for radiation workers.

2. A dose not exceeding 0.5 rem/year whole-body exposure for individual members of the general population.

3. An average dose to the general population not exceeding 0.17 rem/year (170 mrem/year) whole-body exposure.

It is also recommended that actual exposures be kept as low as practical.

In recent years there has been criticism of these radiation standards from a number of sources. Two scientists at the AEC's Lawrence Radiation Laboratory at Livermore, Calif., Dr. John W. Gofman, Director of the Bio-Medical Division and Associate Director of the Laboratory, and Dr. Arthur R. Tamplin of the Bio-Medical Division, have been in the forefront of the fight to have the radiation protection standards lowered by a factor of 10 so that man-made exposures to the general population do not average more than 0.017 rem (17 millirems) per year [6, 13]. Their argument has been based on an examination of the records related to human exposure to ionizing radiation, together with conservative assumptions about the extrapolation of known data to low levels of radiation exposure.

Studies of human exposure to radiation have dealt with the Japanese survivors of the Hiroshima and Nagasaki bombings [7, 8], radiologists who suffered heavy exposures during the first part of the 20th century before the long-term effects of radiation were recognized, children subjected to diagnostic X rays *in utero*, patients treated with radiation for various nonmalignant diseases (those treated for malignant diseases cannot be included in studies seeking later malignancies), uranium miners, and victims of radiation accidents [23]. These studies suggest that leukemia and other forms of cancer occur more often in persons exposed to greater amounts of radiation. The rate increases are superimposed on natural rates of leukemia and other cancers, which lead to about 350,000 cancer deaths annually in the U.S. For each rad of exposure, there is an almost constant fractional increase in the incidence of each type of cancer, according to Gofman and Tamplin's review of the data. Some of these are shown in Table 15-8. For large enough radiation exposures to be able to detect a statistically significant increase in leukemia rates, for example, each rad of exposure appears to correspond to an increase in usual leukemia rates of about 2 or 3%, and an exposure of about 30 to 60 rads will double the usual rate. The spontaneous leukemia rate is about 60×10^{-6}/year; i.e., about 60 out of every million persons contract leukemia each year.

TABLE 15-8. Best estimates of doubling dose of radiation for human cancers and the increase in incidence rate per rad of exposure, from data of Gofman and Tamplin [13, p. 661].

Type of cancer	Doubling dose (rads)	Percent increase in incidence rate per rad
Leukemia	30–60	1.6–3.3
Thyroid cancer:		
Adults	100	1
Young persons	5–10	10–20
Lung cancer	175	0.6
Breast cancer	100	1
Stomach cancer	230	0.7
Pancreas cancer	125	0.8
Bone cancer	40	2.5
Cancer of lymphatic and other hematopoietic organs	70	1.4
Carcinomatosis of miscellaneous origin	60	1.7

Gofman and Tamplin believe that all forms of human cancer can probably be induced by ionizing radiation and that the doubling dose averages about 50 rads for every form, i.e., that on the average the increase in the spontaneous incidence rate is about 2% per rad [6]. They also argue that there is no reason to suppose that the effects of radiation are not linear at all doses, even very small doses. Figure 15-3 shows two possibilities for the dose-effect relationship, the dose being measured in rads of whole-body exposure, for example, and the effect in cases of induced cancer. A linear relationship as shown in Figure 15-3(a) appears to hold for sufficiently large doses but the effects for low doses are difficult to measure. Figure 15-3(b) shows a dose-effect relationship with a threshold, i.e., a dose below which there is no effect (or at least no irreversible effect). There are many scientists who believe that radiation thresholds exist for somatic damage [13], and mammalian systems do appear to be able to repair radiation damage that is not too extensive. As Gofman and Tamplin have argued, however, the absence of clinical symptoms does not rule out the possibility that increased cancer rates will be observed later in life. Appearance of cancer many years after

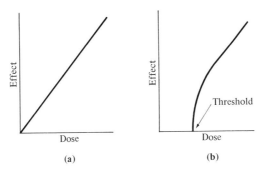

FIG. 15-3. Possible simple dose-effect relationships for radiation exposures (or, in fact, for exposures to any type of pollution). **(a)** Linear dose-effect curve. **(b)** Threshold dose-effect curve.

exposures is a well-known phenomenon (see Chapter 1), but it is not known whether or not small radiation exposures can have similar effects. If the linear relationship holds, exposure to natural radiation of 0.1 rad/year for 30 years would amount to 3 rads, and if 1 rad corresponds to 2% of the spontaneous cancer rate, we might suppose that 6% of the spontaneous cancers at age 30 are due to natural radiation. It is impossible to determine whether or not this is the case. It is also impossible to determine whether or not regions of greater natural radiation have greater cancer rates because the effects are too small to be detected by statistical tests even when they involve millions of persons.

If people in the U.S. were exposed to 0.17 rad/year from man-made radiation sources (as permitted by the FRC guidelines), the 30-year dose (which Gofman and Tamplin believe to be the most meaningful length of time) would be 5 rads—perhaps sufficient to increase cancer mortality by 10%. For the U.S. this would amount to about 35,000 extra deaths annually, or even more when one takes into account the fact that the effects are much greater when the radiation exposures are to small children [6]. It is the size of this number—somewhat less than the number of automotive fatalities but several times the number of murders—that has led Gofman and Tamplin to suggest a lowering of the guidelines by at least a factor of 10, to 0.017 rad/year.

There is not enough information to know whether or not Gofman and Tamplin are overestimating the seriousness of radiation exposures since the effects of long-term, low-level exposures to radiation and other pollutants are simply not known today. Under the circumstances the lowered radiation standard they have proposed is an eminently sensible one.

At present, of course, the actual exposures of the population to man-made radiation from the nuclear industry are much smaller than Gofman and Tamplin's suggested standard, apart from medical X rays. While no one can quarrel with high radiation exposures in the treatment of diseases that might otherwise soon lead to death, the use of X rays in medicine and dentistry for diagnostic purposes may still be dangerously high, as medical experts often warn [24]. Today they constitute a much greater source of radiation than the nuclear industry and there is no prospect that this will change in the near future. Medical radiation uses are much older than the nuclear industry and have never been regulated by the AEC, whose jurisdiction does not extend to these uses. Regulation has been left up to state and local governments and has generally lagged behind the introduction of new technical apparatus.

In 1970 the federal government's power in regard to radiation standards was transferred to the new Environmental Protection Agency (see Chapter 20). The EPA has been urged to undertake a complete review of radiation standards and protection guidelines, and changes could be in the offing.

In concluding this section, mention should be made of the arguments presented by Dr. Ernest Sternglass to show that Sr^{90} fallout from atmospheric weapons tests in the 1950's and early 1960's caused a pause in the decline of infant mortality rates and may have led to an excess of 500,000 infant deaths in the U.S. [25]. These arguments have been widely criticized by other scientists [26], and Arthur Tamplin has argued that the correct number, based on the data given earlier in this section, is probably 4000 [27].

NUCLEAR POWER PLANTS

The importance of the growing nuclear power industry, and the possibility that within a few decades most of the electric power generation in this country will be nuclear, makes it essential in any discussion of environmental pollution to devote some attention to the present and future problems of nuclear power plants. Information about their thermal pollution problems was presented in Chapter 14, and Chapter 16 will contain a comparison of nuclear power plants with fossil-fuel plants.

The four areas of concern (apart from thermal pollution) in the present-day operation of nuclear power plants are radiation releases from the plants themselves and the nuclear fuel reprocessing plants, the possibility of accident, the disposal of radioactive wastes, and the radiation problems associated with the mining of the nuclear fuel.

The reactor core of a nuclear power plant, in which the heat is produced by fission of U^{235} or other fissionable nuclei, may contain 50 to 100 metric tons of fuel in the form of uranium dioxide (UO_2). The individual fuel rods are perhaps 1 cm in diameter and 3.5 m long, and there may be 40,000 such rods in the core. With all the rods in place, a chain reaction can occur but the fission rate, and hence the heat-generation rate, is controlled by control rods of materials (such as cadmium) that absorb neutrons. The fuel rods have a cladding of stainless steel or a zirconium alloy, and despite

FIG. 15-4. Cutaway view of a nuclear steam-supply system based on a pressurized-water reactor. Note the size of the man entering the door at the right. (*Courtesy of the U.S. Atomic Energy Commission.*)

the care with which they are inspected for flaws there are imperfections in the cladding through which fission products can leak out and contaminate the coolant. Radioactive materials in the coolant can be reduced by purification processes. Some, like Kr^{85} and tritium, are gases. Radioactive gases can be held up for a while and later permitted to vent into the atmosphere in minute concentrations. Every year or so the spent fuel must be removed from the reactor core and replaced; the spent fuel is allowed to sit for several months while much of the short-lived radioactivity dies down, and then it is taken to a reprocessing plant.

Accidental releases of radiation are prevented by the use of a large containment vessel around the reactor and steam generators; the container is typically reinforced concrete 1 m thick with a vaporproof steel liner. Intentional releases are permitted in accordance with AEC specifications in the operating license for the plant, and they apply to specific isotopes and take such things as possible biological concentration into account. In practice most of the large nuclear power plants release less than 1% of the radiation permitted.

Although by design and in practice the radiation releases lead to radiation exposures of no more than 2 mrem/year at the plant boundaries and much less some distance away, there could be problems developing in the future if nuclear power capacity increases hundreds or thousands of times as expected. The releases of Kr^{85} and tritium (H^3) have attracted particular attention since both have half-lives of over a decade and may build up

(a)

(b)

(c)

FIG. 15-5. Views of nuclear power plants. **(a)** The 575 MW(e) plant of the Connecticut Yankee Atomic Power Company at Haddam Neck, Conn., on the Connecticut River. (*Courtesy of Connecticut Yankee Atomic Power Company.*) **(b)** The Consumers Power Company's Big Rock Point nuclear power plant at Charlevoix on Lake Michigan. Power output is 75 MW(e). (*Courtesy of Consumers Power Company.*) **(c)** The Yankee nuclear power plant at Rowe, Mass. (*Courtesy of Yankee Atomic Electric Company.*)

steadily in the environment. It is believed that the Kr^{85} in the environment, if present release practices continue, could lead to 2 mrem/year exposure to the general population (of the world) by the year 2000 [15]. Although this is small compared to natural radiation and to medical radiation, it is significant according to the arguments of Gofman and Tamplin. Far greater efforts to prevent the worldwide contamination by such substances should be made, even if the costs of producing electric power are increased in the process. The Oak Ridge Gaseous Diffusion Plant of the AEC has been developing a promising process to remove radioactive krypton and xenon gases from contaminated gas streams using a fluorocarbon solvent, and 99.9% removal appears likely [28].

Tritium is also a problem but not as great a one. By the year 2000, it is estimated that the exposure of the population to tritium from nuclear power plants might be 0.02 mrem/year [15]. It will take many decades for tritium produced by nuclear power plants to equal that produced by the fallout from weapons testing in the years before the 1963 test ban treaty. Fusion ("hydrogen") bombs produced about 6.7 megacuries of tritium per megaton of fusion yield, and all the U.S. and U.S.S.R. tests added about 1700 megacuries to the atmosphere, which formerly had only about 70 megacuries from equilibrium between radioactive decay of tritium and its production by cosmic radiation (through interaction with oxygen and nitrogen atoms in the upper atmosphere). Much of the bomb-produced tritium was released in the stratosphere and is being gradually introduced into the troposphere, in which its effective half-life (before being incorporated into the lithosphere) is only 35 to 40 days. Tritium produced by nuclear plants, if their growth is correctly projected, will not equal the natural tritium concentration until about the end of the 20th century, and even then the residual tritium from weapons testing fallout will still be greater. The concern with tritium would be greater if it could be biologically concentrated but no evidence exists that biological systems distinguish between tritium and ordinary hydrogen atoms [13].

Safety has been an important consideration of the nuclear energy program since its inception, and numerous safeguards have been instituted to protect against various possible accidents. Nevertheless serious accidents are still a possibility. A nuclear explosion similar to that in a fission ("atomic") bomb is not possible because there is no way to bring sufficient fissionable material together, not even by sabotage [29]. A number of minor accidents have occurred in nuclear power plants with no radiation releases to the environment, and in 1966 an obstruction in the coolant system at the Enrico Fermi breeder reactor near Detroit led to meltdown of some of the fuel, again with no radiation releases. The possibility of a loss of coolant in the reactor core is the most worrisome problem facing nuclear engineers, and it is far from certain that the hazards are absent in presently designed plants [30]. Such a loss of coolant would permit the heat release from fission to remain in the core and melt it down; melting through the containment vessel would eventually occur and radioactive materials would be released.

To date there have been several hundred reactor-years experience with nuclear power plants and several thousand reactor-years experience with other reactors used for research and training, all without a major accident endangering the public. The odds against a major accident are impossible to calculate, although one author favorable to the nuclear power industry has suggested that they are at least 1 billion to 1 for each reactor each year [31]. Whatever the odds, there are some simple precautions that can be taken to reduce the chances or effects of a major accident, such as avoiding construction in areas prone to earthquakes or in the vicinity of major population centers or building the plants underground [32].

Nuclear power plants and other users of radioactive materials eventually produce highly radioactive wastes that require special handling and disposal. In the past, many of these wastes have been stored in liquid form. Estimates of the volume of liquid wastes to be stored in the future have actually decreased despite the increases in projected nuclear capacity because greater concentrations of wastes have proved feasible. The ultimate disposal problem has not been solved. The AEC is presently storing radioactive wastes in large underground tanks but is considering solidifying them and encasing them in shielded cylinders for permanent (i.e., for several centuries') storage

FIG. 15-6. A view inside the Carey Salt Company Mine in Lyons, Kan., scene of a radioactive waste disposal experiment conducted by the Oak Ridge National Laboratory. Holes in the salt mine floor 12 ft (3.7 m) deep and lined with stainless steel are used to contain radioactive fuel assemblies. (*Courtesy of the U.S. Atomic Energy Commission.*)

in underground salt mines. Salt mines are often very stable and dry, so that problems of earthquakes and groundwater contamination are minimized; some salt mines, however, are not continuous, and groundwater contamination is then possible. In addition, the high thermal conductivity of the salt will permit the heat produced in radioactive decay to be dissipated readily. The AEC has tentatively chosen the Carey Salt Company Mine in Lyons, Kans., for the Federal Wastes Repository, although this choice has been criticized by the Kansas State Geological Survey [33]. The land requirements will be very small compared to that for disposal of other solid wastes—about 1.1 hectares (2.8 acres) per year by the year 2000 [15].

There are pollution problems involved in the mining of nuclear fuel. Uranium miners in the U.S. and other countries have been subject to veritable epidemics of lung cancer caused by exposure to radioactive radon gas and its decay products [6]. This problem can be reduced by proper ventilation of the mines, and strict exposure controls were promulgated by the Secretary of Labor after the 1967 hearings of the Joint Committee on Atomic Energy on radiation exposure of uranium miners. Radiation dangers are also present near the piles of tailings left over from the grinding of radioactive ores. For many years these tailings were given away to builders in Colorado, with the result that many buildings in Grand Junction, Colo., and other towns have dangerous radiation levels from the presence of radon and its decay products [34].

The pollution problems of nuclear power plants and the allied mining and fuel reprocessing industries can be controlled with proper emphasis on reducing radiation exposures, even though some financial cost will result. This is no different from the situation with other types of electric power generation or other industries in the U.S.

REFERENCES

1. Davis, Edward P., "The Study of the Infant's Body and of the Pregnant Womb by Röntgen Rays." *Amer. Jour. Med. Sciences* **111**:263–270 (1896).
2. Cipollaro, A. C., and M. B. Einhorn, "The Use of X-Ray for the Treatment of Hypertrichosis is Dangerous." *Jour. Amer. Med. Assn.* **135**:350–353 (1947).
3. Frigerio, Norman A., *Your Body and Radiation.* Oak Ridge, Tenn.: U.S. Atomic Energy Commission, 1969. Available free from USAEC, P.O. Box 62, Oak Ridge, Tenn. 37830.
4. Pollard, Ernest C., "The Biological Action of Ionizing Radiation." *American Scientist* **57**:206–236 (1969).
5. Brannigan, F. L., "Radiation in Perspective." *Nuclear Safety* **5**:226–228 (1964).
6. Tamplin, Arthur R., and John W. Gofman, *"Population Control" through Nuclear Pollution.* Chicago, Ill.: Nelson-Hall, 1970.
7. Miller, Robert W., "Delayed Radiation Effects in Atomic-Bomb Survivors." *Science* **166**:569–574 (1969).
8. Boffey, Philip M., "Hiroshima/Nagasaki. Atomic Bomb Casualty Commission Perseveres in Sensitive Studies." *Science* **168**:679–683 (1970).
9. Kastner, Jacob, *The Natural Radiation Environment.* Oak Ridge, Tenn.: U.S. Atomic Energy Commission, 1968. Available free from USAEC, P.O. Box 62, Oak Ridge, Tenn. 37830.
10. Livesey, Derek L., *Atomic and Nuclear Physics.* Waltham, Mass.: Blaisdell Pub. Co., 1966. pp. 497–500.
11. Adams, J. A. S., and W. M. Lowder (eds.), *The Natural Radiation Environment.* Chicago, Ill.: The University of Chicago Press, 1964.
12. Asimov, Isaac, and Theodosius Dobzhansky, *The Genetic Effects of Radiation.* Oak Ridge, Tenn.: U.S. Atomic Energy Commission, 1966. Available free from USAEC, P.O. Box 62, Oak Ridge, Tenn. 37830.
13. *Environmental Effects of Producing Electric Power.* Hearings before the Joint Committee on Atomic Energy, October and November 1969. Part I. Washington, D.C.: U.S. Govt. Printing Office, 1969.
14. Gillette, Robert, "Reactor Emissions: AEC Guidelines Move Toward Critics' Position." *Science* **172**:1215–1216 (1971).
15. *Selected Materials on Environmental Effects of Producing Electric Power.* Washington, D.C.: U.S. Govt. Printing Office, 1969.
16. Sagan, L. A., "Human Radiation Exposures from Nuclear Power Reactors." *Arch. Environ. Health* **22**:487–492 (1971).
17. Comar, C. L., *Fallout from Nuclear Tests.* Oak Ridge, Tenn.: U.S. Atomic Energy Commission, 1967. Available free from USAEC, P.O. Box 62, Oak Ridge, Tenn. 37830.

18. Palmer, B. D., "Interhemispheric Transport of Atmospheric Fission Debris from French Nuclear Tests." *Science* **164**:951–952 (1969).
19. Murthy, G. K., and J. E. Campbell, *Profile of Long-Lived Radionuclides in Milk*. Washington, D.C.: U.S. Dept. of Health, Education, and Welfare, 1964. Public Health Service Publication No. 999-FP-2.
20. The monthly issues of *Radiological Health Data and Reports* (currently published by the Environmental Protection Agency) contain the results of monitoring of air, water, milk, and other food for radioactive isotopes. The Pasteurized Milk Network monitors milk in dozens of U.S. cities for Sr^{90}, Cs^{137}, I^{131}, and other radioisotopes.
21. Gerber, C. R., et al., *Plowshare*. Oak Ridge, Tenn.: U.S. Atomic Energy Commission, 1966. Available free from USAEC, P.O. Box 62, Oak Ridge, Tenn. 37830.
22. Mamuro, T., and T. Matsunami, "Plutonium-238 in Fallout." *Science* **163**:465–467 (1969).
23. Holcomb, Robert W., "Radiation Risk: A Scientific Problem?" *Science* **167**:853–855 (1970).
24. Hahn, D. R., and D. E. Van Farowe, "Misuse and Abuse of Diagnostic X-Ray." *Amer. Jour. Public Health* **60**:250–254 (1970).
25. Sternglass, E. J., "Infant Mortality and Nuclear Tests." *Bull. Atomic Scientists* **25**(4):18–20 (April 1969); Sternglass, E. J., *Low-Level Radiation*. New York: Ballantine Books, Inc., 1972.
26. Sagan, L. A., et al., "The Infant Mortality Controversy: Sternglass and His Critics." *Bull. Atomic Scientists* **25**(8):26–32 (October 1969).
27. Tamplin, Arthur R., "Fetal and Infant Mortality and the Environment." *Bull. Atomic Scientists* **25**(10):23–29 (December 1969).
28. Merriman, J. R., et al., *Removal of Radioactive Krypton and Xenon from Contaminated Off-Gas Stream*. Oak Ridge, Tenn.: Union Carbide Corporation, Nuclear Division, Oak Ridge Gaseous Diffusion Plant, 1970. Report No. K-L-6257.
29. Lapp, Ralph E., "The Four Big Fears About Nuclear Power." *N.Y. Times Magazine*, February 7, 1971, pp. 16–35.
30. Gillette, Robert, "Nuclear Reactor Safety: A Skeleton at the Feast?" *Science* **172**:918–919 (1971).
31. Starr, Chauncey, "Radiation in Perspective." *Nuclear Safety* **5**(4):325–335 (1964).
32. Rogers, Franklyn C., "Underground Nuclear Power Plants." *Bull. Atomic Scientists* **27**(8):38–41, 51 (October 1971).
33. Holden, Constance, "Nuclear Waste: Kansans Riled by AEC Plans for Atom Dump." *Science* **172**:249–250 (1971).
34. Ripley, Anthony, "Radiation Hazard Found in 82 Colorado Buildings." *N.Y. Times*, February 17, 1970, p. 70.

16

ELECTRIC POWER GENERATION

*Of the great number of industries which the enlightened genius of the
nineteenth century has called into existence, few, if any, can show
such a remarkable record as the electrical industry, and none other
such a rapid growth and development. . . . Electric light and electric
power have already produced more changes in mechanical servants
and conveniences of human life than have ever been caused by the
use of any other force which has been subjected to the service
of man. No backward step has been taken; there has been no check
to its onward triumphal progress.* —"Electricity in the United
States." *The Engineer* **75**:23 (January 13, 1893).

At the time these words were written, nearly 80 years ago, the electrical
industry was less than 20 years old, and despite the "rapid growth and
development" was perhaps only producing 1/1000 as much electric power as
today. To the "mechanical servants and conveniences of human life" have
since been added an incredible array of new electrical devices—fans, razors,
radios, dryers, ranges, irons, toasters, blenders, scissors, water heaters, food
freezers, refrigerators, automatic ice-making machines, toothbrushes, dehu-
midifiers, can openers, hair curlers, dishwashers, television sets, stereo equipment, air conditioners, frying pans, washing machines, vacuum cleaners,
knives, central home heating, tape recorders, clocks, ...

Electricity may be cheap as far as out-of-pocket expenses are concerned, but it is accompanied by environmental effects that constitute social
costs to all of us. As the data presented earlier in this book have shown,
electric power generation is presently responsible for about:

80% of the U.S. cooling water demand and the resulting thermal
pollution.

50% of the sulfur oxides emissions.

20% of the nitrogen oxides emissions.

20% of the particulate matter emissions.

Small amounts of radiation, hydrocarbons, and carbon monoxide.

The following sections of this chapter will discuss the history of electric power generation in the U.S., the pollution produced by steam-electric

power plants, present and future alternate methods of generating electricity with particular emphasis on their ability to reduce environmental pollution, and some miscellaneous problems occurring in the use of electricity.

HISTORY OF ELECTRIC POWER GENERATION

The phenomenal growth of the electrical industry in the late 19th century, referred to in the quotation opening this chapter, has continued throughout the 20th century, as shown in Figure 2-4. Table 16-1 shows the actual electric energy production and capacity at 5-year intervals since 1902, the only decrease occurring early in the Great Depression of the 1930's; the steady annual increases in electric power production have not been interrupted by recent depressions and recessions, the last year-to-year decrease occurring in 1946.

TABLE 16-1. Electric power production (in billions of kilowatt-hours), electric power capacity (in millions of kilowatts), and the average utilization of capacity in the U.S. at 5-year intervals [1].

Year	Production	Capacity	Average utilization (%)
1902	6	3.0	23
1907	14	6.8	23
1912	25	11.0	26
1917	43	15.5	32
1922	61	21.3	33
1927	101	34.6	33
1932	99	42.8	26
1937	146	44.4	38
1942	233	57.2	46
1947	307	65.2	54
1952	463	97.3	54
1957	716	146	56
1962	947	210	51
1967	1317	288	52
1972	ca 1800	ca 410	ca 50

The U.S. increase in electric power production has been matched by the worldwide increase; during the 12-year period from 1955 to 1967 world production increased 142% from 1539 billion to 3751 billion kWh, while U.S. production rose 110% from 629 billion to 1317 billion kWh [2].

Over four-fifths of U.S. electricity is produced by the combustion of fossil fuels (coal, oil, gas), most of the rest being produced by water power (hydroelectricity). In 1970 the approximate percentages were [3] 82.5% from fossil-fuel combustion, 16% from hydroelectricity, and 1.5% from nuclear power. Other countries show very different percentages [2]. Canada generates about 80% of its electricity as hydroelectricity, and the Scandinavian countries of Sweden and Norway have even higher percentages of hydro-electricity. The United Kingdom, on the other hand, derives less than 3% from hydroelectricity but already well over 10% from nuclear power. There are also some minor sources of electric power in use around the world— tidal power and geothermal power, for example. In 1967, geothermal electric power production from steam and hot water wells amounted to 4.0 billion kWh worldwide—2.6 billion kWh in Italy, 1.1 billion kWh in New Zealand, and 0.3 billion kWh in the U.S. [4].

TABLE 16-2. Energy consumption in the U.S. for selected years and projections for 1980 [5].

	1937	1947	1957	1965	1980 (est.)
Total (10^{12} kWh)	6.7	9.7	12.2	15.8	25.8
Electricity (10^{12} kWh)	0.15	0.31	0.72	1.16	2.7
Electricity/total (%)	2.2	3.1	5.9	7.4	10.4
Sources of total energy (%)					
Coal	55	48	27	23	18–22
Oil	30	34	44	43	41
Natural gas	11	14	25	30	28
Hydroelectric power	4	4	4	4	4
Nuclear power	0	0	0	0.1	5–9
Total	100	100	100	100	100

The fundamental sources of energy in the U.S. today are the fossil fuels (coal, oil, and gas), hydroelectricity, and nuclear, apart from some minor sources such as wood and geothermal power. Table 16-2 lists the U.S. energy consumption and its breakdown into fundamental sources for several different years, together with Federal Power Commission (FPC) projections to 1980. The electric power production—most of it derived from fossil fuels— is also given, and it is seen to be less than 10% of our total energy consumption. In 1965, however, when 1.16×10^{15} kWh of electricity were produced the total energy *input* into electric power production was 3.2×10^{15} kWh [5]. Thus, while electricity produced and consumed amounted to only 7.4% of total U.S. energy consumption, its production required 20% of our energy consumption, the remaining 12.6% being waste heat capable of causing thermal pollution.

Consumption of electricity is roughly 50% by industry, 30% by households and 20% by commercial establishments. The industrial share has been decreasing over the years, while the residential share has been rising.

POLLUTION FROM STEAM-ELECTRIC GENERATING PLANTS

Many environmental pollution problems that arise from electric power production are particular to steam-electric generating plants, whether they use fossil fuels or nuclear energy as the fundamental source of energy. Hydroelectric plants do not have these problems as they convert the mechanical energy of moving water directly into electricity. Hydroelectric plants have often been criticized because of their effects on the ecology and on scenic values, however. A proposed hydroelectric plant in Norway's Jotunheim Mountains will drain water away from the Eastern Mardola waterfall, which at over 500 m in height is the sixth highest in the world, and great opposition has been encountered from persons who prefer to retain the natural scenery. In the U.S., plans for pumped-storage hydroelectric plants have sometimes encountered opposition; these are plants associated with huge reservoirs that can be filled with water during periods of low demand and emptied during peak demand periods to increase the peak electric power capacity. The plans for one such plant in the lower Berkshires along the Massachusetts-Connecticut border called for 920 hectares of reservoirs with water levels varying by as much as 45 m.

The air pollutant emissions from fossil-fuel plants depend on the type of fuel used. The emission rates and total annual emissions for typical 1000 MW(e) plants using coal, fuel oil, and natural gas have been calculated by Terrill, et al. [6] and are summarized in the discussion below and in Table 16-3.

TABLE 16-3. Typical pollutant emissions from different types of fossil-fuel-fired electric generating plants, expressed as thousands of metric tons annually for a 1000-MW(e) plant operating at capacity [6].

Pollutant	Coal-fired	Oil-fired	Gas-fired
Aldehydes	0.005	0.117	0.031
Carbon monoxide	0.52	0.008	—
Hydrocarbons	0.21	0.67	—
Nitrogen oxides	21	21.7	12
Sulfur oxides	139	52.6	0.012
Particulates	4.5	0.73	0.46

1. *Coal-fired plant.* A 1000-MW(e) plant will burn about 2.1 million metric tons of semibituminous coal annually (1 metric ton every 15 s). Assuming that the coal has a sulfur content of 3.5% (15% of which remains in the ash) and an ash content of 9% and that there is no pollution control equipment except for fly ash removal equipment having 97.5% efficiency, the following emission rates (in kilograms per metric ton of coal burned) may be expected:

    0.0025 kg aldehydes
    0.25 kg carbon monoxide
    0.1 kg hydrocarbons
    10 kg nitrogen oxides
    66.5 kg sulfur oxides
    2.25 kg particulates

There will also be nearly 90 kg of fly and bottom ash to dispose of for every metric ton of coal burned. Coal also contains trace quantities of two naturally occurring radioactive materials, U^{238} (about 1.1 ppm) and Th^{232} (about 2 ppm), as well as their decay products. As a consequence the coal-burning plant described above would release to the atmosphere (in the emitted particulates) about 10.8 millicuries of Ra^{228} and 17.2 millicuries of Ra^{226} annually [7]. In terms of the hazards to humans, these releases would constitute a more serious problem than the radiation releases from nuclear power plants [7].

2. *Oil-fired plant.* A 1000-MW(e) oil-fired electric power plant would require about 1.7 million m^3 of oil annually (1 m^3 every 18 s). Assuming a sulfur content of 1.6% and an ash content of 0.05%, and assuming no pollution control equipment, the emission rates of air pollutants would be approximately as follows per metric ton of oil burned [6]:

> 0.07 kg aldehydes
> 0.005 kg carbon monoxide
> 0.4 kg hydrocarbons
> 13 kg nitrogen oxides
> 31.4 kg sulfur oxides
> 0.44 kg particulates

Oil-fired plants also emit some radioactive elements, including a total of about 0.5 millicuries annually of Ra^{226} and Ra^{228} in the case of the plant described above [7].

3. *Gas-fired plant.* A 1000-MW(e) electric plant operating on natural gas might burn about 1.9 billion m^3 (68 billion ft^3, defined at standard conditions of 60°F and 14.73 lb/in^2 pressure) annually. Assuming no pollution control equipment, the emission rates in kilograms per million cubic meters of gas will be approximately [6]:

> 16 kg aldehydes
> negligible quantities of carbon
>   monoxide and hydrocarbons
> 6240 kg nitrogen oxides
> 6.4 kg sulfur oxides
> 240 kg particulates

From the summary of total annual releases in Table 16-3 the superiority of natural gas over either of the other two fossil fuels is immediately apparent. Natural gas may have some radioactive gases present in small amounts but information is lacking about them. It is unfortunate, as discussed in Chapter 2, that natural gas supplies are running low. Some consideration has been given to the use of new natural gas supplies released by nuclear explosives through Plowshare programs but radioactive contamination of the gas with Kr^{85} and other radioisotopes may occur.

Some control methods are available to alleviate the air pollution problems of fossil-fired plants, as detailed in Chapter 5. Electrostatic precipitation works well to control fly ash problems, but a 1000-MW(e) coal-fired plant operating near capacity will require 2.4 hectares (6 acres) of land annually for fly ash disposal if the ash is piled to a depth of 7.5 m, so the solid waste disposal problem is not trivial. Sulfur dioxide removal methods are being developed but no clearly economical method has yet emerged. Many electric utilities have reduced the sulfur oxide problem by going to the use of low-sulfur fuel, which is especially important during smog episodes. There is plenty of low-sulfur coal in the U.S., as shown in Table 16-4, but 90% of it is west of the Mississippi River, distant from the great metropolitan areas of the Atlantic seaboard and the Great Lakes [5].

TABLE 16-4. U.S. coal reserves (in terms of heat content) by sulfur content [5].

Type	Low S (0–1%)	Medium S (1–3%)	High S (>3%)	Total (%)
Bituminous	17.3	15.6	25.1	58.0
Subbituminous	22.4	0.1	—	22.5
Lignite	16.6	1.7	—	18.3
Anthracite	1.2	—	—	1.2
Total	57.5	17.4	25.1	100.0

Air pollution problems from fossil-fuel plants can be reduced somewhat by the use of tall stacks or the construction of mine-mouth generating plants. Tall stacks 150 m high or more can help disperse the pollutants and thereby lower pollution concentrations at ground level, but they obviously do not decrease the amount of air pollution and may simply transfer the problem to a new locale; in addition they are likely to be rather unaesthetic. Tall stacks can be especially desirable during smog episodes since the stack may rise above a low inversion or may have a hot plume that rises above the inversion. Mine-mouth generating plants, which have been built or planned in western Pennsylvania, southern Illinois, western Kentucky, northwestern New Mexico, and elsewhere, are located near coal mines rather than in or near metropolitan areas. They avoid coal transportation costs and have been made possible by recent advances in the technology of extra-high-voltage transmission that have reduced the costs of transmission. These plants still have pollution but may annoy and affect the health of fewer people. Unfortunately the people whose health is affected may not be the people benefiting from the plant. A number of large electric plants have been built near the Four Corners in the Southwest, and American Indians and others in the area have voiced complaints about being subjected to air pollution so people far away can have air conditioners and other electrical appliances.

The conversion of coal into gas or liquid fuels essentially free of sulfur and fly ash is technologically possible today at costs almost competitive with

FIG. 16-1. The world's tallest chimney—381 m high (taller than the Empire State Building)—was built in 1970 for the Copper Cliff (Ontario) smelter of the International Nickel Company of Canada by the Canadian Kellogg Company, Ltd. The diameter tapers from 35.5 m at the base to 15.8 m at the top; 39,000 metric tons of concrete and 900 metric tons of reinforcing steel were used in constructing the chimney. The structure was unscathed by a freak storm with 145 km/h (90 mi/h) winds that hit the Sudbury area on August 20, the day before the chimney was topped out. (*Courtesy of the M. W. Kellogg Company, a Division of Pullman Incorporated.*)

the cleaner fuels [5]. Research is continuing at several places in the U.S. The natural gas industry is especially interested in gasification since it may need an economical replacement for natural gas in the near future. El Paso Natural Gas Company is already investing $250 million in a coal gasification facility to be operational in 1976.

TABLE 16-5. Electric power statistics and projections [8].

	1950	*1968*	*1980*	*2000*
U.S. population (millions)	152	202	235	320
Power capacity (10^3 MW)	85	290	600	1352
Kilowatt capacity per person	0.6	1.4	2.5	4.25
Kilowatt-hours per year per person	2,000	6,500	11,500	25,000
Nuclear capacity (10^3 MW)	0	2.7	150	941
Nuclear fraction of total (%)	0	1	25	69

The pollution problems of electric power generation are important now and may become much greater in the near future if the growth of electric production continues its historical and current trend of doubling every 10 years. Table 16-5 gives AEC historical and projected data on electric power capacity and consumption [8].

The summary of pollution problems from nuclear and fossil-fuel power plants is as follows:

1. Thermal pollution is currently a more important problem with nuclear plants in operation or under construction, but the efficien-

cies of nuclear plants should equal those of new fossil-fuel plants within a few years.

2. Ordinary air pollution is much worse with fossil-fuel plants than with nuclear plants when plants of the same generating capacity are compared, but emissions from the former could be controlled much more than they are today. This air pollution is undoubtedly responsible for adverse health effects in a large part of the U.S. population.

3. Radiation from nuclear power plants is presently quite trivial compared to numerous other radiation sources and is probably actually less important than radiation from the fly ash of coal-burning plants of the same capacity. Proliferation of nuclear power plants could lead to serious problems in several decades, however.

4. The potential for a serious accident involving thousands of persons is probably greater for nuclear plants but the probability of such an accident appears to be small, and air pollution from fossil-fuel plants may be killing as many persons every few years as would be killed in such an accident.

5. Nuclear power plants involve solid waste disposal problems for dangerous radioactive wastes, while coal-burning plants have large amounts of ash to dispose of.

6. Fossil-fuel plants release carbon dioxide to the atmosphere, while nuclear plants do not. A buildup of CO_2 in the atmosphere could have serious long-term consequences (see Chapter 4).

Any plans for the construction of a new power plant involve the balancing of these risks from the different types of plants. These risks need to be carefully balanced against the benefits, and it may very well be that the best solution would be to try to reduce the fantastic growth in our use of electricity.

PRESENT ALTERNATIVE METHODS OF GENERATING ELECTRICITY

There are a number of ways of generating electricity other than the steam-electric plants based on nuclear power or fossil-fuel combustion but none offers any immediate hope of satisfying the electric power requirements of the U.S. or the world [9].

Hydroelectric power is already an important factor in the U.S. and many other countries. In some countries (such as Norway and Sweden) or particular localities within a country, it may satisfy the electric power requirements quite well. The worldwide scope of water power, however, is not sufficient to generate all the world's electric power. The world's present electric power capacity is approximately 10^{12} watts, and water power is estimated [9] to provide no more than 2.86×10^{12} watts if completely developed, and even then much of it would be too far from potential users. The United States now has a total electric power capacity of about 0.4×10^{12} watts but a potential hydroelectric power capacity of only 0.16×10^{12} watts, of which about one-third is presently developed.

Solar radiation falling on the earth amounts to 1.8×10^{17} watts (see Chapter 4), many times the power used by the human population. Although it can be used for some purposes, such as heating water or greenhouses, it is so diffuse (never more than about 1 kW/m^2) that it has not yet been exploited on a large scale. Solar power has been under active investigation for a number of years and there is renewed hope that it can be utilized to produce significant amounts of power [10, 11].

Another possibility is tidal power, which is obtained by harnessing the energy of ebbing (falling) tides. There are several power plants around the world making use of it, notably at La Rance in France. Recoverable tidal power worldwide is probably only about 1.3×10^9 watts, however, much less than the 10^{12} watts used today [9].

Geothermal energy from wells drilled into reservoirs of superheated water or steam in volcanic areas is now used for about 10^9 watts around the world but is probably capable of supplying at most 6×10^{10} watts worldwide [9]. Although this is insufficient to be the answer to the world's electric power requirements, it may be the answer to the needs of certain localities. Members of the faculty of the Institute of Geophysics and Planetary Physics at the University of California at Riverside are presently investigating the development of geothermal resources in the Imperial Valley of California for the production of electricity and the desalination of seawater [12, 13].

There are several other minor possibilities for the production of electricity, such as the combustion of municipal refuse or the use of wind power, but they cannot generate sufficient power to meet the needs of an advanced economy.

At present, then, only fossil-fuel combustion, which can supply our energy needs only a few more centuries at best, and nuclear energy of various types can meet the world's energy (including electrical energy) needs over a period of decades or longer [9]. Of course, there always remains the possibility of a major scientific breakthrough—the development of economical ways to harness solar energy for large-scale power purposes or the discovery of a new source of energy hitherto unsuspected—but commercial development would require years or decades following the breakthrough, and the world cannot wait in hope for such an event.

Fossil-fuel plants could be built that reduced pollution drastically from current levels. They would need to use coal after a few more decades at most since petroleum and natural gas supplies as such would not last indefinitely, and they would produce serious thermal pollution problems even with the best possible advances in increasing the efficiency of steam-electric plants (see the next section). Research on obtaining "clean power from coal" [14] is highly desirable.

Nuclear power plants of the present types, which require U^{235} for fissioning, cannot supply the world's electric power needs for very long since uranium reserves are limited and since U^{235} constitutes only 0.7% of the naturally occurring uranium (U^{238}, which is not fissionable, constitutes the other 99.3%). Known economically recoverable uranium reserves will last several decades if nuclear power plants increase in number at the projected rate, and new discoveries of uranium or new technologies might extend this period somewhat but it is quite possible that uranium shortages could develop in the first half of the 21st century.

FUTURE METHODS OF GENERATING ELECTRICITY

Hope for the future of electric power generation centers around increasing the efficiencies of power plants and developing new fuel resources.

The present steam-electric plants, fossil-fuel or nuclear, are unlikely to attain thermal efficiencies much above 40%. There are several schemes for combining them with other concepts for generating electricity, however, which may lead to significantly higher efficiencies.

Particular attention has been devoted to the possibility of direct conversion of heat and other forms of energy into electricity, i.e., conversion that does not involve mechanical motion [15]. The ordinary flashlight battery produces electricity from chemical energy by direct conversion, for example. When heat is not involved (as in batteries and fuel cells), direct conversion is not limited by Carnot efficiencies (see Chapter 14). Most of the direct conversion methods being investigated for large-scale power production—thermoelectricity, thermionic conversion, magnetohydrodynamics—involve heat at some stage, however, and thus are limited by Carnot efficiencies; these methods offer the hope of high efficiencies only because of high-temperature heat sources.

Thermoelectricity is the flow of electric current in a circuit composed of wires of two different materials when the two junctions are kept at different temperatures [see Figure 16-2(a)]. This property was discovered in Germany by T. L. Seebeck in 1821 and is known as the Seebeck effect. The electric power is produced directly from heat by this method. Although thermoelectric power production using metals has never proved to be practical, some useful devices for special applications have been constructed in recent years using semiconductors [15]. Thermal efficiencies are no more than 10% at present, however.

Thermionic conversion [see Figure 16-2(b)] involves the boiling off of electrons from a hot metal surface ("thermionic emission") and their collection on a cooler metal surface with a different work function so that a potential difference can be created and electric power generated. Soviet scientists have claimed a recent breakthrough permitting the construction of a thermionic converter with a capacity of several kilowatts [16] but again the thermal efficiencies are still low in most of these devices.

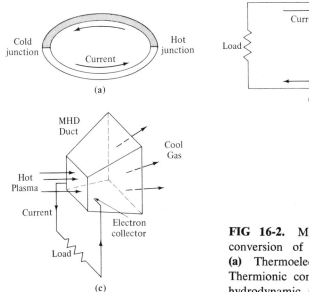

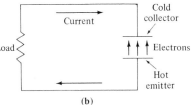

FIG 16-2. Methods for the direct conversion of heat into electricity. **(a)** Thermoelectric conversion. **(b)** Thermionic conversion. **(c)** Magnetohydrodynamic (MHD) conversion.

The most promising method under investigation today is MHD— magnetohydrodynamic conversion. This method involves several steps. First, a gas is heated until it is a plasma, a highly ionized and electrically conducting gas. The plasma is passed through a nozzle to convert the random motion of heat into directed kinetic energy. It is then passed through a magnetic field, which produces a Lorentz force on the electrons and directs them to one side, where they are collected. Electric power is produced by the expanding plasma. The cooled gas exhausted can be used as a heat source for an ordinary steam-electric generating plant. In this way the MHD generator serves as a "topping unit" that converts 15 or 20% of the heat into electricity, and the overall efficiency of the complete power plant might be as high as 60%. MHD is depicted in Figure 16-2(c).

Even if these or other ideas for increasing the efficiency of electric power production are developed successfully, there remains the problem of finding suitable energy sources to meet future needs. The most promising possibilities for the near future are breeder reactors [17–19] and fusion power [20].

The case for the development of breeder reactors (or "catalytic nuclear burners"), which the AEC is pursuing, has been presented by Weinberg and Hammond [18]. Breeder reactors are those in which the breeding of fissionable material from nonfissionable material is accomplished. About 99.3% of natural uranium is in the form of nonfissionable U^{238}, which is a "fertile" material that can be converted into the fissionable isotope Pu^{239}:

$$U^{238} + n \longrightarrow U^{239} \xrightarrow{\beta^-} Np^{239} \xrightarrow{\beta^-} Pu^{239}$$

It is also possible to convert Th^{232} (the only naturally occurring isotope of thorium) into fissionable U^{233}:

$$Th^{232} + n \longrightarrow Th^{233} \xrightarrow{\beta^-} Pa^{233} \xrightarrow{\beta^-} U^{233}$$

The known reserves of uranium can be used to produce power because of the isotope U^{235}, which is needed in present-day nuclear power plants. Breeder reactors that make possible the use of all the uranium—U^{238} and U^{235}—would greatly extend the life of our uranium reserves. The use of thorium would be even more important. With the successful development of breeder reactors, which seems likely within the next decade or two, the energy available from the uranium and thorium in U.S. rocks at minable depths containing at least 50 g combined uranium and thorium per metric ton of rock would be hundreds or thousands of times that available by the combustion of all our fossil-fuel reserves and would last for many millennia [9].

Breeder reactors may solve the energy source problem for a very long time and produce very cheap electricity besides [18], but they will have environmental problems comparable to those of modern nuclear power plants —radiation from fission products, thermal pollution about equal to that of the best fossil-fired plants, radioactive waste disposal, and the possibility of accident or sabotage. There has also been concern expressed about the possibility that sufficient plutonium (about 5 kg) could be stolen to permit the construction of atom bombs [21].

Fusion power would be much more desirable from the point of view of an adequate energy supply and from the point of view of reducing environmental pollution [20]. A hydrogen bomb is an example of uncontrolled nuclear fusion, in which light nuclei combine together ("fuse") to form larger nuclei with a corresponding release of energy. Fusion is the opposite of fission, in which very heavy nuclei release energy by splitting into smaller nuclei. For 2 decades scientists in a number of countries, notably the U.S. and U.S.S.R., have sought a way to produce controlled nuclear fusion, i.e., to generate power by fusion reactions without having an explosion. Although controlled fusion reactions have taken place, their energy input has always exceeded their energy output, and the research that is taking place is directed at reversing this situation.

Typical fusion reactions involving pairs of deuterium nuclei are the following, which will occur with about equal probability:

$$D^2 + D^2 \rightarrow He^3 + n + 3.27 \text{ MeV}$$
$$D^2 + D^2 \rightarrow T^3 + H^1 + 4.0 \text{ MeV}$$

The first reaction, for example, fuses the two deuterium nuclei into a He^3 nucleus plus an extra neutron, with 3.27 MeV (5.1×10^{-13} joules) of energy being released as heat in the form of kinetic energies of the He^3 nucleus and the neutron. Another possible reaction involving the fusion of a deuterium nucleus with a tritium nucleus is

$$D^2 + T^3 \rightarrow He^4 + n + 17.6 \text{ MeV}$$

Since tritium is produced by the second of the deuterium-deuterium reactions given above, an overall reaction that is theoretically possible is

$$5D^2 \rightarrow He^4 + He^3 + H^1 + 2n + 24.8 \text{ MeV}$$

The actual reactions that will be chosen for a fusion reactor will depend on a number of technical considerations and may not involve just deuterium. The emphasis on deuterium is due to the availability of deuterium in seawater. The abundance of D relative to H (ordinary hydrogen) in seawater is about 1/6500 so that 1 m^3 of seawater will have about 10^{25} deuterium atoms of total mass 34 g. If the last reaction given above were to occur, there would be available 5 MeV energy per deuterium atom. The deuterium in 1 m^3 of seawater would have a potential to release 8×10^{12} joules, equivalent to the combustion energy of about 270 metric tons of coal or 1360 barrels of crude oil, and the total supply of deuterium in all the world's oceans would be perhaps 50 million times as much as the world's total initial supply of fossil fuels on the basis of energy content [9]. Recoverable energy is likely to be only a small fraction of this but still a substantial amount.

The difficulty of extracting fusion energy arises from the fact that fusion of deuterium or other nuclei can occur only when the nuclei are very close together (about 10^{-15} m apart), and the electrostatic Coulomb repulsion of the positive nuclei will oppose their coming that close together. Fusion will occur at very high temperatures (typically 50 to 100 million K), however, since the kinetic energy of the nuclei will then permit the overcoming of the Coulomb repulsion. It is also necessary to have a dense enough plasma of the fusion fuel and maintain it for a sufficiently long period of time. Formidable problems are involved in meeting these conditions, the most obvious being to find a container in which the fusion can take place. No materials capable of withstanding temperatures of millions of degrees exist, of course, but a number of magnetic configurations have been devised and tested to maintain the plasmas—the Soviet "tokamak," the Princeton "stellarator," the Los Alamos "scyllac," etc. [20]. Although no scientist knows whether controlled fusion will prove feasible, or when, there is general optimism that it will be proved feasible before the end of the 20th century.

The environmental characteristics of fusion reactors are of particular interest to the present discussion, and they appear to be a great improvement over modern methods of electric power generation. The following characteristics have been noted [20]:

1. Like nuclear fission plants, they do not consume fossil-fuel resources, which are valuable for many other uses, nor do they

release carbon dioxide to the atmosphere (the possible long-term consequences of such release are discussed in Chapter 4).

2. They do not produce radioactive fission wastes. The only radioactive isotope being considered today for use as a fusion fuel or that might be a product of fusion reactions is tritium (H^3), which would generally be a desirable fuel to recover. No fusion plant would ever need to contain a large amount of tritium, and the dangers of radiation would be much less than from present-day nuclear fission power plants. It should be noted, however, that fusion reactors are likely to produce intense neutron fluxes that would constitute a formidable problem.

3. There would never be enough fuel present for any sort of runaway accident to occur so the complete safety of the public would be assured. There would be no concern for safety in the event of sabotage or natural disaster.

4. Thermal pollution could be drastically reduced by direct conversion of the heat into electricity. The thermal efficiency of fusion reactors might approach 96% if combined with an efficient thermal electric plant [22]. Such a high efficiency could result from a choice of a fusion reaction in which all or most of the energy went into charged particles (i.e., anything except a neutron) that could be passed through magnetic fields and collected on appropriately situated electrodes [20, 22]. A 96% efficient method of generating electric power would waste only 4% of the heat, in contrast to the 60 to 65% wasted today, and the waste heat/electric power ratio would be only 4% instead of 150 to 200%. In practice fusion reactor efficiencies may never approach these very high values but it is expected that they would exceed current steam-electric plant efficiencies by a wide margin.

Fusion power is many years off if it is ever developed, but its environmental advantages appear to be so great that the adequate funding of fusion research should be an important priority in pollution control.

CONCLUSION

By any standards the environmental problems associated with electric power production are already great and threaten to become overwhelming within our lifetime if strong corrective measures are not taken. The present discussion has not even touched on a number of related concerns. Aboveground transmission lines are never very aesthetic, for example, but underground transmission presents formidable scientific and economic problems [23]. The choice of a site for a power plant is of immense importance to the utility and its customers and to the plant's neighbors (who are not always the customers), and it involves consideration of the geology, seismology, meteorology, hydrology and transportation facilities of the area, the density and distribution of the population, the effects of noise from the plant and its operations, etc., not to mention the basic problem of land use [24]. Many minor environmental problems show up from time to time, also, such as the fish kills produced at Consolidated Edison plants on the Hudson River in connection with the cooling water supply [25]. A major problem that should never be forgotten is the illness and death produced in coal and uranium miners over the years in the process of obtaining the fuel for electric power production.

Energy of all types—not just electrical energy—is basic to an advanced, technological society, but there are many social costs (see Chapter 19) associated with energy production, distribution, and consumption that are not being taken into account today. A national debate concerning energy appears to have commenced in the U.S. [26, 27]; it is of vital importance to the health and well-being of everyone.

An annotated bibliography on energy resources, production, and environmental effects may be found in Reference 28.

REFERENCES

1. *Statistical Abstract of the United States*. Washington, D.C.: U.S. Govt. Printing Office. Published annually.
2. Federal Power Commission, *World Power Data*. Washington, D.C.: U.S. Govt. Printing Office. Published annually.
3. Federal Power Commission, *Electric Power Statistics*. Washington, D.C.: U.S. Govt. Printing Office. Published monthly.
4. *World Energy Supplies, 1964–1967*. New York: United Nations Dept. of Economic and Social Affairs. Statistical Papers Series J.
5. Federal Power Commission, *Air Pollution and the Regulated Electric Power and Natural Gas Industries*. Washington, D.C.: U.S. Federal Power Commission, 1968.
6. Terrill, James G., et al., "Environmental Aspects of Nuclear and Conventional Powerplants." In *Selected Materials on Environmental Effects of Producing Electric Power*. Washington, D.C.: U.S. Govt. Printing Office, 1969.
7. Eisenbud, Merril, and Henry C. Petrow, "Radioactivity in the Atmospheric Effluents of Power Plants That Use Fossil Fuels." *Science* **144**:288–289 (1964).
8. *Selected Materials on Environmental Effects of Producing Electric Power*. Washington, D.C.: U.S. Govt. Printing Office, 1969.
9. Hubbert, M. King, "Energy Resources." In National Academy of Sciences— National Research Council Committee on Resources and Man, *Resources and Man*. San Francisco: W. H. Freeman and Co., 1969.
10. Hammond, Allen L., "Solar Energy: A Feasible Source of Power?" *Science* **172**:660 (1971).

11. Summers, Claude M., "The Conversion of Energy." *Scientific American* **224**(3):148–160 (September 1971).

12. Rex, Robert W., *Investigation of Geothermal Resources in the Imperial Valley and Their Potential Value for Desalination of Water and Electricity Production*. Riverside, Calif.: The Institute of Geophysics and Planetary Physics, University of California, June 1, 1970.

13. Rex, Robert W., "Geothermal Energy—The Neglected Energy Option." *Bull. Atomic Scientists* **27**(8):52–56 (October 1971).

14. Squires, Arthur M., "Clean Power from Coal." *Science* **169**:821–828 (1970).

15. Corliss, William R., *Direct Conversion of Energy*. Oak Ridge, Tenn.: U.S. Atomic Energy Commission, 1968. Available free from USAEC, P.O. Box 62, Oak Ridge, Tenn. 37830.

16. Shabad, Theodore, "Soviet Develops Direct Converter for Atomic Power." *N.Y. Times,* March 26, 1971, p. 1.

17. Seaborg, Glenn T., and Justin L. Bloom, "Fast Breeder Reactors." *Scientific American* **223**(5):13–21 (November 1970).

18. Weinberg, Alvin M., and R. Philip Hammond, "Limits to the Use of Energy." *American Scientist* **58**:412–418 (1970).

19. Mitchell, Walter, III, and Stanley E. Turner, *Breeder Reactors*. Oak Ridge, Tenn.: U.S. Atomic Energy Commission, 1971. Available free from USAEC, P.O. Box 62, Oak Ridge, Tenn. 37830.

20. Gough, William C., and Bernard J. Eastlund, "The Prospects of Fusion Power." *Scientific American* **224**(2):50–64 (February 1971).

21. Shapley, Deborah, "Plutonium: Reactor Proliferation Threatens a Nuclear Black Market." *Science* **172**:143–146 (1971).

22. Post, Richard F., "Fusion Power: Direct Conversion and the Reduction of Waste Heat." USAEC Microfiche TID-25414.

23. Rose, P. H., "Underground Power Transmission." *Science* **170**:267–273 (1970).

24. Carter, Luther J., "Land Use: Congress Taking Up Conflict over Power Plants." *Science* **170**:718–719 (1970).

25. Millones, Peter, "Con Edison Plant Closed 3 Days to Stop Fish-Kill." *N.Y. Times,* February 3, 1970, pp. 37, 71.

26. The September 1971 issue of *Scientific American* is devoted to the topic "Energy and Power" and contains a number of relevant articles.

27. The September, October, and November 1971 issues of the *Bulletin of the Atomic Scientists* contain articles on the energy crisis.

28. Romer, R. H., "Resource Letter ERPEE-1 on Energy: Resources, Production, and Environmental Effects." *Amer. Jour. Phys.* **40**:805–829 (1972).

17
FOODS, DRUGS, AND COSMETICS

A Butcher that selleth Swines Flesh meazled, or Flesh dead of the Murrain [plague], or that buyeth Flesh of Jews, and selleth the same unto Christians, after he shall be convict thereof, for the first Time, he shall be grievously amerced [fined]; the Second Time he shall suffer Judgement of the Pillory; and the Third Time he shall be imprisoned and make Fine; and the Fourth Time he shall forswear the Town.
—The Judgement of the Pillory, ca. 1267 A.D. [1].

The Judgement of the Pillory, which dates from the fifty-first year of the reign of Henry III of England, might be considered an early consumer protection statute, marked by an anti-Semitic bias, and it shows the importance attached to the quality of meat. In the early days of U.S. history, when each family grew its own food or obtained it from neighbors, it was relatively easy to know the history of the food and the likely sources of its contamination. The growth of urban populations led to the expansion of the food industry, however, especially in the years after the Civil War. Adulteration of food was so common and sanitation so rare that as early as 1880 Peter Collier of the Agriculture Department's Bureau of Chemistry was urging the enactment of a national food (and drug) act. During the period from 1880 to 1905, a total of 103 bills were introduced in vain in Congress. Dr. Harvey W. Wiley, who became head of the Bureau of Chemistry in 1883, campaigned indefatigably for many years for the Federal Food and Drug Act that was finally passed in 1906 after the public had become aroused through the efforts of muckrakers like Upton Sinclair, author of *The Jungle*.

The 20th century has seen the growth of the food, drug, and cosmetic industries and the appearance of a wide variety of new problems. In this chapter these problems will be discussed to show that they constitute environmental pollution as defined in Chapter 1 and are closely related to some of the other problems treated in other chapters. One section deals with the long and interesting history of federal legislation in this field.

FOODS

The laws of the United States protect consumers against adulterated foods, defined as foods containing poisonous or deleterious substances, including natural substances if present in amounts that can be injurious to health; containing filthy, putrid, or decomposed substances; containing ingredients that conceal the use of inferior or decomposed substances; containing products of diseased animals; packaged in a container having poisonous or deleterious substances that may be injurious to health; or containing unsafe amounts of pesticides, food additives, or color additives. As might be expected, this law and its interpretation have been central to a vast number of important court cases in consumer law [2].

Incidents of intentional substitution of inferior foods or "watering down" of foods in one way or another do not often come to light in the U.S., although regulations have had to be instituted to prevent gradual degradation of foods, such as the greater and greater fat content of hot dogs or the decreasing fruit content of jams and fruit pies. No great scandals comparable to those exposed by the early 20th century muckrakers or to the modern Italian ox-blood wines have occurred in the U.S. in recent years.

Food additives are substances other than basic foodstuffs that are present in food as a result of production, processing, storage, or packaging [3]. Many of these additives occur intentionally and have been added by the food industry in order to make their products more attractive or less susceptible to deterioration. The food chemical industry has grown very rapidly in the period since World War II as the food industry has emphasized new products, convenience foods, low-calorie foods, ethnic foods, frozen foods, and many other innovations [4]. The list of different types of additives and their purposes is given in Table 17-1.

Until 1958, the Food and Drug Administration (FDA), which has been responsible for enforcing the provisions of the various Food and Drug Acts, was not allowed to ban any additive unless it could prove it to be injurious to health. Today the manufacturer of a proposed food additive must satisfy the FDA of the safety of the product, and approval can be withdrawn anytime new information casts doubt on the safety of the additive.

In past years a number of additives have had to be banned [4]. Monochloroacetic acid used as a preservative in wines, soft drinks, and salad dressings was banned in the early 1940's after it caused large numbers of persons to become ill. Nitrogen trichloride ("Agene"), a bleaching and maturing agent in flour, was banned in 1949 after it was discovered that it reacted with methionine in the flour to produce a toxic compound, methionine sulfoximine, capable of producing convulsions in laboratory animals (especially dogs, which are more sensitive to the compound). The artificial sweetener dulcin (4-ethoxyphenylurea) was banned in 1950 when it was discovered to be carcinogenic, and the natural flavoring material safrole, extracted from sassafras bark and used in root beer, was banned in 1960 for the same reason. Coumarin, a flavoring ingredient, and the tonka bean from which it was obtained were banned in 1954 after being found to cause liver damage in rats and dogs. Cobaltous salts (acetate, chloride, sulfate) were used in the U.S. from 1963 to 1966 to improve the stability of the foam in beer and then discontinued when they were implicated in the unexplained deaths of several dozen heavy beer drinkers in the U.S. and Canada in 1966.

TABLE 17-1. Food additives in use in the U.S. [3, 4].

Acidulants add an appealing tartness to soft drinks, fruit juices, jams, sherbets, and other foods. Citric acid, once obtained from citric fruits but now produced by fermentation, accounts for 60% of U.S. use and phosphoric acid for 25%, followed by fumaric, malic, adipic, tartaric, lactic, and succinic acids.

Anticaking compounds keep powders and crystalline materials such as table salt, baking powder, vanilla powder, garlic salt, malted milk powders, and nondairy coffee creamers from absorbing moisture and caking. These include calcium silicate, magnesium silicate, tricalcium phosphate, sodium aluminosilicate, silica gel, and aluminum calcium silicate, all of which preferentially absorb any moisture that gets into the food.

Antioxidants retard the oxidative breakdown of fats and oils and thus prevent objectionable flavors and odors due to rancidity in shortenings, cooking oils, potato chips, breakfast cereals, salted nuts, dehydrated soups, and other foods. Gum guaiac and lecithin were once used for this purpose but the main antioxidants in use today are BHA (butylated hydroxyanisole), BHT (butylated hydroxytoluene), and propyl gallate, with ascorbic and erythorbic acids also finding some use.

Antistaling agents ("antifirming agents" or 'bread emulsifiers") are used in bread to retard staling, which involves the increased crystallization of starch. These are generally monoglycerides and diglycerides.

Artificial sweeteners are nonnutritive compounds added to foods (especially those advertised as low-calorie foods) to produce a sweet taste. A few years ago the sodium and calcium salts of cyclamic acid ("cyclamates") were the most common artificial sweeteners but they have since been banned (see text), and saccharin (discovered in 1879) is the only one presently permitted in food.

Bleaching and maturing agents such as benzoyl peroxide, chlorine, or nitrosyl chloride are added to flour to promote proper bleaching and aging.

Colors are used in a wide variety of foods—butter, cheeses, soft drinks, ice cream, candies, sherbets, cake mixes, breakfast cereals, jellies, etc. Some colors are natural compounds used for many years (even centuries)—annatto, saffron, cochineal, caramel, and turmeric. There are also a number of FD&C (Food, Drug, and Cosmetic) certified colors of synthetic origin—Red No. 40, Yellow No. 5, etc. Since 1956 a number of previously certified colors have been banned or permitted only under special restrictions.

Dough conditioners produce a bread dough that is drier and more extensible and thus easier to machine. Calcium stearyl-2-lactylate and sodium stearyl fumarate are used for this purpose.

Emulsifiers are surface-active agents used to disperse one liquid in another (usually oil in water); they are used in nondairy creamers, confectionary coatings, ice cream, margarine, candy, cakes, desert toppings, soft drinks (to disperse flavorings), and other foods. Natural emulsifiers include lecithin (from egg yolks; still the major emulsifier in mayonnaise), saponins, and lipoproteins. Synthetic emulsifiers include glycerol esters of fatty acids (the monoglycerides and diglycerides), sorbitan monostearate, and various polysorbates.

Firming agents such as calcium chloride, calcium citrate, and monocalcium and dicalcium phosphates are used to increase firmness of canned vegetables and fruits—tomatoes, potatoes, sliced apples, etc.

TABLE 17-1 (continued).

Flavors in use in the U.S. number well over 1000. Natural flavors include cinnamon, cloves, pepper, fruit extracts, and many types of spices (and even salt and sugar, which are generally regarded as basic food ingredients rather than additives). Synthetic flavorings include isoamyl acetate, ethyl butyrate, benzaldehyde, and methyl salicylate. In addition, there are a number of "flavor enhancers" such as MSG (monosodium glutamate), disodium 5'-inosinate, disodium 5'-guanylate, and maltol that do not add a flavor but bring out the flavor of food ingredients already present. MSG intensifies the flavor of high-protein foods such as meat and fish and is popular among Orientals.

Leavening agents such as phosphate, cream of tartar (potassium acid tartrate), and sodium aluminum sulfate ("soda alum") are used to release carbon dioxide from sodium bicarbonate ("baking soda") or other compounds in baked products.

Miscellaneous additives are used to cure meats, reduce foaming in cooking and fermentation, ripen bananas, prevent sprouting of potatoes, etc.

Moisture-retaining agents prevent the escape of moisture from foods and thus have an effect opposite to that of anticaking compounds. Examples are sorbitol in shredded coconut and candies, glycerol in marshmallows, and propylene glycol in candies.

Nutritional supplements include vitamins (A in cheeses, skim milk, and baby foods, B in wheat flour, C in fruit juices and drinks, D in milk, etc.), amino acids (such as lysine in some breads and cereals), iodides in salt, etc.

Preservatives control molds, yeasts, rope bacteria, and other organisms. Wood smoke was used in prehistoric times to preserve meat. Molds in bread, cheese, jellies, cakes, and dried fruits are inhibited by calcium and sodium propionates, sorbates, sodium diacetate, sodium benzoate, and other compounds. Yeasts can be controlled by ethyl formate, methyl formate, ethylene oxide, and propylene oxide. Browning of fruits and vegetables is inhibited by sodium sulfite, ascorbic acid, sulfur dioxide, and citric acid. At one time, antibiotics were used in chilled, uncooked poultry and seafood.

Sequestrants improve the color, flavor, and texture of foods by tying up trace metals (iron, copper, zinc, etc.). Leading sequestrants are citric acid, sodium hexametaphosphate, sodium tripolyphosphate, and the calcium disodium and disodium dihydrogen salts of ethylenediaminetetraacetic acid (EDTA).

Stabilizers and thickeners are used in ice cream, chocolate milk, icings, cheese spreads, salad dressings, pie fillings, syrups, gravies, and other foods to provide the proper texture and consistency. Important additives of these types are gum arabic, guar gum, sodium carboxymethylcellulose, gelatin, carrageenan, carob bean gum, agar-agar, and methylcellulose.

There are some food additives that Americans do not use but are common in other countries. The Japanese prefer their rice to be treated with talc, which is usually contaminated with asbestos, a known carcinogen, and it has been suggested [5] that the unusually high incidence of stomach cancer in Japan may be due to this food additive.

Some additives have been partially but not wholly banned. The synthetic female sex hormone diethylstilbestrol (DES) was used for about 10 years in the feed of young male chickens to increase weight gain and then banned after it was found that small residues remained after slaughter, because DES is known to have carcinogenic properties. Its use was still permitted in beef cattle until shortly before slaughter because no residues could then be detected. In October 1971 the FDA increased the mandatory waiting period between the last DES application and slaughter from 48 h to 7 days after DES was detected in a few beef liver samples. DES continued to be found, however, and in August 1972 the FDA banned DES from cattle feed beginning January 1, 1973. In 1964 the coal-tar tint Red No. 4 was banned when it was found to injure the bladders and adrenal glands of dogs but its use was permitted in maraschino cherries since the maraschino cherry industry had no satisfactory substitute.

When the Food, Drug, and Cosmetic Act was revised in 1958, a number of food additives that had been in general use without having doubts cast on their safety were placed on the GRAS ("generally recognized as safe") list of approved substances. A number of substances have since been removed from the list.

One of these was the artificial sweetener cyclamate (the sodium or calcium salt), shown in Figure 17-1. The sweetening nature of cyclamates was discovered in 1937 by Michael Sveda, a graduate student at the University of Illinois, and they were first marketed as a nonnutritive sweetener in 1950 by Abbott Laboratories. In 1962 the National Academy of Sciences expressed concern over the greatly increased use of cyclamates in beverages and diet foods. By 1969 evidence had accumulated showing that cyclamate, largely through its derivative cyclohexylamine, could cause birth defects, chromosome damage, and bladder cancer in various animals [6, 7]. The FDA at first banned cyclamates, acting on the Delaney clause of the 1958 Food and Drug Act, which stipulates that no material that is capable of causing cancer in appropriate tests (even in large doses) in animals or man may ever be permitted in any food. The FDA later decided to permit cyclamates to be consumed upon a physician's advice with warning, as by diabetics, but then in 1970 it reversed this stand and instituted a complete ban on cyclamates. The Department of Health, Education, and Welfare, over FDA objections, acceded to the requests of the National Soft Drink Association and permitted the continued use for 3 years of returnable soft drink bottles labeled "sugar-free" even though the beverage now contained sugar—thereby misleading some diabetics who missed a small warning placed on the bottle cap.

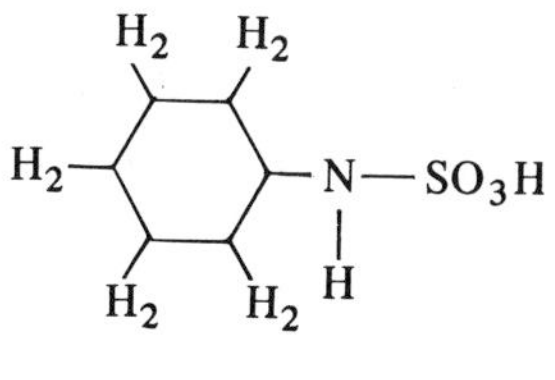

FIG. 17-1. N-cyclohexyl sulfamate ("cyclamate").

The food additives that can cause difficulties are sometimes natural food products rather than synthetics. The noted "Chinese restaurant syndrome"—burning sensations, facial pressure, and chest pain—is regarded as caused by the flavor enhancer MSG because of its heavy use (typically 5 g per serving) in wonton soup and other foods served in Chinese restaurants [8]. MSG has also been reported to induce brain lesions in monkeys [9] and mice [10] when fed in large quantities. These results led to strong criticism of the former use of MSG in baby foods, which has now been discontinued.

When the Food and Drug Act was revised in 1958, the FDA requested authority to ban additives that, although apparently safe, were of no benefit to the consumer. Examples cited were chrome yellow used to color coffee beans to make them appear to be of better value, sulfite compounds used to make stale meat redder, copper used to color canned peas and make them seem to be less mature, and fluorine compounds used in wine and beer as a substitute for pasteurization to stop fermentation. The FDA today requires additives to be of some use but a great deal of subjective judgment is involved —one person might regard a pleasing color as desirable and another might regard it as completely immaterial.

Accidental additives are those that find their way into food unintentionally—pesticide residues, fertilizers, feed adjuvants (antibiotics, hormones, tranquilizers) and drugs, packaging materials, and radioactive substances (such as Cs^{137} and Sr^{90} from atmospheric weapons testing found in milk). Pesticide residues in food are subject to tolerances (maximum concentrations) set by the Environmental Protection Agency. Some pesticides have zero tolerances and must not be present in detectable amounts, just as cyclamates can no longer be present in food. The concept of zero tolerance has often been criticized on scientific grounds because it presumes that certain substances are "poisons per se." Actually all substances are poisonous in large enough quantities, and different substances differ in toxicity rather than in their poisonous or nonpoisonous nature.

From time to time accidental additives cause great harm and even death but their danger appears to be rather small in comparison with the many technological hazards of our environment. Accidental substitution of toxic substances for foods or food additives is probably as serious a problem; a person recently died because a meat tenderizer jar had accidentally been filled with pure sodium nitrite.

The Food Protection Committee of the National Academy of Sciences-National Research Council has urged that no intentional or incidental food additive be permitted until its safety has been demonstrated beyond reasonable doubt as judged by competent experts [3], and the FDA tries to act in accordance with this opinion. There are many degrees of standards possible, however, and society must ultimately be the judge of how conservative or how liberal its standards are to be and how much it is willing to pay (in testing costs) to ensure itself against unsafe additives. The Delaney clause of the Food, Drug, and Cosmetic Act, which prohibits *any* amounts of carcinogenic substances in food, represents an extremely conservative standard; it has often been attacked by scientists and professional groups as too restrictive and too reminiscent of the old "poison per se" concept.

One type of food contamination whose undesirability is not questioned is that which leads to "food poisoning." At one time, food poisoning was referred to as "ptomaine poisoning" because it was erroneously attributed to the presence of ptomaines, organic compounds isolated from putrid meat. The actual causes are those listed in Table 17-2. Most of the problems giving rise to food poisoning can be avoided by proper sanitation, cooking, and refrigeration. Laws requiring the pasteurization of milk and the ripening of cheese have been of great value. Health departments prohibit known disease carriers from holding jobs in which they might contaminate food; the New York health department, for example, is keeping 200 typhoid carriers under surveillance.

TABLE 17-2. Important sources of food poisoning [11].

---

1. *Infections*—bacterial, viral or rickettsial—such as typhoid fever, salmonella food poisoning, streptococcal sore throat, scarlet fever, bacillary dysentery, tuberculosis, and diphtheria. The most common bacterial infections are due to *Salmonella, Clostridium perfringens,* and *Streptococcus. Salmonella* organisms, which are natural inhabitants of the intestinal tracts of humans and animals, lead to salmonellosis, usually within 12 to 36 h of eating contaminated food; symptoms are fever, headache, diarrhea, and vomiting, lasting 1 to 8 days. Salmonellosis probably affects several million people in the U.S. each year, especially through poultry, egg, and meat products.
2. *Bacterial toxins*, especially botulism (from the toxin of *Clostridium botulinum*) and staphylococcal intoxication. Botulism occurs most commonly in sausage, fish, and improperly canned (generally home-canned) foods; symptoms set in after 1 or 2 days and involve the central nervous system: swallowing difficulty, double vision, respiration difficulties, inability to speak properly; death occurs in over 50% of the cases. Staphylococcal intoxication is most common in milk and egg products; symptoms occur after 2 or 3 h and include nausea, vomiting, diarrhea, cramps, and sometimes fever.
3. *Parasites*, such as tapeworm in beef, pork, and fish, trichinosis in pork, or amebiasis in raw vegetables and fruits.

Heat can destroy some bacteria and their toxins. *Salmonellae* are destroyed by 10 min heating at 60°C (140°F). *Staphylococcus* is easily destroyed by heat but the toxin is more resistant and requires several hours boiling for destruction. Spores of the *Clostridium botulinum* bacillus are resistant to heat but the toxin is quite sensitive to heat and can be destroyed by boiling for 10 to 20 min. Refrigeration of foods usually helps prevent food poisoning outbreaks since the organisms do not multiply at cold temperatures.

Food poisoning occurs surprisingly often at large catered affairs because much of the food is prepared hours in advance and bacteria have time to multiply if the temperature is favorable. The number of food poisoning outbreaks annually in the U.S. is not known. In 1967 there were about 1700 outbreaks (involving 115,000 persons and 17 deaths) reported to public health officials but these are believed to be only a very small fraction of the total that occurred [12].

No discussion of the possible environmental dangers resulting from the use of food would be complete without mention of the fact that there are many natural toxicants that are found in foods [13, 14]. These include oxalates in rhubarb and spinach, glycosides capable of yielding hydrogen cyanide in lima beans and almonds, and pressor amines in certain fruits. In addition there are poisonous plants mistaken for edible plants—the poisonous mushrooms are an important example. Some foods can also be contaminated by dangerous natural substances, such as the carcinogenic aflatoxin molds occurring on peanuts and other crops.

The dangers of natural toxicants are sufficiently great that some plants generally regarded as edible would be illegal if they were manufactured products. Our ignorance of the effects of natural toxicants is probably as unfortunate as our ignorance of the effects of synthetic additives.

Food in the United States is, on the whole, quite safe when one appreciates the immense problems that could (and did, in the past) result in the absence of standards of safety and cleanliness. Nevertheless, several dozen deaths each year are due to food poisoning and accidental additives, and we cannot be sure of the long-term safety of many of the intentional additives in current use. The FDA protects the public to some degree, ordering the recall of dozens of food products each week—frozen pizzas made with mushrooms containing botulism toxin, jars of sliced pimientos containing glass particles, candy infested with rodent remains, canned vichyssoise containing botulism toxin, egg noodles contaminated by salmonella organisms, etc. Food monitoring procedures in the U.S. are nevertheless not completely adequate, as shown by such incidents as the recent discoveries in tuna and swordfish of mercury concentrations exceeding the Public Health Service limits for food.

DRUGS

From a legal standpoint, drugs can be classified into three groups:

1. "Quack drugs" which have no therapeutic value or which have significant dangers for users.
2. Therapeutic drugs used in medical practice.
3. Illegal drugs whose use or sale is prohibited by law.

Quack drugs were very common before passage of the 1938 Food and Drug Act and some products generally considered of no value are still on the market. In the early 20th century some extraordinarily dangerous products were being marketed and the FDA was legally unable to prevent their sale [15].

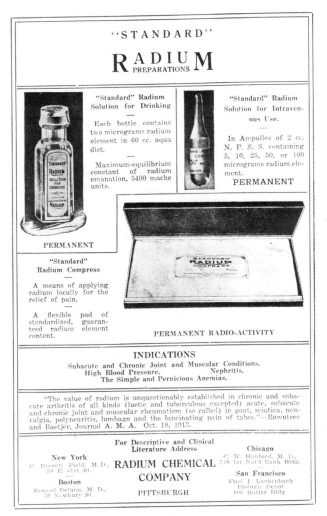

FIG. 17-2. An old advertisement for radium preparations dating back to a time when their great dangers were not fully recognized. (*Courtesy of the U.S. Atomic Energy Commission.*)

An obesity cure called *Marmola* contained thyroid extract and bladderwrack (a seaweed rich in iodine) that speeded up the metabolic rate to a dangerous extent. The radium water *Radithor* caused disintegration of the jawbone and a slow, tortuous death, claiming the wealthy Pittsburgh steel manufacturer Eben M. Byers as one of its victims. *Bromo-seltzer* was dangerous to heavy users because of its ingredients acetanilid (which could produce fatalities) and sodium bromide (which led to bromide intoxication). A number of weight reducers (*Slim, Formula 17,* and others) contained excessive quantities of dinitrophenol, for which the lethal adult dose is only 1 to 3 g orally.

Dinitrophenol speeds up the metabolic rate, increasing oxygen consumption, body temperature, and heart rate, without increasing circulation and respiration to a comparable extent.

Many other medicines were claimed in advertising to cure large numbers of ailments, even though manufacturers kept within the law by omitting the claims from their labels. *Fleischmann's Yeast* was advertised as being good for indigestion, stomach troubles, headaches, constipation, bad breath, skin troubles, and various other ills. *Crazy Water* from Mineral Wells, Tex., which was rich in sodium sulfate, was supposed to cure rheumatism, arthritis, high blood pressure, diabetes, and other illnesses. *B. & M. Liniment*, made from turpentine, ammonia, eggs, and water, was advertised to cure tuberculosis, and the FDA spent $75,000 over 10 years just to force removal of fraudulent claims from its label. *Cancer and Scrofula Syrup* was mainly potassium iodide with some plant drugs. Many of these medicines were quite harmless, of course, but they did keep cancer and tuberculosis victims from seeking proper medical attention until it was too late [15].

Therapeutic drugs all involve the balancing of benefits (from their therapeutic effects) and risks (from overdoses or from side effects). Some nonprescription drugs can be rather dangerous. Aspirin (acetylsalicylic acid) is widely used for alleviating pain, headache, and neuralgia but it is easily taken in overdose, resulting in gastric irritation, perspiration, severe dehydration, thirst, fever, and lethargy [16]. It results in the death of several hundred persons annually in the United States and in Great Britain. Oral contraceptives have been criticized in recent years as being responsible for increased incidence of blood clotting, breast cancer, diabetes, sterility, and other side effects, but all the evidence is not yet in.

Sometimes the risks associated with a drug are not recognized until it is in general use. This was the situation in West Germany with the tranquilizer, thalidomide, which possessed an unsuspected teratogenicity; pregnant women who took thalidomide later often gave birth to children with congenital deformities of the limbs. Although never approved for use in the United States, some 2 million tablets had been distributed in the U.S. to 1200 doctors [17]. The tragic consequences of the thalidomide incident were instrumental in convincing the U.S. Congress of the need for more effective drug testing procedures, which were required by the 1962 amendments to the Food, Drug and Cosmetic Act. These amendments permitted the FDA to withhold approval of new drug products until the manufacturer proved their safety and effectiveness. The last decade has witnessed important developments in programs for the toxicological evaluation of experimental drugs [18].

The testing of drugs is a long and expensive process, involving the use of animals before human experimentation. The stringency of the legal requirements when new drugs are used on human subjects has led to the testing of some drugs outside the United States. Dr. Albert Sabin's oral polio vaccine, which was developed at the University of Cincinnati College of Medicine, was actually field tested by physicians in the U.S.S.R. [17].

The 1962 amendments to the Food, Drug and Cosmetic Act also gave the FDA authority to examine the effectiveness of all drugs marketed between 1938 and 1962, a formidable task beyond the manpower resources of the FDA. In 1966 the National Academy of Sciences-National Research Council agreed to provide the manpower to carry out this work. Thirty different panels, each responsible for particular categories of diseases, reported back in 1969, concluding that the claims made for many of the therapeutic preparations were not supported by substantial evidence. On November 27, 1970 the FDA released a comprehensive list of 369 drug products that the NAS-NRC panel and the FDA had decided were either ineffective or unduly hazardous (see Reference 19 for the complete list). These accounted for about 12% of the 3000 products that came on the market between 1938 and 1962. Included were both prescription and nonprescription drugs, as well as some mouthwashes, nasal sprays, nose drops, lozenges, and toothpastes [19]. Some of the consumer products would have been allowed on the market except for the advertising claims made for them, such as the claims for reduced tooth decay from the use of the mouthwashes and toothpastes.

The FDA over the past few years has initiated actions to remove these products from the market. The most hotly contested action involved the antibiotic combination sold by The Upjohn Company under the brand name *Panalba* [20]. *Panalba*, which contained the two antibiotics tetracycline and novobiocin, became a sales hit in the late 1950's, when it was introduced, and was estimated to have annual sales of close to $20 million when The Upjohn Company decided to oppose the FDA's ban. The NAS-NRC panel had found the risks associated with novobiocin to be greater than the benefits because of the availability of safer and more efficacious drugs and was concerned with the rapid and frequent emergence of bacterial strains (especially staphylococci) resistant to novobiocin. In early 1970 the FDA was successful in forcing *Panalba* off the market, clearing the way for action against other antibiotic combinations [21]. These combinations, such as "pen-sulfas" (penicillin and sulfa) and "pen-streps" (penicillin and streptomycin), were regarded as dangerous by the NAS-NRC panel for a number of reasons: They promoted bacterial resistance to several antibiotics at once, they compounded the problems associated with allergic reactions and other side effects, and they were no more effective than the individual antibiotics taken separately. On the other hand, numerous physicians have protested the withdrawal of combination drugs, and some will undoubtedly remain on the market.

A related problem that is quite disturbing is the extensive use of antibiotics such as tetracycline and penicillin in animal feeds (for pigs, cattle, fowl) and the resulting emergence of drug-resistant strains of a number of pathogenic bacteria [22]. These strains are now responsible for a considerable proportion of the diseases produced by the bacteria; the problem of controlling disease has been complicated, with important implications for human health. In 1969, Great Britain decided to institute strict controls on the use of antibiotics in animal feed, permitting the continued use of antibiotics that promote animal growth but have little or no medicinal value to humans and forbidding the use of antibiotics of therapeutic value to humans except when prescribed to cure or prevent animal disease.

The illegal drugs are a very different problem, with important social and political implications [23]. Table 17-3 lists several categories of drugs subject to considerable abuse in the U.S., not all of which are actually illegal. Their overuse constitutes a more-or-less voluntary type of "internal pollution" that is very much a public health problem (there are over 1000 narcotics deaths annually in New York alone). Most of these drugs are on a quite different legal footing, however, from cigarette smoking and excessive alcohol consumption, both of which are also important public health problems. Of special importance in the 20th century has been the growth in the use and abuse of hallucinogens of plant origin [24] or synthetic manufacture.

TABLE 17-3. Common drugs constituting public health problems.

A. *Stimulants*
   1. Amphetamines—synthetic drugs such as dextroamphetamine and methamphetamine.
   2. Cocaine—a white crystalline alkaloid from leaves of the coca bush; formerly used as a local anesthetic.

B. *Depressants* (narcotics)
   1. Barbiturates—synthetic drugs such as phenobarbital, pentobarbital, secobarbital; used as sedatives.
   2. Opiates—natural drugs obtained from the fruit of the opium poppy: heroin, morphine, codeine (methylmorphine), paregoric, etc.
   3. Synthetic narcotics—demerol, methadone, etc.

C. *Hallucinogens* (mind-affecting drugs)
   1. Marijuana—a mild hallucinogen obtained from the Indian hemp plant.
   2. Mescaline—the active ingredient of the peyote cactus.
   3. Psilocybin—obtained from *Psilocybe mexicana* and other mushrooms.
   4. LSD (lysergic acid diethylamide)—synthesized from an alkaloid of ergot (a rye fungus).

D. *Deliriants* (Toxic vapors such as toluene, a solvent in airplane glue.)

COSMETICS

Cosmetics, rarely noted today for any dangerous properties, have been responsible for serious problems in the past. During the early part of the 20th century, before the passage of the 1938 Food, Drug, and Cosmetic Act, many dangerous cosmetics flooded the market [15]. *Lash-Lure*, a synthetic aniline dye of the paraphenylenediamine group for use on eyebrows and lashes, led to many cases of blindness, eye injury, and even death. Hair tonics often contained arsenic, and skin whiteners and freckle removers contained mercury. Hair dyes with silver salts led to silver poisoning (argyrism) with its permanent bluish-black discoloration of the skin. Probably the most amazing cosmetic was *Koremlu Cream*, a depilatory containing thallium acetate (which incidentally sold for $10 per 2-oz jar even though the ingredients were worth $0.35); it led to many severe cases of thallium poisoning, of which extensive loss of hair is only a symptom. Most of the cosmetics of the day were harmless, of course, but the abuses made corrective legislation imperative.

FEDERAL FOOD, DRUG, AND COSMETIC LEGISLATION

The following paragraphs give a brief summary of the most important food, drug, and cosmetic legislation at the federal level in the U.S., insofar as it concerns the topics touched on in this chapter and elsewhere in the book. There have been other laws dealing with meat and seafood inspection, butter, narcotic drugs (especially in recent years), and other products, but only the general laws are considered here.

The 1906 Federal Food and Drug Act ("Wiley Act") dealt largely with misbranded and improperly labeled foods and drugs (and cosmetics only to the extent that they were drugs), protecting the consumer against adulterated foods, foods labeled falsely or misleadingly, and foods containing harmful ingredients. The law applied only to products in interstate commerce. A major loophole as far as drugs and patent medicines were concerned was the legality of false advertising as long as the label contained no false claims. False claims on the label were actually not prohibited until the Sherley Amendment was passed in 1912, and even it prohibited "false and fraudulent" claims so that the FDA could prosecute only when it could prove both falsehood and fraudulent intent. The Gould Amendment of 1913 required packaged products to list their net weight.

In 1938 a totally new Federal Food, Drug, and Cosmetic Act (PL 75-717) was passed, and in its amended form it is still the basic law applying today. Impetus for the passage of this law came from the inability of the FDA to prosecute many dangerous products or to prohibit false advertising. A noted tragedy that occurred in 1937 contributed to its successful passage. The S. E. Massengill Company marketed an "Elixir of Sulfanilamide" (which one was supposed to drink for improved health) in order to cash in on the wonder drug craze, but it used the highly toxic diethylene glycol as the solvent. The product did not need and did not have FDA approval and had not been tested for toxicity. About 107 persons died from using the preparation, and the FDA was able to seize the product only under the labeling provisions of the old Food and Drug Act because the name "elixir" falsely implied the use of alcohol as the solvent.

The 1938 Act gave the Secretary of Agriculture the authority to set up definitions and standards of identity for foods. It prohibited misbranding of products and required foods, drugs, and cosmetics to carry labels listing ingredients. It forbade false or misleading advertisements (giving enforcement powers to the Federal Trade Commission) or food additives intended to promote deception. Only coal-tar colors certified as safe by the FDA were permitted in food. Contaminated food or food produced in unsanitary conditions were prohibited.

Although the 1938 act gave the Secretary of Agriculture the authority to set tolerances for accidental additives such as pesticides, the procedure given was too cumbersome to carry out in practice. In 1954 the Pesticide ("Miller") Amendments gave the FDA the authority to set and enforce safe tolerances for pesticide residues.

This provision was greatly broadened by the 1958 Food Additives Amendment (PL 85-929), which made the FDA responsible for evaluating the safety of new food additives before marketing. Both intentional and accidental food additives were regulated by this amendment. The amendment also contained the noted Delaney Clause forbidding the presence in any amount in foods of any known carcinogen. Although this clause has been attacked as an unscientific revival of the old "poison per se" doctrine, consumer groups have urged that the U.S. consumer be further protected by prohibition of additives with mutagenic and teratogenic properties [25].

In 1960 the Color Additives Amendment (PL 86-618) gave the FDA authority to set tolerances for color additives in food.

In 1962, following hearings held by Senator Estes Kefauver, the Drug Amendments (PL 87-781) were passed. They provided for the review of all new drug applications by the FDA, with marketing of a new drug permitted only after the FDA was satisfied with the safety and efficacy of the drug and had given formal approval. The FDA was also granted authority over prescription drug advertising.

REFERENCES

1. *The Statutes of the Realm,* Vol 1, p. 203. London, 1822.
2. Keeton, Page, and Marshal S. Shapo, *Products and the Consumer: Defective and Dangerous Products.* Mineola, N.Y.: The Foundation Press, Inc., 1970.
3. Food Protection Committee, *The Use of Chemicals in Food Production, Processing, Storage, and Distribution.* Washington, D.C.: National Academy of Sciences-National Research Council, 1961.
4. Sanders, Howard J., "Food Additives." *Chemical and Engineering News* **44**(42):100–120 (October 10, 1966) and **44**(43):108–128 (October 17, 1966).
5. Merliss, R. R., "Talc-Treated Rice and Japanese Stomach Cancer." *Science* **173**:1141–1142 (1971).
6. Bryan, G. T., and E. Erturk, "Production of Mouse Urinary Bladder Carcinomas by Sodium Cyclamate." *Science* **167**:996–998 (1970).
7. Price, J. M., et al., "Bladder Tumors in Rats Fed Cyclohexylamine or High Doses of a Mixture of Cyclamate and Saccharin." *Science* **167**:1131–1132 (1970).

8. Schaumburg, H. H., et al., "Monosodium L-Glutamate: Its Pharmacology and Role in Chinese Restaurant Syndrome." *Science* **163:**826–828 (1969).
9. Olney, J. W., and L. G. Sharpe, "Brain Lesions in an Infant Rhesus Monkey Treated with Monosodium Glutamate." *Science* **166:**386–388 (1969).
10. Arees, E. A., and J. Mayer, "Monosodium Glutamate-Induced Brain Lesions: Electron Microscopic Examination." *Science* **170:**549–550 (1970).
11. Hobbs, Betty C., *Food Poisoning and Food Hygiene,* 2nd ed. London: Edward Arnold Ltd., 1968.
12. Woodward, W. E., et al., "Foodborne Disease Surveillance in the United States, 1966 and 1967." *Amer. Jour. Public Health* **60:**130–137 (1970).
13. Food Protection Committee, *Toxicants Occurring Naturally in Foods.* Washington, D.C.: National Academy of Sciences-National Research Council, 1966.
14. Liener, Irvin E. (ed.), *Toxic Constituents of Plant Foodstuffs.* New York: Academic Press, 1969. Food Science and Technology, Vol. 6.
15. Lamb, Ruth D., *American Chamber of Horrors.* New York: Farrar & Rinehart, Inc., 1936.
16. Deichmann, W. B., and H. W. Gerarde, *Toxicology of Drugs and Chemicals.* New York: Academic Press, 1969.
17. Pearson, Michael, *The Million-Dollar Bugs.* New York: G. P. Putnam's Sons, 1969.
18. Zbinden, Gerhard, "Drug Safety: Experimental Programs." *Science* **164:**643–647 (1969).
19. Schmeck, Harold M., Jr., "F.D.A. Lists 369 Drugs as Ineffective or Perilous." *N.Y. Times,* November 28, 1970, pp. 1, 35.
20. Mintz, Morton, "FDA and Panalba: A Conflict of Commercial, Therapeutic Goals?" *Science* **165:**875–881 (1969).
21. Gruchow, Nancy, "FDA Wins Round in Panalba Fight." *Science* **167:**1710 (1970).
22. Smith, H. W., "Drugs in Animal Feeds." In J. Rose (ed.), *Technological Injury.* London: Gordon and Breach, 1969. pp. 79–91.
23. Many of these social and political implications are discussed in the official report of the National Commission on Marihuana and Drug Abuse. See *Marihuana. A Signal of Misunderstanding.* New York: The New American Library, Inc., 1972.
24. Schultes, R. E., "Hallucinogens of Plant Origin." *Science* **163:**245–254 (1969).
25. Turner, James S. (Project Director), *The Chemical Feast.* The Ralph Nader Study Group Report on Food Protection and the Food and Drug Administration. New York: Grossman Publishers, 1970.

18

POLLUTION IN FOREIGN COUNTRIES

In Kohln, a town of monks and bones,
And pavements fanged with murderous stones,
And rags, and hags, and hideous wenches;
I counted two and seventy stenches,
All well defined, and several stinks!
Ye Nymphs that reign o'er sewers and sinks,
The river Rhine, it is well known,
Doth wash your city of Cologne;
But tell me, Nymphs, what power divine
Shall henceforth wash the river Rhine?
—Samuel Taylor Coleridge, *Cologne* (1828).

Pollution problems are not limited to the United States, and despite occasional claims that the capitalist economic system with its emphasis on profits tends to promote environmental deterioration, these problems occur in other countries with very different economic systems. Environmental problems are clearly associated with increasing population, increasing industrial development, and increasing consumption of goods, regardless of the economic system.

The other chapters in this book have concentrated largely on pollution problems in the U.S., with occasional discussions of those in Great Britain. This chapter touches on some representative national problems in several other countries and on a few international problems. Information about pollution in the world today is not readily available, partly because widespread recognition of the importance of environmental problems is a very recent phenomenon. As a consequence, no complete discussion of pollution in foreign countries is possible at present.

NATIONAL PROBLEMS

CANADA. Canada has many of the same problems as the U.S.—air pollution from automobiles and paper and pulp mills, water pollution by mercury and oil, etc. In fact, the 1970 mercury pollution concern in North America was triggered by the Canadian discovery of mercury-contaminated fish in Lake Ontario. Mercury levels in excess of the 0.5 ppm limit for human food were even discovered among beluga whales in Hudson Bay, and a whale factory in Rupert Inlet had to stop canning whale meat; in this case the mercury appeared to come from the mouths of the rivers in James Bay and of the Nelson River system in Manitoba.

An oil spill in June, 1970, from the Fort McMurray plant of the Great Canadian Oil Sands Ltd. drifted into Lake Athabasca and forced closure of the commercial fishing season on the lake. The fishing industry supported about 100 wage earners in the Fort Chipewyan, Alberta, community of 1400 Indians and Metis, who thus lost their livelihood. (Newspaper articles nevertheless reported that the oil spill had had no effect on humans!) Environmental disruption of a different sort had originally led the Indians to become fishermen. Previously they had been trappers but the trapping industry had been dealt a fatal blow when the Bennett Dam was completed hundreds of miles away at Hudson Hope, British Columbia, lowering the water table in the marshy muskrat and beaver breeding areas below the dam.

On the national scene, the Canadian government has taken steps to correct pollution problems by eliminating the use of phosphates in detergents, by restricting DDT use and setting stricter standards for pesticide residues in food, and by charting a water cleanup program to cost several billion dollars over the decade of 1970 to 1980. The serious water and air pollution problems around the Great Lakes and connecting boundary waters between Canada and the U.S. (see Chapter 10) will require U.S. cooperation, however.

Canada has also begun to face up to the need to preserve the rare natural habitat of its Arctic regions and to protect these regions from pollution and despoilment, especially since American oil companies have been considering making use of the Northwest Passage to transport oil from Alaska. The U.S. State Department, however, has already rejected Canada's claims over Arctic waters [1].

JAPAN. Rapid postwar industrial growth in Japan (see Figure 2-6, for example) has brought pollution problems with it, especially since pollution abatement has generally been ignored in Japan's drive to become the third-ranking industrial nation in the world. The trend toward environmental degradation may soon be reversed, however, since several noted smog episodes have alarmed the public and led to intensive newspaper and television discussion of pollution [2]. Every opposition party has availed itself of the opportunity to denounce the Sato government for the increase in pollution, especially since the government leaders have many financial ties with pollution-creating industries.

Tokyo has often shown signs of both classical and photochemical smog, and gauze masks (which offer protection against airborne pathogens) and oxygen vending machines are a common sight. Much of the air pollution arises from the use of heavy fuel oil containing a high sulfur content from the Persian Gulf. The role of automobiles has become more important in recent years, and oxidant levels during a 4-day smog episode in July 1970 reached 0.4 ppm [3].

Japanese water pollution is quite similar to that in some parts of the U.S. The best-known case of mercury poisoning in the world was that which occurred over a period of many years at Minamata Bay, where several dozen deaths and numerous cases of mental and physical retardation were produced in persons eating fish contaminated with mercury originating with the wastes of the Nippon Nitrogen Company. At Tagonoura, slimy organic wastes from paper plants have filled in half the harbor, which once was 8.5 m deep; fishing interests were successful in preventing the wastes from being dumped into nearby Suruga Bay.

Cadmium, which has recently come under scrutiny in the U.S., has also proved to be a problem. Near the Nippon Mining Company's zinc refinery on the coast of the Sea of Japan, cadmium has poisoned 130 hectares of farm and pasture land by becoming impregnated into the soil, leading to the condemnation of rice from the land and milk from cows grazing on the land. In 1971, in a case that brought widespread attention, a number of victims of a bone disease produced by cadmium successfully sued the company for damages.

PEOPLE'S REPUBLIC OF CHINA. Mainland China has taken important strides in upgrading the quality of its environment, which has been affected for centuries by China's vast population. The 20th century brought a desperate need for improvement in sanitation, public health, water conservation, land reclamation, and reforestation [4]. The Communist government has emphasized preventive medicine and environmental sanitation during the past 2 decades, partly because of the scarcity of medical personnel and facilities. Urban sewerage and sewage treatment plants have been greatly expanded, flies and mosquitoes have been virtually eliminated, with a consequent reduction in diseases for which these insects served as vectors, and the old habit of expectorating on the streets has disappeared [4]. An important remaining problem (important enough for Chairman Mao Tse-Tung to have written a poem about it) is the prevalence of schistosomiasis ("snail fever"), aggravated by unsanitary use of human excrement as fertilizer [5].

Many polluted streams and lakes that received human and animal wastes for centuries have been cleaned up, with the organic wastes returned to the land; even urban wastes are being collected and used for irrigation and fertilization. Similarly, industrial pollution has been reduced by emphasis on recovering and reusing the "four wastes"—waste materials, waste water, waste gas, and waste heat [4].

The concern of the Chinese Communists with environmental pollution appears to have been due to two important Communist policies: the improvement of health and sanitation and the program of frugality and economy. The Chinese record of environmental concern is far from perfect, however. At the 1972 U.N. Conference on the Human Environment, the Chinese representative argued in favor of population expansion, increased industrial development without fear of the consequences, and continued nuclear weapons testing, in addition to claiming that resources were inexhaustible.

U.S.S.R. Environmental problems in the U.S.S.R. are as great as those of the U.S. despite the fact that state ownership of factories theoretically implies that the public interest will always be protected and that social as well as private costs will be taken into account [6]. Communist economists, such as Oscar Lange in Poland, have asserted that the communist system is environmentally superior to the capitalist system because it does take social costs (see Chapter 19) into account, but this claim has not been borne out in practice, presumably because the main emphasis has been on industrial development and the fulfillment of production quotas.

A number of prominent Soviet writers, including some with a reputation for supporting the Soviet government even when other writers have been critical of it, have spoken out harshly about environmental degradation. One notable example was the address of Mikhail Sholokov to the 23rd Congress of the Communist Party in 1966 [7]. Sholokov's address directly criticized the Soviet Minister of Fish Industry, Comrade Ishkov, for allowing industrial water pollution to kill fish in the Volga, Lake Baikal and its basin, the Don, and the Azov Basin.

Soviet water pollution is as serious as U.S. water pollution. In the largest Soviet Republic, the Russian Federated Socialist Republic, 65% of the factories discharge untreated wastes into waterways; mines, oil wells, and ships add more wastes. Domestic water wastes, like industrial wastes, are generally not treated and many cities do not even have sewer systems. Adequate drinking water supplies are often lacking, and in 1960 62% of the urban residences in the U.S.S.R. did not even have running water [6].

The water level in the Aral and Caspian Seas has been dropping due to diversion of water for crop irrigation and the building of hydroelectric plants and dams on rivers supplying these seas. In 20 years the level of the Caspian Sea has fallen 2.5 m, eliminating one-third of the spawning areas of sturgeon. The sturgeon has also been affected by oil on the sea, and caviar exports by the U.S.S.R. have dropped sharply; caviar is difficult to buy inside the Soviet Union except at restaurants.

One of the most widely discussed Soviet problems is the pollution of Lake Baikal, the world's oldest and deepest freshwater lake. Two pulp and paper mills were built on its shores in 1966 and immediately produced extensive air and water pollution. This pollution was criticized by conservationists and other citizens who feared the destruction of the lake and its unique aquatic life (such as freshwater seals) might result. In addition, the depletion of the forest lands around the lake is leading to flooding and water erosion, and fears have been expressed that the area may become part of the Gobi Desert [6]. In 1969, in the face of protests about the degradation of the lake, stringent antipollution measures were promulgated, but it remains to be seen whether or not they will prove successful. In the meantime, the situation has been used by the Soviet film director Sergei A. Gerasimov in a movie, "At the Lake," which focuses on man's dependence on nature and the need for conservation [8].

Classical air pollution arising from fossil-fuel combustion by industries and electric generating plants is quite common in the U.S.S.R. and although the country is less industrialized than the U.S., it has also placed less emphasis on pollution control devices such as electrostatic precipitators. Photochemical air pollution is not yet much of a problem because of the scarcity of motor vehicles. The lead aerosol problem is nonexistent because the U.S.S.R. is the only major country in the world that does not use tetraethyl lead in gasoline [6].

Soviet biologists have also spoken out against the dangers of pesticides, and Soviet newspapers have carried reports of wildlife kills following aerial spraying of pesticides. An article [9] in *Komsomolskaya Pravda* by the Moscow conservationist Vladimir Peskov in April 1970 was followed in May by a letter from the Ministry of Agriculture stating tersely that DDT would be forbidden for use on food and fodder crops.

TABLE 18-1. U.S.S.R. air quality standards [10].

Pollutant	Peak maximum	24-h average
Carbon monoxide	6.0 mg/m^3 (6.2 ppm)	1.0 mg/m^3 (1.0 ppm)
Nitrogen oxides	0.3 mg/m^3 (0.15 ppm)	0.1 mg/m^3 (0.05 ppm)
Soot	0.15 mg/m^3	0.05 mg/m^3
Sulfur dioxide	0.5 mg/m^3 (0.23 ppm)	0.15 mg/m^3 (0.07 ppm)

The Soviet Union has set air quality standards that are to be striven for over a long period of time [10]. Some examples are shown in Table 18-1; these are similar to those set in the U.S. by the Environmental Protection Agency (given in Table 20-2) though some are stricter. The Soviet standards have been chosen by requiring that no reflexive reaction of the receptors of the respiratory organs including the nose be detectable from exposures to pollutants. U.S.S.R. scientists also consider lead to be much more toxic than U.S. scientists do.

INTERNATIONAL PROBLEMS AND PROGRAMS

In some parts of the world environmental problems are clearly international; i.e., they extend across national boundaries. One important example is the pollution of the Great Lakes and other boundary waters between the U.S. and Canada, discussed at the beginning of Chapter 10. The existence of this pollution has been recognized since before World War I but has become more severe over the years.

Some of the most marked examples of international pollution occur in Europe. Scandinavian countries experience a very acidic rain believed to be produced by industrial sources in Britain and western Europe. The Rhine River becomes progressively more and more polluted by organic and industrial wastes as it winds its way from Switzerland and Germany to France and the Netherlands. The accidental spilling of an insecticide into the river in West Germany in June 1969 led to the killing of thousands of fish and the contamination of drinking water supplies in the Netherlands. Air pollution from the heavily industrialized Ruhr Valley in West Germany affects Belgium and the Netherlands as well as Germany. Pollution of the North Sea originates in and contaminates several different countries.

Excluding the U.S.S.R., Europe has well over 450 million persons in an area of 5 million km^2, compared to 210 million persons in an area of 9 million km^2 in the United States. The greater population concentration, the industrial development of European nations, and the large number of political (and cultural) units all combine to make environmental problems extensive and difficult to control by legislation. The European Economic Community (EEC) and other multinational groups have been studying ways of instituting international protection of the environment, and some positive steps to prevent international pollution have been taken on a voluntary basis.

On a global scale, international efforts to study and combat environmental pollution are just starting, most of them under the auspices of United Nations organizations. For example, the Intergovernmental Oceanographic Commission of UNESCO is beginning to investigate marine pollution on a worldwide scale. In addition, a great deal of valuable information has been gathered in recent years through cooperative undertakings such as the International Geophysical Year.

The World Meteorological Organization (WMO) has begun a global Air Pollution Monitoring Programme that will monitor certain air pollutants at several base line stations in areas unaffected by local pollution sources and at a greater number of regional stations. As a minimum program the WMO began measuring atmospheric turbidity and the chemical constituents of precipitation; it intends later to add to its program sulfur oxides, nitrogen oxides, lead and other heavy metals, fluorine, and other substances. About 5 base line and 25 regional stations became operational in 1970.

The 1970's seem certain to witness the growth of international efforts to control pollution, especially since the global effects of environmental pollution are becoming recognized as serious problems. A major step in this direction was the convening of the U.N. Conference on the Human Environment at Stockholm in June 1972 [11]. Some 114 countries—the U.S.S.R. and East Germany were the major absentees—were represented at the conference, which resolved to set up a U.N. environmental organization, the Governing Council for Environmental Programmes, for which only $100 million in funds would be provided over the first 5 years. It also resolved to establish an international convention on marine dumping and recommended a 10-year moratorium on whaling. The conference also issued a Declaration on the Human Environment with 26 principles [12].

DECLARATION ON THE HUMAN ENVIRONMENT

1. Man has the fundamental right to freedom, equality and adequate conditions of life, in an environment of a quality which permits a life of dignity and well-being, and bears a solemn responsibility to protect and improve the environment for present and future generations. In this respect, policies promoting or perpetuating apartheid, racial segregation, discrimination, colonial and other forms of oppression and foreign domination stand condemned and must be eliminated.

2. The natural resources of the earth including the air, water, land, flora and fauna and especially representative samples of natural ecosystems must be safeguarded for the benefit of present and future generations through careful planning or management as appropriate.

3. The capacity of the earth to produce vital renewable resources must be maintained and wherever practicable restored or improved.

4. Man has a special responsibility to safeguard and wisely manage the heritage of wildlife and its habitat which are now gravely imperiled by a combination of adverse factors. Nature conservation including wildlife must therefore receive importance in planning for economic developments.

5. The nonrenewable resources of the earth must be employed in such a way as to guard against the danger of their future exhaustion and to insure that benefits from such employment are shared by all mankind.

6. The discharge of toxic substances or of other substances and the release of heat, in such quantities or concentrations as to exceed the capacity of the environment to render them harmless, must be halted in order to insure that serious or irreversible damage is not inflicted upon ecosystems. The just struggle of the peoples of all countries against pollution should be supported.

7. States shall take all possible steps to prevent pollution of the seas by substances that are liable to create hazards to human health, to harm living resources and marine life, to damage amenities or to interfere with other legitimate uses of the sea.

8. Economic and social development is essential for insuring a favorable living and working environment for man and for creating conditions on earth that are necessary for the improvement of the quality of life.

9. Environmental deficiencies generated by the conditions of underdevelopment and natural disasters pose grave problems and can be remedied by accelerated development through the transfer of substantial quantities of financial and technological assistance as a supplement to the domestic effort of the developing countries and such timely assistance as may be required.

10. For the developing countries, stability of prices and adequate earnings for primary commodities and raw material are essential to environment management since economic factors as well as ecological processes must be taken into account.

11. The environmental policies of all states should enhance and not adversely affect the present or future development potential of developing countries, nor should they hamper the attainment of better living conditions for all, and appropriate steps should be taken by states and international organizations with a view to reaching agreement on meeting the possible national and international economic consequences resulting from the application of environmental measures.

12. Resources should be made available to preserve and improve the environment, taking into account the circumstances and particular requirements of developing countries and any costs which may emanate from their incorporating environmental safeguards into their development planning and the need for making available to them, upon their request, additional international technical and financial assistance for this purpose.

13. In order to achieve a more rational management of resources and thus to improve the environment, states should adopt an integrated and coordinated approach to their development planning so as to insure that development is compatible with the need to protect and improve the human environment for the benefit of their population.

14. Rational planning constitutes an essential tool for reconciling any conflict between the needs of development and the need to protect and improve the environment.

15. Planning must be applied to human settlements and urbanization with a view to avoiding adverse effects on the environment and obtaining maximum social, economic and environmental benefits for all. In this respect projects which are designed for colonialist and racist domination must be abandoned.

16. Demographic policies, which are without prejudice to basic human rights and which are deemed appropriate by governments concerned, should be applied in those regions where the rate of population growth or excessive population concentrations are likely to have adverse effects in the environment or development, or where low population density may prevent improvement of the human environment and impede development.

17. Appropriate national institutions must be entrusted with the task of planning, managing or controlling the environmental resources of states with the view to enhancing environmental quality.

18. Science and technology, as part of their contribution to economic and social development, must be applied to the identification, avoidance and control of environmental risks and the solution of environmental problems and for the common good of mankind.

19.   Education in environmental matters, for the younger generation as well as adults, giving due consideration to the underprivileged, is essential in order to broaden the basis for an enlightened opinion and responsible conduct by individuals, enterprises and communities in protecting and improving the environment in its full human dimension. It is also essential that mass media of communications avoid contributing to the deterioration of the environment, but, on the contrary, disseminate information of an educational nature on the need to protect and improve the environment in order to enable man to develop in every respect.

20.   Scientific research and development in the context of environmental problems, both national and multinational, must be promoted in all countries, especially the developing countries. In this connection, the free flow of up-to-date scientific information and experience must be supported and assisted, to facilitate the solution of environmental problems: environmental technologies should be made available to developing countries on terms which would encourage their wide dissemination without constituting an economic burden on the developing countries.

21.   States have, in accordance with the Charter of the United Nations and the principles of international law, the sovereign right to exploit their own resources pursuant to their own environmental policies, and the responsibility to insure that activities within their jurisdiction or control do not cause damage to the environment of other states or of areas beyond the limits of national jurisdiction.

22.   States shall cooperate to develop further the international law regarding liability and compensation for the victim of pollution and other environmental damage caused by activities within the jurisdiction or control of such states to areas beyond their jurisdiction.

23.   Without prejudice to such general principles as may be agreed upon by the international community, or to the criteria and minimum levels which will have to be determined nationally, it will be essential in all classes to consider the systems of values prevailing in each country, and the extent of the applicability of standards which are valid for the most advanced countries but which may be inappropriate and of unwarranted social cost for the developing countries.

24.   International matters concerning the protection and improvement of the environment should be handled in a cooperative spirit by all countries, big or small, on an equal footing. Cooperation through multilateral or bilateral arrangements or other appropriate means is essential to prevent, eliminate or reduce and effectively control adverse environmental effects resulting from activities conducted in all spheres, in such a way that due account is taken of the sovereignty and interests of all states.

25.   States shall insure that international organizations play a coordinated, efficient and dynamic role for the protection and improvement of the environment.

26. Man and his environment must be spared the effects of nuclear weapons and all other means of mass destruction. States must strive to reach prompt agreement, in the relevant international organs, on the elimination and complete destruction of such weapons.

REFERENCES

1. Szulc, Tad, "U.S. Rejects Canadians' Claim to Wide Rights in Arctic Seas." *N.Y. Times,* April 10, 1970, p. 13.
2. Oka, Takashi, "A Japanese Youth, 15, Lectures Sato on Pollution." *N.Y. Times,* September 22, 1970, p. 4.
3. "Tokyo Smarting from 4-Day Smog." *N.Y. Times,* July 27, 1970, p. 19.
4. Orleans, L. A., and R. P. Suttmeier, "The Mao Ethic and Environmental Quality." *Science* **170:**1173–1176 (1970).
5. Reston, Sally, "Mao's Poem Urges Drive on Virulent Snail Fever." *N.Y. Times,* July 30, 1971, p. 7.
6. Goldman, Marshall I., "The Convergence of Environmental Disruption." *Science* **170:**30–42 (1970). Goldman, Marshall I., *The Spoils of Progress. Environmental Pollution in the Soviet Union.* Cambridge, Mass.: The MIT Press, 1972.
7. Goldman, Marshall I. (ed.), *Controlling Pollution: The Economics of a Cleaner America.* Englewood Cliffs, N.J.: Prentice-Hall, Inc., 1967.
8. Gwertzman, Bernard, "New Russian Movie Discusses Industry's Environmental Role." *N.Y. Times,* March 2, 1970, p. 6.
9. "Russian Reports Ecology Threat." *N.Y. Times,* April 27, 1970, p. 10.
10. Ryazanov, V. A., "Sensory Physiology as Basis for Air Quality Standards. The Approach Used in the Soviet Union." *Arch. Environ. Health* **5:**480–494 (1962).
11. Hawkes, Nigel, "Stockholm: Politicking, Confusion, but Some Agreements Reached." *Science* **176:**1308–1310 (1972).
12. "Text of the Environmental Principles." *N.Y. Times,* June 17, 1972, p. 2.

19

ECONOMIC AND LEGAL QUESTIONS

One broad principle well established in the law of noise, both in this country and England, curiously illustrates the serious bent of our Anglo-Saxon nature, and that is the sharp distinction drawn between money-making noises and those which are made in the pursuit of pleasure. The law is tender to a steam engine or a boiler-maker, and will allow them to disturb a whole neighborhood with impunity, but it is severe upon a brass band or a game of skittles.   —Boston Advertiser, February 27, 1886 [1].

The economic and legal questions posed by the problems of pollution and pollution control are intertwined and are of great importance to society. Pollution has its costs but they have generally not been readily expressible in terms of the usual pricing mechanisms of a capitalist society and have been neglected (by noncapitalist societies, too). Economics textbooks until recently have generally ignored the concept of social costs or external diseconomies and have shown little concern for pollution and other by-products of a technological society, with most of them stating that the science of economics is concerned with "scarce" or "economic" goods and not with "free" or "noneconomic" goods such as air, water, or sunshine! (Other disciplines, physics and chemistry, for example, have no better record with respect to the treatment of environmental problems.)

In this chapter will be discussed the concept of social costs, the problem of assessing benefits and costs, the legal aspects of pollution, the various types of remedies that might be used to alleviate the problems of pollution, and some rough estimates of the cost of a clean environment.

ECONOMICS OF POLLUTION

The harmful effects of pollution, many of which have been discussed in earlier chapters, constitute examples of social costs, which have been defined by Kapp [2] as "all those harmful consequences and damages which third persons or the community sustain as a result of the productive process, and for which private entrepreneurs are not easily held accountable."

As long as the unpleasant or damaging effects of an action taken by a person affect only him, it can be presumed that he will be able to make a rational decision as to whether or not the benefits that accrue to him from that action offset the harmful effects. This is not true to society as a whole if the undesirable effects fall upon someone who is not party to the decision. A corporation might decide to use a river as a dump for its wastes without making any effort at treating the wastes, and as a result a landowner downstream might find himself unable to use the river for drinking water or swimming or other purposes. There would be no economic reason for the corporation to treat the wastes or avoid dumping the wastes (it might be bound legally or morally to do so, however). Such a situation would be an "external diseconomy" since the diseconomy is external to the decision-making unit. It also constitutes a social cost to society as a whole. Of course, if an analysis of the costs of preventing the pollution showed that the pollution abatement costs exceeded the harm to the downstream landowner, we might have to admit that the situation corresponded to the most efficient alternative for society as a whole, even though it would not appear to be the "fairest" one. The landowner might decide to reduce the costs to himself by building a treatment plant for himself or for the corporation if he really felt that this lowered his overall costs, and again an efficient but unfair solution would have been reached.

If the laws of the land are properly formulated to correspond with the usual notion of fairness, they would provide for the costs to the landowner to be borne by the corporation; the corporation and the landowner might disagree over the costs but the courts would provide for a just resolution of the case. The most beneficial solution for the corporation and for society as a whole might be for the corporation to buy out the landowner, which is another way of saying that the fact that the landowner was where he was had as much to do with the fact that the wastes were harmful as the fact that the corporation decided to locate where it did. It should be pointed out that just as excessively weak antipollution laws may be unfair to the landowner, excessively rigid antipollution laws may be unfair to the corporation, and society is the loser in either case [3].

There are many examples of social costs that can be cited. Kapp [2] has considered in detail examples such as

Air pollution.

Water pollution.

Impairment of the physical and mental health of workers, now often corrected by occupational safety and health acts, workmen's compensation acts, and social insurance.

Destruction of self-renewable animal and fishing resources by commercial exploitation.

Waste of nonrenewable energy and mineral resources.

Water and wind erosion of soils.

Diminution or degradation of groundwater stores.

Deforestation.

Unemployment.

Obsolescence of capital equipment and human skills by technological change.

"Planned obsolescence" of consumer goods.

Unnecessary advertising (as opposed to informational advertising).

Wasteful duplication and uneconomic utilization of transportation facilities.

Secrecy and duplication in scientific research.

Patent protection of technical innovations.

Inadequate land use policies.

If one chooses some particular industry or process and lists the social costs associated with it, the list can sometimes be surprisingly long. The social costs associated with electric power generation from combustion of coal include [4]

Depletion of an irreplaceable natural resource (coal).

Impairment of the health of coal miners due to inhalation of coal dust.

Deaths of coal miners in accidents.

"Aesthetic pollution" due to culm piles or strip mines.

Smoke from underground coal fires.

Earth subsidence from underground tunneling.

Water pollution from runoff from coal piles and from erosion at coal mines.

Water pollution from acid mine drainage.

Air pollution (especially sulfur oxides, particulate matter, and nitrogen oxides) from coal combustion.

Solid waste disposal problem from ash collection.

Water pollution from chemicals used in the power plant.

Land use impairment by rail lines from coal mines to the power plant.

Thermal pollution of water or atmosphere.

Noise from the power plant and from the transportation of coal to the plant.

This probably does not exhaust the list of possible social costs. It should be recognized that some of these costs are now being borne in part by the producers of electricity.

Economists over the centuries—Adam Smith, the socialists, and others —seem to have been quite conscious of the existence of social costs, but to an outsider it appears that the science of economics has concentrated its attention on those topics which can be readily considered in terms of market prices to the exclusion of those (such as air and water pollution) which cannot be so expressed. Too often the costs and benefits to the entrepreneur have been identified as the total costs and benefits to society. All the sciences probably tend to overconcentrate on those problems they know how to handle, gaining valuable knowledge in the process but neglecting more complex but equally important problems.

It is possible to use the framework of cost-benefit analysis to reach some sensible conclusions about social costs [5]. The benefits of a manufacturing process can be determined by the value of the manufactured product to the individuals who decide to purchase it. The costs include the usual costs of the manufacturing process (labor, fuel, raw materials, capital investment, interest, etc.) plus the social costs. The social costs include public expenditures to avoid pollution from the manufacturing process (such as the construction of drinking water treatment plants), private expenditures to avoid pollution damage or to correct it (medical bills, cleaning bills, travel expenses to unpolluted recreational areas), and the actual pollution damages that are suffered rather than avoided or corrected (the latter are clearly very difficult to measure quantitatively). The manufacturing costs may include some treatment or abatement costs, but to the extent that other social costs remain, the manufacturer is being subsidized. Society must somehow weigh the benefits and the costs and decide whether or not the manufacturing process is desirable; if costs exceed benefits, it is clearly undesirable; but even if benefits exceed costs, there may be more beneficial alternatives.

The laws directed against pollution can have important effects. A simple and highly idealized example to demonstrate this is presented in Table 19-1. Suppose a product is clearly desirable; i.e., its benefits far exceed its costs regardless of the strictness of the antipollution laws. Suppose further that it can be manufactured by either of two processes, A or B, the ordinary manufacturing costs being only $80 per unit for process A but $100 per unit for process B, while the social costs are $40 per unit for A and zero for B. (Still another process C will be discussed below.) In our hearts, we would feel process B to be superior unless pollution control equipment can be used with process A to bring total manufacturing and social costs below $100 per unit. Suppose however that the costs (counting capital and operating costs) of equipment to reduce social costs to zero would amount to $30 per unit.

In the absence of antipollution laws, a manufacturer might very well choose process A since its manufacturing costs are less, even though its total costs to society are greater. With laws insisting upon zero social costs, i.e., no external diseconomies, the manufacturer will be led to choose process B since its manufacturing costs will be less than those of process A, which will now include the costs of the pollution control equipment. Still another possibility is that the laws specify zero social costs but provide for a 50% subsidy (in the form of grants or tax rebates, for example) for the cost of pollution

TABLE 19-1. Idealized costs to the manufacturer and to society for three different manufacturing processes assuming different types of laws. The social costs in the case of regulation with a subsidy are the costs of the subsidy made by the government.

	Process A	*Process B*	*Process C*
No regulation			
Manufacturing costs	$ 80	$100	$ 60
Social costs	40	0	20
Total costs to society	**$120**	**$100**	**$ 80**
Regulation requiring zero social costs			
Manufacturing costs	$110	$100	$120
Social costs	0	0	0
Total costs to society	**$110**	**$100**	**$120**
Regulation with 50% subsidy			
Manufacturing costs	$ 95	$100	$ 90
Social costs	15	0	30
Total costs to society	**$110**	**$100**	**$120**

control equipment. For process A, half the latter cost, or $15 per unit, would be borne by society and thus amount to a social cost, while the other half becomes part of the manufacturing costs. With the subsidy, the manufacturer would choose process A, with its $95 per unit manufacturing costs, over process B, even though the latter has lower total costs to society.

This concocted example demonstrates the important role that legislation plays in economic considerations and incidentally demonstrates that subsidy plans might encourage manufacturers to choose the less desirable alternative.

Pure regulation without subsidies can also force an undesirable alternative to be chosen. It would discriminate against process C (shown in Table 19-1), which is characterized by low expenses but high costs for avoiding social costs. If such a process C were available, the preferred alternative to society would be to permit the use of process C and to suffer its social costs since the total would be only $80 per unit. "Fairness" could be ensured by levying a direct charge of $20 per unit against the manufacturer to make him pay compensation for what would otherwise be social costs.

If process A were the only possible process, society would benefit by requiring pollution control equipment since the total costs would be only $110 per unit compared to $120 per unit without it. The total costs would not depend on whether subsidies existed, and the subsidies might be quite desirable if they speeded up the use of pollution control equipment.

LEGAL ASPECTS OF POLLUTION

In the absence of specific antipollution legislation, individuals or corporations harmed by pollution from other individuals or corporations have had to rely on the law of public nuisance to obtain abatement of the problem [6, 7]. Unfortunately, the practice of suing polluters in court has never worked well against those polluters who harmed a large region, causing a little damage to everyone, since no one individual feels it is worth his time and money to sue. In principle, government officials could have sued to abate public health nuisances but the information presented in this book demonstrates how inadequate this remedy has been. In recent years there has been interest in the institution of "class suits" in which one individual sues a corporation to require payment of damages to himself and to all others similarly injured by faulty products or pollution [6, 8].

Because of the special importance of water to human civilization, an extensive body of water law has developed in the United States and other countries [9]. It is of interest to examine the two basic systems of water law in the United States.

According to the *riparian doctrine*, a riparian owner (one who owns land contiguous to a stream) automatically owns the right to use the water from the stream on the riparian tract of land, whether he makes use of it or not, but he cannot use it off the land or sell it to another for use off the land. The use may follow either the natural flow doctrine, according to which the riparian landowner may not impair (pollute) or diminish the flow of the stream to the detriment of any other riparian landowner, or the reasonable use doctrine, according to which any riparian landowner may make reasonable use of the flow, taking into account the needs and uses of the other landowners. As one might expect, the courts have an important role in interpreting these doctrines. The riparian doctrine has been firmly established in the eastern United States, where the availability of water has never been a major problem.

According to the *appropriation doctrine*, land ownership is irrelevant to water rights. A water right is obtained simply by appropriating water for a beneficial use, and it is lost by termination of the use. The use may be on or off the riparian land and may even be outside the watershed. When several users are competing for water rights, the rule is "first come, first served," although this applies only to the use of water in amounts previously used, which cannot be increased at the expense of someone who came later. In the arid parts of the western United States, the appropriation doctrine found favor because it was more suitable for an arid region in which water was scarce. Numerous problems and legal battles have occurred over the years, however, including the "Owens Valley War" in which the city of Los Angeles sought the rights to the water of the Owens Valley [10].

Legislation can reduce the social costs of pollution, and several methods are available to accomplish this purpose. Existing or suggested laws are generally one of three types: regulation, subsidization, or direct charges.

REGULATION. Under the regulatory system, manufacturers and individuals who do not meet standards (such as temperature or BOD limits for effluent waste waters) are fined or otherwise penalized. The choice of a control method is left up to private enterprise. The standards could be uniform standards, which are easy to administer but probably less efficient economically, or they could be "point-by-point" standards, which vary with location and situation.

SUBSIDIZATION. Tax credits, loans, grants, and other types of subsidies can also be used to encourage pollution control, and a number of states have recently enacted legislation providing for subsidies. The federal government has encouraged the construction of sewage treatment facilities by municipalities by providing grants, a type of subsidy that involves only governmental units. Subsidies to industry have often been referred to as bribes to polluters, but if they lead to an improvement in environmental quality, they may be quite desirable. Subsidies are usually only a supplement and are unlikely to lead to a reduction in pollution in the absence of other antipollution legislation.

DIRECT CHARGES. Direct charges levied against polluters will encourage them to find ways to avoid pollution and encourage other companies to come up with control equipment to sell to them. A system of direct charges has worked comparatively well in the heavily industrialized Ruhr Valley of Germany for many years. Between 1904 and 1958 a number of cooperative associations (*Genossenschaften*) were set up as legal authorities to deal with the problems of waste disposal, water supply, flood control, and land drainage [11]. Charges were leveled against polluters roughly in proportion to the amount of dilution necessary to make the wastes nontoxic to fish. In some cases the companies have taken steps to avoid the wastes and thus avoid the charges, while in other cases the authorities have used the proceeds from the charges to treat the wastes from a number of factories in an economical fashion. The *Genossenschaften* are pretty much controlled by the industries and the pollution problems are far from eliminated, but this procedure has kept the problems from being as serious as they are in similarly industrialized areas of the U.S. and Japan.

J. H. Dales has suggested [5] a slightly different approach, which he believes would work well in certain situations and would be very inexpensive to administer. A government body, with the help of expert opinion, would classify wastes in terms of some standard unit (much as toxicity to fish is used in the Ruhr Valley) and then sell a certain number of pollution rights on the open market. Each right would permit the owner to dump one unit of wastes into a river. The number of rights sold for each watershed or other area chosen would provide for a net water quality that is acceptable (or the number of rights sold could be decreased a little each year until such a point were reached). This would provide for maximum efficiency: The companies that could remedy their pollution problems for the least cost would do so since the other companies would bid the rights higher than the former companies would be willing to pay. Of course, a pollution right would not have to be used, and conservation groups might purchase some to ensure water quality higher than the government was willing to specify. The revenue from the sale of pollution rights could also be used to improve environmental quality.

In the U.S. there has been some use made of regulation and of subsidies, and some individuals, notably Senator William Proxmire of Wisconsin, have argued in favor of direct charges. Direct charges have been essentially nonexistent in the U.S., although it can be argued that sewer charges are really a form of direct charge against a polluter.

It will be the public and its elected representatives who will have to decide how great the social costs of pollution are and what the best methods of maximizing net benefits are. Costs and benefits cannot be too narrowly defined, either, or else undesirable policies may result. For example, it is essential to take future generations into account. An analysis from a present-day point of view may very well lead to the use of some manufacturing process that depletes a nonrenewable natural resource without enabling it to be recycled, and future generations would have the right to curse us for making the decision to use it. Social costs to the future are especially difficult to determine because technological innovations may later greatly reduce the apparent cost, or new and important uses for the resource may later greatly increase the cost.

All the possible remedies to environmental problems have the potentiality to become "licenses to pollute." Society should regard any pollution as basically undesirable and adopt the point of view that any pollution that is permitted to exist must be justified in some reasonable fashion. Today we try not to permit the use of new food additives or therapeutic drugs unless they are proved to be of net benefit to society, and we should not allow the existence of any polluting action unless it too can be proved to be of net benefit to society. Today, unfortunately, we still operate with the opposite philosophy—that any pollution is permitted to exist unless it is proved to be of great harm to man or his environment.

THE COST OF A CLEAN ENVIRONMENT

The exact cost of a clean environment is impossible to establish, in large part because of the difficulty of defining the "clean environment." The exact costs of meeting specified air or water pollution standards, such as the national ambient air quality standards of the Environmental Protection Agency given in the next chapter, are not known exactly either.

The Council on Environmental Quality, however, has estimated [12] the annual costs (including capital, operating, and maintenance costs) to meet federal air and water quality standards by 1975. Table 19-2 lists the actual 1970 costs and the required annual costs by 1975. The total for air and water

TABLE 19-2. Annual pollution abatement costs in the U.S. [12]. The 1970 figures are actual figures; the 1975 figures are estimates if federal air and water quality standards are to be met. Capital, operating, and maintenance costs are all included.

	1970	1975
Air	$0.5 billion	$4.7 billion
Water	3.1 billion	5.8 billion
Solid wastes	5.7 billion	7.8 billion
Total	$9.3 billion	$18.3 billion

pollution abatement and solid waste management would amount to $18.3 billion annually by 1975, or perhaps $85 annually per person. This is approximately twice the actual 1970 cost.

Even if one allows for the costs of reducing other types of pollution, such as noise and radiation, it is very difficult to imagine an extensive campaign against environmental pollution costing as much as 3% of the gross national product, or about $30 billion annually. The 1972 report [13] of the Council on Environmental Quality has included some of these other costs and estimated that pollution control expenditures in the 1970's should total $287 billion, or 2.2% of GNP in the same period. This sum would appear largely as taxes and increased prices of products and would probably not even be noticed by a public accustomed to years of inflation. Although one often hears claims that pollution control will be enormously expensive, it actually appears to amount to about 1 year's productivity increase. This means that if the people of the U.S. would be willing to forgo one average year's rise in the standard of living as it is customarily measured (neglecting social costs), a vastly improved environment—with all its benefits—would be possible.

REFERENCES

1. Quoted in the *New York Times,* March 1, 1886, p. 2.
2. Kapp, Karl William, *The Social Costs of Private Enterprise.* Cambridge, Mass.: Harvard University Press, 1950.
3. Stepp, J. M., and H. H. Macaulay, *The Pollution Problem.* Washington, D.C.: American Enterprise Institute for Public Policy Research, 1968.
4. Abrahamson, Dean E., *Environmental Cost of Electric Power.* New York: Scientists' Institute for Public Information, 1970.
5. Dales, John Harkness, *Pollution, Property & Prices.* Toronto, Ont.: University of Toronto Press, 1968.
6. Dolan, Edwin G., *TANSTAAFL (There ain't no such thing as a free lunch). The Economic Strategy for Environmental Crisis.* New York: Holt, Rinehart and Winston, Inc., 1971.

7. Crocker, Thomas D., and A. J. Rogers, III, *Environmental Economics*. Hinsdale, Ill.: The Dryden Press, 1971.
8. Sax, Joseph L., *Defending the Environment: A Strategy for Citizen Action*. New York: Alfred A. Knopf, Inc., 1971.
9. Sax, Joseph L., *Water Law, Planning, and Policy: Cases and Commentary*. Indianapolis, Ind.: The Bobbs-Merrill Co., Inc., 1968.
10. Nadeau, R. A., *The Water Seekers*. Garden City, N.Y.: Doubleday & Co., Inc., 1950.
11. Goldman, Marshall I. (ed.), *Controlling Pollution: The Economics of a Cleaner America*. Englewood Cliffs, N.J.: Prentice-Hall, Inc., 1967.
12. *Environmental Quality. The Second Annual Report of the Council on Environmental Quality*. Washington, D.C.: U.S. Govt. Printing Office, August, 1971.
13. *Environmental Quality. The Third Annual Report of the Council on Environmental Quality*. Washington, D.C.: U.S. Govt. Printing Office, August, 1972.

20

LEGISLATION

*It shall not be lawful to throw, discharge, or deposit, or cause, suffer,
or procure to be thrown, discharged, or deposited either from or
out of any ship, barge, or other floating craft of any kind, or from the
shore, wharf, manufacturing establishment, or mill of any kind,
any refuse matter of any kind or description whatever other than
that flowing from streets and sewers and passing therefrom in a liquid
state, into any navigable water of the United States, or into any
tributary of any navigable water from which the same shall float or
be washed into such navigable water . . . . Every person and every
corporation that shall violate . . . the provisions . . . of this Act
shall be guilty of a misdemeanor, and on conviction therefore shall be
punished by a fine not exceeding twenty-five hundred dollars nor
less than five hundred dollars, or by imprisonment (in the case of a
natural person) for not less than thirty days nor more than one year,
or by both such fine and imprisonment, in the discretion of the court;
one-half of said fine to be paid to the person or persons giving
information which shall lead to conviction.* —"Refuse Act of 1899"
(River and Harbor Act of March 3, 1899, Chapter 425, 30 Stat. 1151).

These strong provisions of the Refuse Act of 1899, which was intended to
protect navigation, have been revived in recent years through the efforts of
Rep. Henry Reuss of Wisconsin, chairman of the House Subcommittee on
Conservation and Natural Resources [1, 2]. The Act provides (in another
part) for discharges only under permit from the U.S. Army Corps of Engineers, but until recently few permits have ever been requested and the Act has
clearly been violated many times over the years. Environmental groups have
begun to use the provisions of the Refuse Act successfully, encouraged by the
fact that it authorizes *qui tam* actions—actions in the name of the United
States that permit collection of half of the penalty. In November 1970, for
example, a New York concrete company was fined $25,000 ($500 on each
of 50 counts) for discharging waste concrete into the East River in violation
of the Refuse Act, and a Manhattan social worker and her son, who had
notified authorities of the violations, received $12,500.

This chapter will review the history and present status of federal legislation dealing with environmental pollution, occasionally mentioning state or
local legislation of note. Legislation affecting air and water pollution, solid

316

wastes, and pesticides are dealt with in this chapter, while that affecting radiation has been touched on in Chapter 15 and food, drug, and cosmetic legislation is discussed in Chapter 17. The last section of the chapter discusses the new Environmental Protection Agency, which was organized in 1970.

AIR POLLUTION

In contrast to water pollution discussed in the next section, air pollution was not dealt with in federal legislation until 1955, and even then only at the level of research and technical assistance to the states [3, 4]. The major public laws dealing with air pollution and their most important provisions are listed in Table 20-1.

Federal air pollution legislation has always regarded the state and local governments as having the primary responsibility for dealing with air pollution problems, although by the time the Clean Air Act amendments of 1970 were passed, this power had become limited to implementation of plans to meet the national ambient air quality standards promulgated by the Environmental Protection Agency. The first vigorous federal attack on air pollution problems was provided by the Clean Air Act of 1963, which is still the fundamental federal law but which has led to only a few court actions directed at abating interstate air pollution—one example being the Bishop Processing Company case mentioned in Chapter 5. Federal standards for motor vehicle emissions (discussed in Chapter 6) began with the 1968 model year and will become quite strict by 1975 and 1976.

Responsibility for air pollution control was originally vested in the U.S. Public Health Service. Its Division of Air Pollution became the National Center for Air Pollution Control in 1967 and then the National Air Pollution Control Administration (NAPCA) in 1968—always in the Department of Health, Education, and Welfare. This organization had the responsibility for the coordinated attack on air pollution on a regional basis provided by the 1967 Air Quality Act. The 48 contiguous states were divided into eight atmospheric areas based upon long-term meteorological factors, and these were subdivided into air quality control regions (AQCR's) such as the metropolitan areas of New York, Philadelphia, Chicago, Denver, Los Angeles, and Washington, D.C. The AQCR's were charged with setting air quality standards based upon the air quality criteria published by NAPCA and then with devising implementation plans to meet these standards, making use of the control technique documents published by NAPCA [5].

This rather cumbersome procedure has been revamped completely by the Clean Air Act Amendments of 1970, which provide for national ambient air quality standards to be set by the Environmental Protection Agency by April, 1971, for the major air pollutants, and later standards for other air pollutants deemed to be hazardous. The air quality control regions—designated over the whole nation—must submit implementation plans by January

TABLE 20-1. Summary of federal air pollution legislation.

1955   PL 84-159: Untitled. Provided temporary authority and $5 million annually for 5 years for federal program of research in air pollution and technical assistance to state and local governments.

1959   Extension of 1955 act for 4 years more.

1963   PL 88-206: Clean Air Act. Granted permanent authority to federal air pollution control activities and authorized expenditures of $95 million over 3 years. Major provisions: (1) provided for federal grants to state and local air pollution control agencies to establish and improve their control programs; (2) provided for federal action to abate interstate air pollution through a system of hearings, conferences, and court actions; (3) provided for an expanded federal research and development program with particular emphasis on motor vehicle pollution and sulfur oxide emissions from coal and fuel oil combustion.

1965   PL 89-272 (Title I): Clean Air Act amendments. (1) Provided for the promulgation of national standards relating to motor vehicle pollution (initially applied to the 1968 model year); (2) provided for cooperation with Canada and Mexico to abate international air pollution.

1965   PL 89-675: Clean Air Act amendments. Authorized 3-year, $186 million expansion of air pollution program, including funds to operate local control agencies.

1967   PL 90-148: Air Quality Act (amending the Clean Air Act). Enunciated a national policy of air quality enhancement and provided a procedure for designation of air quality control regions and setting of standards by cooperation between federal and state governments. Also provided for registration of fuel additives.

1969   PL 91-137: Clean Air Act amendments. Extended authorization for research on low-emission fuels and motor vehicles.

1970   PL 91-604: Clean Air Act amendments. Provided for the establishment of national ambient air quality standards and their achievement by July 1, 1975 through the implementation plans of air quality control regions and states. Provided for 90% reductions of automotive hydrocarbon and carbon monoxide emissions from 1970 levels by the 1975 model year and 90% reduction in nitrogen oxide emissions from 1971 levels by the 1976 model year (with 1-year extensions if necessary). Provided for studies of aircraft emissions and noise pollution.

1972 and achieve the standards by mid-1975. (A complete calendar of the various steps in this procedure may be found in Reference 6.)

The Environmental Protection Agency announced its standards for major air pollutants on April 30, 1971, and they are given in Table 20-2. As may be seen by reference to the toxicological and epidemiological data discussed in Chapters 3 and 6, they are strict standards. In the light of our present knowledge, at least, it appears that they will protect the health and welfare of the public, probably even asthma and emphysema victims and others who are particularly sensitive to air pollution.

TABLE 20-2. National ambient air quality standards promulgated by the Environmental Protection Agency on April 30, 1971 [7].

Sulfur oxides: 80 μg/m^3 (0.03 ppm) as annual arithmetic mean and 365 μg/m^3 (0.14 ppm) as maximum 24-h concentration not to be exceeded more than once a year.

Suspended particulate matter: 75 μg/m^3 as annual geometric mean and 260 μg/m^3 as maximum 24-h concentration not to be exceeded more than once a year.

Carbon monoxide: 10 mg/m^3 (9 ppm) and 40 mg/m^3 (35 ppm) as maximum 8-h and 1-h concentrations, respectively, not to be exceeded more than once a year.

Hydrocarbons: 160 μg/m^3 (0.24 ppm) as maximum 3-h concentration (6 to 9 A.M.) not to be exceeded more than once a year.

Nitrogen oxides: 100 μg/m^3 (0.05 ppm) as annual arithmetic mean.

Photochemical oxidants: 160 μg/m^3 (0.08 ppm) as maximum 1-h concentration not to be exceeded more than once a year.

State and local governments were active in air pollution control a number of years before the federal government in some cases, and today many states and large urban areas have active air pollution control agencies. The pioneer in these efforts was the state of California, which in 1947 passed the Air Pollution Control District Act providing for countywide control districts; the districts in the Los Angeles and San Francisco Bay areas have been particularly active. The Los Angeles A.P.C.D. was established in 1947 and the very next year began attacking open burning of refuse by municipalities and private individuals. By 1957 it had also banned incinerators having a single combustion chamber. It also tried unsuccessfully to increase the use of natural gas as a fuel in place of fuel oil, encountering great opposition from industry and little cooperation from the Federal Power Commission [8].

New York, which has had laws prohibiting dense smoke, cinders, dust, and gases for many years, did not set up a Department and Board of Air Pollution Control until 1952. In 1966, following pressures from the public and the U.S. Department of Health, Education, and Welfare, the City Council passed a law requiring more extensive use of low-sulfur fuels and the upgrading of incinerators throughout the city. Although the use of low-sulfur fuel oil by Consolidated Edison Company and other industries brought about a sharp reduction in sulfur oxide levels, the campaign against inadequate incinerators soon became bogged down, with the city's public housing authority being a major violator [8].

WATER POLLUTION

Water pollution legislation at the national level dates back to the Refuse Act of 1899, quoted in part at the beginning of this chapter. Some of the important pieces of legislation relating to federal water and oil pollution control are listed in Table 20-3.

Some of the earliest legislation in this area dealt with oil pollution, but oil pollution acts passed in 1924, 1961, and 1966 have accomplished little. Action against large oil spills from tankers in recent years has been difficult to take under existing legislation, in part because of the difficulty of proving "grossly negligent or willful spilling," as provided by the 1966 law. The Water Quality Improvement Act of 1970 provided for liability of $100 per gross ton of displacement up to a total of $14 million for accidental oil spills, however, and no limit in those cases involving willful negligence or misconduct. Passage of these provisions was aided by the support of Rep. William C. Cramer of Florida, ranking Republican on the House Public Works Committee, who had been a strong opponent of tough federal laws making shipowners liable for damages from oil spills until a major oil spill occurred in his home district

---

TABLE 20-3. Summary of federal water pollution legislation [9].

1899   "Refuse Act of 1899" (River and Harbor Act of 1899, Sections 9 through 20). Prohibited discharges into navigable rivers; permitted *qui tam* actions by citizens.

1912   Public Health Service Act authorized investigation of water pollution in relation to human diseases.

1924   PL 68-238: Oil Pollution Act. Prohibited discharge of oil by any means except in emergency or by accident into navigable waters of the U.S.

1948   PL 80-845: Water Pollution Control Act. Provided 5-year authorization to fund research studies, low-interest loans for construction of sewage and waste treatment works, and the Federal Water Pollution Control Advisory Board. Authorized the Department of Justice to bring suits against individuals or firms, but only after notice, hearing, and consent of the state involved.

1952   PL 82-579: Water Pollution Control Act Extension. Extended the Water Pollution Control Act of 1948 for 3 more years.

1956   PL 84-660: Water Pollution Control Act amendments. Provided permanent authority. Provided $50 million annual authorization for grants for construction of sewage treatment works. Provided for abatement of interstate water pollution by federal enforcement through a conference-public hearing-court action procedure.

1961   PL 87-88: Federal Water Pollution Control Act. Permitted the Secretary of Health, Education, and Welfare, through the Department of Justice, to bring court suits to stop pollution of interstate waters without seeking permission of state. Extended pollution abatement procedures to navigable intrastate and coastal waters with permission of state. Authorized seven regional laboratories for research and development in improved methods of sewage treatment and control. Authorized funds for grants to local communities for sewage treatment plants: $80 million in fiscal year 1962, $90 million in 1963, and $100 million each from 1964 to 1967.

TABLE 20-3 (continued)

1961 PL 87-167: Oil Pollution Act. Enacted to implement provisions of the International Convention for the Prevention of the Pollution of the Sea by Oil, 1954.

1965 PL 89-234: Water Quality Act. Enunciated a national policy of water quality enhancement. Established the Federal Water Pollution Control Administration (FWPCA). Provided for the states to adopt water quality standards for interstate waters and plans for implementation and enforcement, to be submitted by June 30, 1967, to the Secretary of Health, Education, and Welfare (later to Secretary of Interior after FWPCA was transferred to the Department of the Interior) for approval as federal standards; authorized Secretary to initiate federal actions to establish standards if the state criteria were inadequate. Authorized grants for research and development to control storm water and combined sewer overflows and authorized $150 million each in fiscal years 1966 and 1967 for sewage treatment plant grants.

1966 PL 89-551: Oil Pollution Act of 1961 amendments. Various minor amendments.

1966 PL 89-753: Clean Water Restoration Act. Provided for project grants for research and development of advanced waste treatment methods for municipal and industrial wastes. Authorized grants of $450 million in fiscal year 1968, $700 million in 1969, $1 billion in 1970, and $1.25 billion in 1971 for construction of treatment plants. Amended the Oil Pollution Act of 1924 by transferring responsibility to Secretary of the Interior and provided for suits against "grossly negligent, or willful spilling, leaking, pumping, pouring, emitting or emptying of oil."

1970 PL 91-224 (Title I): Water Quality Improvement Act. Strengthened federal authority to deal with sewage discharges from vessels, hazardous polluting substances, and pollution from federal and federally related activities. Provided for liability for oil spills from onshore and offshore drilling facilities and from vessels.

(after the *Delian Apollon* ran aground near St. Petersburg, Fla., on February 13, 1970) [10].

The water pollution control legislation that began in 1948 left most of the enforcement power in the hands of state authorities, but the scope of federal control over water quality standards has been broadened over the years and is likely to become more extensive through further legislation, judging from the widespread dissatisfaction with the slow pace at which action is being taken. Although there is some evidence that improvements in water quality are being made, the procedures based on conferences, hearings, and then court action have often taken years, and the federal government has carried very few cases to successful enforcement [3].

Originally, water pollution control was handled by the Division of Water Supply and Pollution Control in the Department of Health, Education, and Welfare. The Federal Water Pollution Control Administration was created in the same department in 1965 and transferred to the Department of the Interior in 1966 by President Johnson's Reorganization Plan No. 2 of 1966. President Nixon's Reorganization Plan No. 3 of 1970 transferred all powers to the Environmental Protection Agency.

The water quality standards provided for by the Water Quality Act of 1965 have been set by the states and most of them have been accepted by the federal government. Disagreements arose between the Secretary of the Interior and several states over federal insistence on a nondegradation clause— which would provide that water quality in streams which already satisfy the proposed standards should not be permitted to degrade—and over federal insistence on secondary treatment of all wastes flowing into interstate waters. The state of Iowa, in particular, has argued that water quality standards are more important than effluent standards and that wastes receiving no treatment or only primary treatment should be permitted when they are minor enough not to affect water quality.

The individual states have their own water pollution laws and commissions, and there have been several interstate compacts organized to abate pollution. New York and New Jersey joined together in 1936 to form the Interstate Sanitary District but the results were not very notable. The eight-state Ohio River Valley Sanitation Commission (ORSANCO) was set up in 1948 and it has proved to be a more successful institution, although the Ohio River still has serious water pollution problems.

SOLID WASTES

Federal programs dealing with solid wastes were nonexistent before 1965 when the Solid Waste Disposal Act was passed by Congress. Table 20-4 lists federal solid wastes legislation of the past few years. To date, funding has

---

TABLE 20-4. Federal legislation relating to solid wastes.

1965   PL 89-272 (Title II): Solid Waste Disposal Act. Began a national research and development and demonstration program on solid wastes and provided financial assistance to interstate, state, and local agencies for planning and establishing solid waste disposal programs. Authorized increasing amounts from $10 million in fiscal year 1966 to $32.5 million in 1969 to be spent by the Department of Health, Education, and Welfare and by the Bureau of Mines in the Department of the Interior.

1968   PL 90-574: Solid Waste Disposal Act amendment. Authorized $32 million for fiscal year 1970.

1970   PL 91-512: Resource Recovery Act. Provided for extended research into new and improved methods to recover, recycle, and reuse wastes and for financial assistance to the states in the construction of solid waste disposal facilities.

---

been mainly for research and development, for training programs, and for financial assistance to the states. The National Solid Wastes Survey, mentioned in Chapter 13, is providing the first nationwide collection of scientific data about solid waste production and disposal.

To date there has been no federal legislation setting standards for the collection or disposal of solid wastes. Several state and local governments have passed such legislation, often to prevent the creation of air or water pollution problems from burning at open dumps or improper location of dumps and landfills. In addition, public health laws have sometimes been used against disposal areas with rodent or fly problems.

PESTICIDES

As shown in Table 20-5, U.S. pesticide legislation is rather old compared to most of the legislation discussed in this chapter. Much of it has dealt with protecting the farmer or other pesticide user against substandard products and the consumer from contamination of his food by pesticide residues that might endanger health.

The necessity of regulating the amounts of pesticide residues found in food was recognized in the early 20th century, and the 1938 Food, Drug, and Cosmetic Act provided for the setting of tolerances but few such toler-

---

TABLE 20-5. Federal legislation relating to pesticides [11].

1910   Federal Insecticide Act. Protected farmers and other users from substandard or fraudulent insecticides and fungicides.

1938   PL 83-518: Federal Food, Drug, and Cosmetic Act. Provided for the establishment of tolerances for pesticide residues in food.

1947   Federal Insecticide, Fungicide, and Rodenticide Act. Superseded 1910 Federal Insecticide Act. Provided for registration of "economic poisons" by the U.S. Department of Agriculture for products marketed in interstate commerce. Provided for seizures of adulterated, misbranded, unregistered, or insufficiently labeled pesticides.

1954   Miller Amendment to the Federal Food, Drug, and Cosmetic Act. Provided for condemnation of raw agricultural commodities containing pesticide residues in excess of tolerances fixed by the Secretary of Health, Education, and Welfare.

1959   Nematocide, Plant Regulator, Defoliant, and Desiccant Amendment. Extended coverage of the 1947 Act to these materials.

1961   Amendments of March 29, 1961 and April 7, 1961 to the 1947 Act.

1964   Amendment of May 12, 1964 to the 1947 Act (PL 88-305). Eliminated clauses that made it legal for a manufacturer to sell an unregisterable product simply by filing a formal protest.

---

ances were actually set because of the difficulty in reconciling divergent points of view. The 1954 Miller Amendment resolved the problem by giving the Secretary of Health, Education, and Welfare the right to set such tolerances based upon information provided by the U.S. Department of Agriculture. The 1958 Food, Drug, and Cosmetic Act contains an important clause applicable to pesticides—the "Delaney Clause" stipulating that no material that is capable of causing cancer in appropriate tests may ever be permitted in any food.

The 1947 Federal Insecticide, Fungicide, and Rodenticide Act and later amendments provided for registration of pesticides by the U.S. Department of Agriculture, which has generally granted registration to any product that is not a public health hazard. Registration was not permitted until a food tolerance had been set under the provisions of the Miller Amendment or until tests had indicated that no residues could result from the proposed use of the pesticide. In 1970 the registration responsibilities were transferred to the new Environmental Protection Agency.

Individual states also have laws regulating pesticide use within the states [11].

GENERAL ENVIRONMENTAL LEGISLATION

Federal environmental legislation of a more general nature has been passed in recent years.

The National Environmental Policy Act of 1969 (PL 91-190) declared a national policy for the environment to encourage productive and enjoyable harmony between man and his environment, to promote efforts that will prevent or eliminate damage to the environment and biosphere and stimulate the health and welfare of man, and to enrich the understanding of the ecological systems and natural resources important to the nation [9]. It established a three-member Council on Environmental Quality in the Executive Office of the President. The Council is charged with making studies and recommendations to the President and preparing an annual Environmental Quality Report [12, 13].

NEPA—as the Act is often referred to—also directed that:

All agencies of the Federal Government shall . . . include in every recommendation or report on proposals for legislation and other major Federal actions significantly affecting the quality of the human environment, a detailed statement by the responsible official on:
(i)   The environmental impact of the proposed action.
(ii)   Any adverse environmental effects which cannot be avoided should the proposal be implemented.
(iii) Alternatives to the proposed action.
(iv)   The relationship between local short-term uses of man's environment and the maintenance and enhancement of long-term productivity.

(v)   Any irreversible and irretrievable commitments of resources which would be involved in the proposed action should it be implemented.

This provision for the filing of environmental impact statements has become a powerful tool for protecting the environment and a basis for legal action by environmental groups in such cases as the trans-Alaska pipeline and the Tennessee-Tombigbee Waterway. The most noted application of NEPA was in the 1971 "Calvert Cliffs" court ruling against the Atomic Energy Commission which forced the AEC to determine for itself the impact of the effluents of nuclear power plants (including thermal effluents which the AEC had long tried to ignore) and weigh the benefits of nuclear power plants against their environmental costs [14]. In its attempts to make public the environmental costs of federal actions and to compel the federal bureaucracy to protect the biosphere, NEPA has made many enemies [15].

The Environmental Quality Improvement Act of 1970 (PL 91-224, Title II) established an Office of Environmental Quality in the Executive Office of the President to provide professional and administrative services to the Council on Environmental Quality [9].

President Nixon's Reorganization Plan No. 3 of 1970, dated July 9, 1970, set up the Environmental Protection Agency and transferred to it the environmental functions of numerous other agencies and about 6000 federal employees. It was put in charge of programs for which about $1.4 billion had been appropriated in fiscal year 1971. The EPA became fact on December 2, 1970, with the confirmation of William D. Ruckelhaus as its first administrator.

ENVIRONMENTAL PROTECTION AGENCY

The Environmental Protection Agency (EPA), in common with other bureaucratic agencies, has a vast number of officials—the administrator, a deputy administrator, numerous assistant administrators, 10 regional administrators (in Boston, New York, Philadelphia, Atlanta, Chicago, Dallas, Kansas City, Denver, San Francisco, and Seattle), etc.

The agency also has six offices responsible for particular environmental programs: water, air, pesticides, radiation, solid waste management, and noise abatement and control.

The *Office of Water Programs* is responsible for (1) nationwide monitoring of water quality; (2) water pollution research, development, and demonstration programs; (3) financial assistance to municipal waste treatment facilities and to state water pollution control agencies; (4) ensuring the implementation of water quality standards provided for by federal legislation; and (5) the establishment and implementation of drinking water standards for systems subject to federal law.

The *Office of Air Programs* is responsible for (1) nationwide monitoring of air quality; (2) air pollution research, development and demonstration programs; (3) financial assistance to air pollution control agencies; and (4) setting nationwide air quality standards and regulating emissions from stationary and mobile sources in accordance with federal law.

The *Office of Pesticides Programs* is responsible for (1) the establishment of tolerance levels for pesticide residues that occur in or on food; (2) the registration of pesticide uses for protection of human safety; (3) the monitoring of pesticide residue levels in foods and in various parts of the environment; (4) research on effects of pesticides on human health, non-target fish and wildlife, and natural ecosystems; and (5) the establishment of guidelines and standards for analytical methods of residue detection.

The *Office of Radiation Programs* is responsible for (1) the development of radiation protection guidelines and environmental radiation standards; and (2) surveillance and monitoring of such guidelines and standards as well as of levels of background environmental radiation.

The *Office of Solid Waste Management Programs* is responsible for (1) the conduct of a research, development, and demonstration program for new and improved methods of solid waste disposal; (2) studies directed toward the conservation of natural resources by reducing the amount of waste and unsalvageable materials and by recovery and utilization of potential resources in solid wastes; and (3) the provision of technical and financial assistance to state, local, and interstate agencies for solid waste management programs.

The *Office of Noise Abatement and Control* is responsible for investigating the causes and sources of noise and determining their effects on the public health and welfare.

REFERENCES

1. *Our Waters and Wetlands: How the Corps of Engineers Can Help Prevent Their Destruction and Pollution.* Twenty-First Report by the Committee on Government Operations of the House of Representatives. Washington, D.C.: U.S. Govt. Printing Office, March 1970.
2. *Qui Tam Actions and the 1899 Refuse Act: Citizen Lawsuits Against Polluters of the Nation's Waterways.* Washington, D.C.: U.S. Govt. Printing Office, September 1970. Committee print of the Conservation and Natural Resources Subcommittee of the Committee on Government Operations.
3. Degler, S. E., and S. C. Bloom, *Federal Pollution Control Programs: Water, Air, and Solid Wastes.* Washington, D.C.: Bureau of National Affairs, 1969.
4. Davies, J. C., III, *The Politics of Pollution.* New York: Pegasus, 1970.
5. These air quality criteria are References 9, 12, 15, 22, and 24 of Chapter 3 and Reference 5 of Chapter 6. The control technique documents are References 10, 16, 18, 19, and 20 of Chapter 5 and Reference 9 of Chapter 6.
6. Miller, S. S., "New Blueprint Emerges for Air Pollution Controls." *Environ. Sci. Tech.* **5:**106–108 (1971).
7. Miller, S. S., "National Air Quality Standards Finalized." *Environ. Sci. Tech.* **5:**503 (1971).

8. Hagevik, George H., *Decision-Making in Air Pollution Control*. New York: Frederick A. Praeger Inc., Pub., 1970.

9. *Laws of the United States Relating to Water Pollution Control and Environmental Quality*. Compiled by the Committee on Public Works of the U.S. House of Representatives. Washington, D.C.: U.S. Govt. Printing Office, July 1970.

10. Kenworthy, E. W., "Rep. Cramer Stiffens Stand on Oil Spillage." *N.Y. Times,* February 17, 1970, p. 69.

11. Bloom, S. C., and S. E. Degler, *Pesticides and Pollution*. Washington, D.C.: Bureau of National Affairs, 1969.

12. *Environmental Quality. The First Annual Report of the Council on Environmental Quality*. Washington, D.C.: U.S. Govt. Printing Office, August 1970.

13. *Environmental Quality. The Second Annual Report of the Council on Environmental Quality*. Washington, D.C.: U.S. Govt. Printing Office, August 1971.

14. Holden, Constance, "Court Decision Jolts AEC." *Science* **173:**799 (1971).

15. Gillette, Robert, "National Environmental Policy Act: Signs of Backlash Are Evident." *Science* **176:**30–33 (1972).

EPILOGUE

In the past, we have had science for intellectual pleasure, and science for the control of nature. We have had science for war. But today, the whole human experience may hang on the question of how fast we now press the development of science for survival.
—John Platt, "What We Must Do." *Science* **166:**1115–1121 (1969).

Several years ago John Platt listed mankind's major problems and the importance and imminence of major crises resulting from them, and he urged a large-scale scientific mobilization to attack them. Environmental degradation ranked high on the list, which included several other problems of equal or greater importance. Mankind must face up to all of these concerns—nuclear annihilation, poverty, racial wars, famines, pollution, the population explosion, political rigidity, etc.

Pollution is of importance because of its adverse effects on the health, safety, and well-being of humans, animals, and plants, including the animals and plants that are important as food for man and those which are not but which play significant roles in natural ecosystems.

Pollution is especially important in localized areas, such as densely populated urban areas, where man-made pollution dwarfs natural pollution in significance. The evidence shows, however, that pollution is becoming more and more an international and even global concern.

The solution to the problem of pollution involves both population control and the avoidance or control of those actions that pollute the environment. There should be no right to pollute but rather the privilege to pollute when and only when society (global society, of course) decides to grant the privilege.

Much research is needed to increase our understanding of the sources and effects of pollutants, and all disciplines have a role to play in this increased understanding. Research and development are also needed to devise better methods of pollution control, although there is already a great deal that can be accomplished to alleviate the problems if society has the will.

All citizens can contribute by emphasizing the improvement of the quality of life through personal actions, collective actions, and political actions. Much of what has been mistaken for improvements in the quality of life in recent years is now recognized by many persons to have been wasteful and waste-producing consumption providing less physical and mental satisfaction than had been anticipated.

The material that has been presented in this book has been intended as an introduction to an important area of environmental concern—pollution. It is likely to be corrected and updated with the passage of time. But if it is not acted upon despite our partial ignorance of the subject, some of the catastrophes and tragedies that have already occurred will be repeated with increasing frequency and increasing intensity.

APPENDIX A. CONVERSION FACTORS

The following alphabetical list gives the conversion factors between units used in this book and those appearing in the references or in other works dealing with environmental pollution. A table at the end lists the prefixes used to denote multiples and submultiples of the basic units, together with the appropriate symbols.

1 acre = 43,560 ft² = 4047 m² = 0.4047 hectares
1 acre-foot = 3.259 × 10⁵ gal = 1.233 × 10³ m³
1 are = 100 m² = 1076 ft²
1 atmosphere (atm) = 14.7 lb/in² = 2116 lb/ft² = 1.013 × 10⁵ N/m²
1 barrel (petroleum) = 42 gal (U.S.) = 34.97 gal (imperial)
  = 158.99 liters = 0.15899 m³
1 British thermal unit (Btu) = 1055 joules = 2.93 × 10⁻⁴ kWh
1 calorie (cal) = 4.184 joules
1 curie = 3.7 × 10¹⁰ disintegrations per second
1 dyne = 10⁻⁵ newton
1 electron-volt (eV) = 1.602 × 10⁻¹⁹ joule
1 erg = 10⁻⁷ joule
1 foot (ft) = 0.3048 m
1 square foot = 0.0929 m²
1 cubic foot = 2.832 × 10⁻² m³ = 28.32 liters
1 cubic foot per second = 102 m³/h = 2.45 × 10³ m³/day
1 gallon (gal) (U.S.) = 3.785 × 10⁻³ m³
1 gallon (Imperial) = 4.546 × 10⁻³ m³
1 hectare = 100 ares = 10⁴ m² = 2.471 acres
1 horsepower (hp) = 7.457 × 10² watts
1 inch (in.) = 2.54 cm = 2.54 × 10⁻² m
1 joule (J) = 9.481 × 10⁻⁴ Btu = 0.2389 cal = 2.778 × 10⁻⁷ kWh
1 kilogram (kg) = 1000 g; it corresponds to 2.205 lb
1 kilometer (km) = 1000 m = 0.6214 mi
1 kilowatt (kW) = 1000 watts = 3413 Btu/h = 1.341 hp
1 kilowatt-hour (kWh) = 3.6 × 10⁶ joules = 3413 Btu
1 liter (l) = 1000 cm³ = 1.000 × 10⁻³ m³ = 0.03531 ft³
1 meter (m) = 3.281 ft
1 square meter = 10.76 ft²
1 cubic meter = 35.31 ft³
1 meter per second = 2.237 mi/h

1 metric ton = 1000 kg; it corresponds to 1.10 short tons
1 "metric ton of coal equivalent" is approximately equal to
   28.5×10^6 Btu or 3×10^{10} joules or 8.34×10^3 kWh
1 micron (μm) = 10^{-6} m
1 mile (mi) = 5280 ft = 1609 m = 1.609 km
1 square mile = 640 acres = 2.590 km^2 = 259 hectares
1 cubic mile = 4.168 km^3
1 mile per hour = 0.447 m/s
1 newton (N) = 10^5 dynes = 0.2248 lb
1 N/m^2 = 9.869×10^{-6} atm = 1.450×10^{-4} lb/in.^{2}
1 poise = 0.1 N-s/m^2; it is the unit of viscosity
1 pound (lb) = 4.448 N; it corresponds to 0.4536 kg
1 rad, rem, roentgen, or rutherford: see Table 15-2, p. 248
1 ton (long) = 2240 lb; it corresponds to 1.02 metric tons
1 ton (short) = 2000 lb; it corresponds to 0.91 metric tons
1 ton per square mile corresponds to 0.35 metric ton/km^2
1 watt (W) = 3.413 Btu/h = 1.341×10^{-3} hp
1 year = 3.15×10^7 s

MULTIPLES AND SUBMULTIPLES OF UNITS

Multiple	Prefix	Symbol
10^{12}	tera-	T
10^9	giga-	G
10^6	mega-	M
10^3	kilo-	k
10^2	hecto-	h
10	deka-	da
10^{-1}	deci-	d
10^{-2}	centi-	c
10^{-3}	milli-	m
10^{-6}	micro-	μ
10^{-9}	nano-	n
10^{-12}	pico-	p
10^{-15}	femto-	f
10^{-18}	atto-	a

APPENDIX B. SOURCES OF FURTHER INFORMATION

BOOKS

All the books in this list are worth consulting about environmental problems; many have already been referenced elsewhere in this book. A number of these works contain a great deal of biological and ecological information and thus complement the physical and chemical emphasis of this book.

A. GENERAL

Baldwin, Malcolm, and James K. Page, Jr. (eds.), *Law and Environment.* New York: Walker Pub. Co., Inc., 1970.

Boughey, Arthur S., *Man and the Environment. An Introduction to Human Ecology and Evolution.* New York: The Macmillan Co., 1971.

Cleaning Our Environment. The Chemical Basis for Action. Washington, D.C.: American Chemical Society, 1969.

Commoner, Barry, *The Closing Circle.* New York: Alfred A. Knopf, Inc., 1971.

Council on Environmental Quality, *Environmental Quality. The Annual Report of the Council on Environmental Quality.* Washington, D.C.: U.S. Govt. Printing Office. Published annually beginning in 1970.

Council on Environmental Quality, *Toxic Substances.* Washington, D.C.: U.S. Govt. Printing Office, 1971.

Detwyler, Thomas R. (ed.), *Man's Impact on Environment.* New York: McGraw-Hill Book Co., 1971.

Dworsky, Leonard, *Pollution.* New York: Chelsea House Pub., 1971. A history of pollution and pollution legislation in the United States.

Ehrenfeld, David W., *Biological Conservation.* New York: Holt, Rinehart and Winston, Inc., 1970.

Ehrlich, Paul R., John P. Holdren, and Richard W. Holm (eds.), *Man and the Ecosphere.* San Francisco, Calif.: W. H. Freeman and Co., 1971. A collection of readings from *Scientific American.*

Harte, John, and Robert H. Socolow (eds.), *Patient Earth.* New York: Holt, Rinehart and Winston, Inc., 1971.

Lund, Herbert F. (ed.), *Industrial Pollution Control Handbook.* New York: McGraw-Hill Book Co., 1971.

Man's Impact on the Global Environment. Cambridge, Mass.: M.I.T. Press,

1970. The Report of the Study of Critical Environmental Problems, held at Williamstown, Mass. in 1970.

Nicholson, Max, *The Environmental Revolution*. New York: McGraw-Hill Book Co., 1970.

Novick, Sheldon, and Dorothy Cottrell (eds.), *Our World in Peril: An Environment Review*. Greenwich, Conn.: Fawcett Pub., Inc., 1971.

Rose, J. (ed.), *Technological Injury*. New York: Gordon and Breach, Inc., 1969.

Shepard, Paul, and Daniel McKinley (eds.), *The Subversive Science. Essays Toward an Ecology of Man*. Boston, Mass.: Houghton Mifflin Co., 1969.

Singer, S. F. (ed.), *Global Effects of Environmental Pollution*. New York: Springer-Verlag New York Inc., 1970.

Strobbe, Maurice A. (ed.), *Understanding Environmental Pollution*. St. Louis, Mo.: C. V. Mosby, 1971.

Wagner, Richard H., *Environment and Man*. New York: W. W. Norton & Co., 1971.

Ward, Barbara, and René Dubos, *Only One Earth*. New York: W. W. Norton & Co., Inc., 1972.

B. Population and Natural Resources

Borgstrom, Georg, *The Hungry Planet*. New York: The Macmillan Co., 1967.

Borgstrom, Georg, *Too Many. A Study of the Earth's Biological Limitations*. New York: The Macmillan Co., 1969.

Ehrlich, Paul R., and Anne H. Ehrlich, *Population Resources Environment. Issues in Human Ecology*, 2nd ed. San Francisco, Calif.: W. H. Freeman and Co., 1972.

Flawn, Peter T., *Mineral Resources*. Chicago, Ill.: Rand McNally & Co., 1966.

Hardin, Garrett (ed.), *Population, Evolution, and Birth Control*, 2nd ed. San Francisco, Calif.: W. H. Freeman and Co., 1969.

National Academy of Sciences-National Research Council, Committee on Resources and Man, Division of Earth Sciences, *Resources and Man*. San Francisco, Calif.: W. H. Freeman and Co., 1969.

United Nations Statistical Office, *Statistical Yearbook*. Published annually.

United States Bureau of Mines, *Mineral Facts and Problems*. Published every 5 years, most recently in 1970.

United States Bureau of Mines, *Minerals Yearbook*. Published annually.

C. Air Pollution

Air Conservation. Washington, D.C.: American Assn. for the Advancement of Science, 1965. AAAS Publication No. 80.

Air Quality Criteria. Separate monographs dealing with carbon monoxide, hydrocarbons, nitrogen oxides, particulate matter, sulfur oxides, and photochemical oxidants published by the Environmental Protection Agency and predecessor.

Bach, Wilfrid, *Atmospheric Pollution*. New York: McGraw-Hill Book Co., Inc., 1972.

Control Techniques. Separate monographs dealing with the control of major air pollutants from stationary and mobile sources, published by the Environmental Protection Agency and predecessor.

Esposito, John C. (Project Director), *Vanishing Air. The Ralph Nader Study Group Report on Air Pollution*. New York: Grossman Pub., 1970.

Junge, C. E., *Air Chemistry and Radioactivity*. New York: Academic Press, 1963.

Stern, A. C. (ed.), *Air Pollution*, 2nd ed. 3 volumes. New York: Academic Press, 1968.

D. Noise

Beranek, Leo (ed.), *Noise and Vibration Control*. New York: McGraw-Hill Book Co., 1971.

Burns, William, *Noise and Man*. London: John Murray (Pub.) Ltd., 1968.

Environmental Protection Agency, *Report to the President and Congress on Noise*. Washington, D.C.: U.S. Govt. Printing Office, 1972.

Harris, Cyril M. (ed.), *Handbook of Noise Control*. New York: McGraw-Hill Book Co., 1957.

Kryter, Karl D., *The Effects of Noise on Man*. New York: Academic Press, 1970.

Welch, Bruce L., and Annemarie S. Welch (eds.), *Physiological Effects of Noise*. New York: Plenum Press, 1970.

E. Water pollution

American Water Works Association, *Water Quality and Treatment. A Handbook of Public Water Supplies*, 3rd ed. New York: McGraw-Hill Book Co., 1971.

Behrman, A. S., *Water is Everybody's Business: The Chemistry of Water Purification*. New York: Doubleday & Co., Inc., 1968.

Besselievre, Edmund B., *The Treatment of Industrial Wastes*. New York: McGraw-Hill Book Co., 1969.

Camp, Thomas R., *Water and Its Impurities*. New York: Reinhold Pub. Co., 1963.

Clark, John W., Warren Viessman, Jr., and Mark Hammer, *Water Supply and Pollution Control*, 2nd ed. Scranton, Pa.: International Textbook Co., 1971.

Council on Environmental Quality, *Ocean Dumping. A National Policy*. Washington, D.C.: U.S. Govt. Printing Office, 1970.

Eckenfelder, W. W., Jr., *Industrial Water Pollution Control*. New York: McGraw-Hill Book Co., 1966.

Eckenfelder, W. W., Jr., *Water Quality Engineering for Practicing Engineers*. New York: Barnes & Noble, Inc., 1970.

Fair, Gordon M., John C. Geyer, and Daniel A. Okun, *Elements of Water Supply and Wastewater Disposal*, 2nd ed. New York: John Wiley & Sons, Inc., 1971.

Grava, Sigurd, *Urban Planning Aspects of Water Pollution Control*. New York: Columbia University Press, 1969.

Hepple, Peter (ed.), *Water Pollution by Oil*. London: Institute of Petroleum, 1971.

Holden, W. S. (ed.), *Water Treatment and Examination*. London: J. & A. Churchill, 1970.

Hoult, David P. (ed.), *Oil on the Sea*. New York: Plenum Press, 1969.

Hynes, H. B. N., *The Biology of Polluted Waters*. Liverpool: Liverpool University Press, 1963.

Nemerow, N. L., *Liquid Waste of Industry. Theories, Practices, and Treatment*. Reading, Mass.: Addison-Wesley Pub. Co., Inc., 1971.

Warren, Charles E., *Biology and Water Pollution Control*. Philadelphia, Pa.: W. B. Saunders Co., 1971.

Water Quality Criteria. Report of the National Technical Advisory Committee to the Secretary of the Interior. Washington, D.C.: U.S. Dept. of the Interior, Federal Water Pollution Control Administration, 1968.

Wilbur, Charles G., *The Biological Aspects of Water Pollution*. Springfield, Ill.: C. C. Thomas, Pub., 1969.

Wolman, Abel, *Water, Health and Society: Selected Papers by Abel Wolman edited by Gilbert F. White*. Bloomington, Ind.: Indiana University Press, 1969.

Zwick, David, and Marcy Benstock, *Water Wasteland. Ralph Nader's Study Group Report on Water Pollution*. New York: Grossman, 1971.

F. AGRICULTURAL POLLUTION

Brady, Nyle C. (ed.), *Agriculture and the Quality of Our Environment*. Washington, D.C.: American Assn. for the Advancement of Science, 1967. AAAS Publication No. 85.

Held, R. Burnell, and Marion Clawson, *Soil Conservation in Perspective*. Baltimore, Md.: The Johns Hopkins Press for Resources for the Future, Inc., 1965.

Loehr, Raymond C., *Pollution Implications of Animal Wastes—A Forward Oriented Review*. Ada, Okla.: Robert S. Kerr Water Research Center, U.S. Dept. of the Interior, 1968.

Management of Farm Animal Wastes. Proceedings, National Symposium on Animal Waste Management, May 1966. St. Joseph, Mich.: American Society of Agricultural Engineers. ASAE Publication No. SP-0366.

Wadleigh, Cecil H., *Wastes in Relation to Agriculture and Forestry*. Washington, D.C.: U.S. Dept. of Agriculture, March 1968. USDA Miscellaneous Publication No. 1065.

Willrich, Ted L., and George E. Smith (ed.), *Agricultural Practices and Water Quality*. Ames, Iowa: Iowa State University Press, 1970.

G. Pesticides

Burges, H. D., and N. W. Hussey (eds.), *Microbial Control of Insects and Mites.* New York: Academic Press, 1971.

De Bach, Paul (ed.), *Biological Control of Insect Pests and Weeds.* New York: Reinhold Pub. Co., 1964.

Jacobson, Martin, and D. G. Crosby, *Naturally Occurring Insecticides.* New York: Marcel Dekker, Inc., 1971.

Jager, K. W., *Aldrin, Dieldrin, Endrin and Telodrin. An Epidemiological and Toxicological Study of Long-Term Occupational Exposure.* New York: Elsevier Pub. Co., 1971.

Kilgore, Wendell W., and Richard L. Doutt (eds.), *Pest Control. Biological, Physical, and Selected Chemical Methods.* New York: Academic Press, 1967.

Kraybill, Herman F. (ed.), "Biological Effects of Pesticides in Mammalian Systems." *Annals N.Y. Acad. Sci.* **160:**1–422 (1969).

Mellanby, Kenneth, *Pesticides and Pollution.* New York: William Collins Sons & Co., Ltd., 1969.

Miller, M. W., and G. C. Berg (eds.), *Chemical Fallout.* Springfield, Ill.: C. C. Thomas, Pub., 1969.

Moore, M. W. (ed.), *Pesticides in the Environment and Their Effects on Wildlife. Jour. Appl. Ecol.,* Supplement to Vol. 3, 1966.

O'Brien, R. D., *Insecticides: Action and Metabolism.* New York: Academic Press, 1967.

Report of the Secretary's Commission on Pesticides and Their Relationship to Environmental Health. Washington, D.C.: U.S. Dept. of Health, Education, and Welfare, December 1969.

Rudd, Robert L., *Pesticides and the Living Landscape.* Madison, Wis.: University of Wisconsin Press, 1964.

H. Thermal pollution

Eisenbud, Merril, and George Gleason (eds.), *Electric Power and Thermal Discharges. Thermal Considerations in the Production of Electric Power.* New York: Gordon and Breach, Inc., 1969.

Krenkel, Peter A., and Frank L. Parker (eds.), *Biological Aspects of Thermal Pollution.* Nashville, Tenn.: Vanderbilt University Press, 1969.

Parker, Frank L., and Peter A. Krenkel (eds.), *Engineering Aspects of Thermal Pollution.* Nashville, Tenn.: Vanderbilt University Press, 1969.

Parker, Frank L., and Peter A. Krenkel, *Physical and Engineering Aspects of Thermal Pollution.* Cleveland, Ohio: Chemical Rubber Company Press, 1970.

I. Radiation

Casaret, Alison P., *Radiation Biology.* Englewood Cliffs, N.J.: Prentice-Hall, Inc., 1968.

Eisenbud, Merril, *Environmental Radioactivity.* New York: McGraw-Hill Book Co., 1963.

Foreman, Harry (ed.), *Nuclear Power and the Public.* Minneapolis, Minn.: University of Minnesota Press, 1970.

Gofman, John W., and Arthur R. Tamplin, *Poisoned Power.* Emmaus, Pa.: Rodale Press, 1971.

Morgan, K. Z., and J. E. Turner, *Principles of Radiation Protection.* New York: John Wiley & Sons, Inc., 1967.

Rees, D. J., *Health Physics.* Cambridge, Mass.: M.I.T. Press, 1967.

Upton, Arthur C., *Radiation Injury.* Chicago, Ill.: University of Chicago Press, 1969.

PERIODICALS

Periodicals and monograph series are grouped into four categories: those dealing almost exclusively with environmental problems mentioned in this book; general scientific publications which carry many articles about pollution; leading medical and public health journals that carry some articles about the health effects of pollution; and miscellaneous scientific and industrial periodicals that sometimes carry articles of interest. Professional scientific journals (as opposed to trade magazines and periodicals directed at lay readers) are marked by an asterisk (*).

A. ENVIRONMENTAL PERIODICALS

Advances in Environmental Sciences and Technology is a series of volumes with articles by leading environmental scientists.

AIR/WATER Pollution Report is a weekly worldwide environmental newsletter.

Archives of Environmental Health is published monthly by the American Medical Association and contains many interesting articles about the health effects of pollutants.

Bulletin of Environmental Contamination and Toxicology provides rapid publication of significant discoveries.

CF Letter is a newsletter published by the Conservation Foundation.

CRC Critical Reviews in Environmental Control is a quarterly journal containing articles by recognized experts in fields relevant to the area of environmental control.

Environment is directed at the lay reader and publishes information about the effects of technology on the environment and about the peaceful and military uses of nuclear energy.

Environmental Education is a quarterly devoted to environmental education.

Environmental Health Series monographs are published by the Environmental Protection Agency and other federal agencies.

Environmental Letters is an international journal for rapid communication of the results of environmental research.

Environmental Pollution is a quarterly international journal concerned mainly with the biological effects of pollution.

Environmental Research is an international journal of biomedical ecology dealing with epidemiology, physiology, toxicology, etc.

Environmental Science and Technology is published monthly by the American Chemical Society and contains news reports; feature articles by environmental scientists from government, industry, and universities; and original research stressing chemical and physical aspects. At $9.00 per year it is a real "best buy."

Environmental Technology and Economics is a technical, economic, and political digest published 26 times annually.

Pollution Abstracts is an extremely valuable journal of abstracts of articles dealing with pollution published in hundreds of different periodicals.

Water, Air and Soil Pollution is a quarterly technical journal.

Atmospheric Environment is an international journal dealing with air pollution, industrial aerodynamics, micrometeorology, and aerosols.

Clean Air is published by the National Society for Clean Air in England.

Journal of the Air Pollution Control Association deals mainly with methods of controlling air pollution but also contains many articles of fundamental importance.

American Water Works Association Journal is a valuable periodical for the study of water pollution.

Environmental and Water Resources Engineering Conference contains the proceedings of an annual conference held at Vanderbilt University.

Journal of the Water Pollution Control Federation deals mainly with water pollution control.

Marine Pollution Bulletin is a monthly journal dealing with the management of the marine environment, its capacity to absorb waste materials, natural fluctuations in marine organisms, and marine productivity.

Water Pollution Abstracts is a monthly abstract journal published by the Great Britain Ministry of Technology.

Water Pollution Control is published by the Institute of Water Pollution Control in England.

Water Research is the journal of the International Association on Water Pollution Research.

SOLID WASTE Report is a biweekly newsletter.

Health Aspects of Pesticides is an abstract bulletin published monthly by the Environmental Protection Agency.

Pesticides Monitoring Journal deals with the monitoring of pesticides in the environment.

Residue Reviews contains review articles by leading scientists about residues of pesticides and other foreign chemicals in foods and feeds.

Health Physics is the official journal of the Health Physics Society; it contains articles on the effects of radiation.

Nuclear Safety is a bimonthly technical progress review prepared for the USAEC Division of Technical Information by the Nuclear Safety Information Center, Oak Ridge, Tenn.

Radiation Research is the official organ of the Radiation Research Society.

Radiological Health Data and Reports is published monthly by the Environmental Protection Agency and reports data on radiation from air, food, milk, and water.

B. GENERAL SCIENTIFIC PUBLICATIONS

American Scientist is a bimonthly publication of the Society of the Sigma Xi; it contains lengthy but quite readable articles by prominent scientists.

Biological Conservation is an international quarterly journal devoted to the scientific protection of all nature.

Bioscience is published by the American Institute of Biological Sciences.

Bulletin of the Atomic Scientists contains many articles on the use and effects of nuclear energy.

Nature is an important British science weekly.

New Scientist and Science Journal is a British weekly for the lay reader.

Science, the weekly publication of the American Association for the Advancement of Science, contains many articles, comments, and research reports on environmental problems and closely follows government actions and legislation. It is "must" reading for anyone following pollution problems in the U.S.

Scientific American is a monthly publication containing frequent articles about environmental concerns; it is directed at the educated lay reader.

C. MEDICAL AND PUBLIC HEALTH JOURNALS

American Journal of Public Health and the Nation's Health
British Journal of Industrial Medicine
British Medical Journal
Journal of Environmental Health
Journal of the American Medical Association
Lancet
New England Journal of Medicine
Public Health Reports

D. OTHER PERIODICALS

Advanced Energy Conversion
Advances in Ecological Research
Applied Acoustics
Audubon
Bulletin of the Ecological Society
Chemical and Engineering News

Chemical Engineering
Compost Science
Ecology Today
Engineering News-Record
FDA Papers
*Geochimica et Cosmochimica Acta
International Pest Control
*Journal of Applied Ecology
*Journal of Applied Meteorology
*Journal of Geophysical Research
*Journal of Sound and Vibration
*Journal of the Acoustical Society of America
*Journal of the Atmospheric Sciences
Municipal Engineer's Journal
*Natural Resources Journal
Nuclear Industry
Nuclear News
Pest Control
Physics Today
Public Works
*Quarterly Journal of the Royal Meteorological Society
*Radiation Botany
*Safety Series (published by the International Atomic Energy Agency, Vienna)
Scrap Age
Sound and Vibration
*Tellus (geophysics)
Water and Pollution Control
Water and Sewage Works
Water and Wastes Engineering
Water and Waste Treatment Journal
Water Newsletter
*Water Resources Research
World Review of Pest Control

ADDRESSES

The following list includes the addresses of governmental, professional, industrial, and private organizations that provide information or express opinions about environmental problems.

Air Pollution Control Association, 4400 Fifth Ave., Pittsburgh, Pa. 15213
American Association for Conservation Information, 1416 Ninth St., Sacramento, Calif. 95814
American Association for the Advancement of Science, 1515 Massachusetts Ave., N.W., Washington, D.C. 20036

American Chemical Society, 1155 Sixteenth St., N.W., Washington, D.C. 20036

American Conservation Association, 30 Rockefeller Plaza, New York, N.Y. 10020

American Gas Association, Inc., Department of Statistics, 1515 Wilson Blvd., Arlington, Va. 22209

American Geological Institute, 2201 M St., N.W., Washington, D.C. 20037

American Institute of Biological Sciences, 3900 Wisconsin Ave., N.W., Washington, D.C. 20016

American Nuclear Society, 244 East Ogden Ave., Hinsdale, Ill. 60521

American Petroleum Institute, Air and Water Conservation Library, 1271 Avenue of the Americas, New York, N.Y. 10020

American Society for Engineering Education, 1 DuPont Circle, N.W., Washington, D.C. 20036

American Water Works Association Research Foundation, 2 Park Ave., New York, N.Y. 10016

Citizens for Clean Air, 40 West 57th St., New York, N.Y. 10019

Citizens League Against Sonic Boom, 19 Appleton St., Cambridge, Mass. 02138

Committee on Environmental Information, 438 N. Skinker Blvd., St. Louis, Mo. 63130

Conservation Education Association, 1250 Connecticut Ave., N.W., Washington, D.C. 20036

Conservation Foundation, 1717 Massachusetts Ave., N.W., Washington, D.C. 20036

Conservation Law Society of America, Mills Tower, 220 Bush St., San Francisco, Calif. 94104

Conservation Library Center, Denver Public Library, 1357 Broadway, Denver, Colo. 80222

Council on Environmental Quality, 722 Jackson Place, N.W., Washington, D.C. 20006

Defenders of Wildlife, 1346 Connecticut Ave., N.W., Washington, D.C. 20036

Edison Electric Institute, 750 Third Ave., New York, N.Y. 10017

Environmental Defense Fund, P.O. Drawer 740, Stony Brook, N.Y. 11790

Environmental Law Institute, 1346 Connecticut Ave., N.W., Washington, D.C. 20036

Environmental Protection Administration (of New York City), Public Information and Education, 2345 Municipal Building, New York, N.Y. 10007

Environmental Protection Agency, Washington, D.C. 20460

Federation of Western Outdoor Clubs, Box 548, Bozeman, Mont. 59715

Friends of the Earth, 529 Commercial St., San Francisco, Calif. 94111

Glass Container Manufacturers Institute, Public Affairs Department, 330 Madison Ave., New York, N.Y. 10017

Institute of Paper Chemistry, Lawrence University, Appleton, Wis. 54911

International Joint Commission, 1717 H St., N.W., Washington, D.C. 20441
   or 151 Slater St., Ottawa 4, Ont., Canada
International Union for Conservation of Nature and Natural Resources, 1110
   Morges, Switzerland
Izaak Walton League of America, 1326 Waukegan Rd., Glenview, Ill. 60025
League of Conservation Voters, 917 15th St., N.W., Washington, D.C. 20005
National Academy of Science, Printing and Publishing Office, 2101 Constitu-
   tion Ave., Washington, D.C. 20418
National Audubon Society, 1130 Fifth Ave., New York, N.Y. 10028
National Environmental Law Institute, Stanford, Calif. 94305
National Parks Association, 1701 18th St., N.W., Washington, D.C. 20009
National Recreation and Park Association, 1700 Pennsylvania Ave., N.W.,
   Washington, D.C. 20006
National Society for Clean Air, 134/137 North St., Brighton, BN1 1RG,
   England
National Tuberculosis and Respiratory Disease Association, 1740 Broadway,
   New York, N.Y. 10019
National Wildlife Federation, 1412 Sixteenth St., N.W., Washington, D.C.
   20036
Natural Resources Defense Council, 1600 20th St., N.W., Washington, D.C.
   20009
Nature Conservancy, 1522 K St., N.W., Washington, D.C. 20005
New England Interstate Water Pollution Control Commission, 73 Tremont
   St., Boston, Mass. 02106
New York State Department of Environmental Conservation, 50 Wolf Rd.,
   Albany, N.Y. 12201
Open Space Institute, 145 East 52nd St., New York, N.Y. 10022
Planned Parenthood—World Population, 515 Madison Ave., New York, N.Y.
   10022
Population Reference Bureau, 1755 Massachusetts Ave., N.W., Washington,
   D.C. 20036
Resources for the Future, 1755 Massachusetts Ave., N.W., Washington, D.C.
   20036
Robert A. Taft Sanitary Engineering Center, Cincinnati, Ohio 45226
Robert S. Kerr Water Reseach Center, Ada, Okla.
Scientists Institute for Public Information, 30 East 68th St., New York, N.Y.
Sierra Club, 1050 Mills Tower, 220 Bush St., San Francisco, Calif. 94104
Smithsonian Institution Center for Short-Lived Phenomena, 60 Garden St.,
   Cambridge, Mass. 02138
Soil Conservation Society of America, 7517 N.E. Ankeny Rd., Ankeny, Iowa
   50021
Superintendent of Documents, U.S. Government Printing Office, Washington,
   D.C. 20402 (provides free distribution of listings of "Selected U.S.
   Government Publications")

U.S. Atomic Energy Commission, Division of Technical Information, Washington, D.C. 20545

U.S. Bureau of Mines, Publications Distribution Branch, 4800 Forbes Ave., Pittsburgh, Pa. 15213

U.S. Department of Health, Education, and Welfare, Public Health Service, Washington, D.C. 20201

U.S. Geological Survey, Water Resources Division, Washington, D.C. 20242

Water Pollution Control Federation, 3900 Wisconsin Ave., N.W., Washington, D.C. 20016

Wilderness Society, 729 15th St., N.W., Washington, D.C. 20005

Wildlife Management Institute, 709 Wire Building, Washington, D.C., 20005

World Wildlife Fund, 910 17th St., Washington, D.C. 20006

Zero Population Growth, 4080 Fabian Way, Palo Alto, Calif. 94303

INDEX